Statistical Mechanics of Nonequilibrium Liquids

THEORETICAL CHEMISTRY

A Series of Monographs

T. E. PEACOCK: Electronic Properties of Aromatic and Heterocyclic Molecules

R. McWEENY and B. T. SUTCLIFFE: Methods of Molecular Quantum Mechanics

J. LINDBERG and Y. ÖHRN: Propagators in Quantum Chemistry

M. S. CHILD: Molecular Collision Theory

C. D. H. CHISHOLM: Group Theoretical Techniques in Quantum Chemistry

E. R. DAVISON: Reduced Density Matrices in Quantum Chemistry

A. C. HURLEY: Electron Correlation in Small Molecules

D. P. CRAIG and T. THIRUNAMACHANDRAN: Molecular Quantum Electrodynamics

GAD FISCHER: Vibronic Coupling

R. McWEENY: Methods of Molecular Quantum Mechanics. Second Edition

D. J. EVANS and G. P. MORRISS: Statistical Mechanics of Nonequilibrium Liquids

Statistical Mechanics of Nonequilibrium Liquids

DENIS J. EVANS

Research School of Chemistry, Australian National University, Canberra, ACT, Australia

GARY P. MORRISS

School of Physics, University of New South Wales, Kensington, NSW, Australia

ACADEMIC PRESS
Harcourt Brace Jovanovich, Publishers
London San Diego New York
Boston Sydney Tokyo

ACADEMIC PRESS LIMITED
24/28 Oval Road
London NW1 7DX

United States Edition published by
ACADEMIC PRESS INC.
San Diego, CA 92101

British Library Cataloguing in Publication Data

Is available

ISBN 0-12-244090-0

Typeset in 10/12pt Times by P & R Typesetters Ltd, Salisbury, Wiltshire, UK
Printed and bound in Great Britain by T. J. Press Ltd, Padstow, Cornwall

Preface

During the 1980s there have been many new developments regarding the nonequilibrium statistical mechanics of dense classical systems. These developments have had a major impact on the computer simulation methods used to model nonequilibrium fluids. Some of these new algorithms are discussed in the recent book by Allen and Tildesley, Computer Simulation of Liquids. However that book was never intended to provide a detailed statistical mechanical backdrop to the new computer algorithms. As the authors commented in their preface, their main purpose was to provide a working knowledge of computer simulation techniques. The present volume is, in part, an attempt to provide a pedagogical discussion of the statistical mechanical environment of these algorithms.

There is a symbiotic relationship between nonequilibrium statistical mechanics on the one hand and the theory and practice of computer simulation on the other. Sometimes, the initiative for progress has been with the pragmatic requirements of computer simulation and at other times, the initiative has been with the fundamental theory of nonequilibrium processes. Although progress has been rapid, the number of participants who have been involved in exposition and development, rather than with application, has been relatively small.

The formal theory is often illustrated with examples involving shear flow in liquids. Since a central theme of this volume is the nonlinear response of systems, this book could be described as a text on Theoretical Rheology. However our choice of rheology as a testbed for the theory is merely a reflection of personal interest. The statistical mechanical theory that is outlined in this book is capable of far wider application.

All but two pages of this book is concerned with atomic rather than molecular fluids. This restriction is one of economy. The main purpose of this text is best served by choosing simple applications.

Many people deserve thanks for their help in developing and writing this book. Firstly we must thank our wives, Val and Jan, for putting up with our absences, our irritability and our exhaustion. We would also like to thank Dr David MacGowan for reading sections of the manuscript. Thanks must also go to Mrs Marie Lawrence for help with indexing. Finally special thanks must go to Professors Cohen, Hanley and Hoover for incessant argument and interest.

D. J. Evans and G. P. Morriss

Biographies

Denis Evans was born in Sydney, Australia in 1951. He obtained first class honours in astrophysics at Sydney University in 1972 and gained his Ph.D. in statistical mechanics from the Australian National University (ANU) in 1975. After postdoctoral appointments at Oxford, Cornell and ANU and a Fulbright Fellowship to the National Bureau of Standards he became a Fellow in the Research School of Chemistry at ANU in 1982. He has won a number of awards including the Rennie Medal for Chemistry (1983), the Young Distinguished Chemist Award of the Federation of Asian Chemical Societies (1989) and the Frederick White Prize of the Australian Academy of Science (1990). In 1989 he was appointed as Professor of Theoretical Chemistry at ANU and currently serves as Academic Director of the ANU Supercomputer Facility.

Gary Morriss was born in Singleton, Australia in 1951. He obtained first class honours in mathematics/physics at Newcastle University in 1976 and gained his Ph.D. in statistical mechanics from Melbourne University in 1980. After postdoctoral appointments at Cornell and ANU, he became a Research Fellow and later Senior Research Fellow at ANU. In 1989 he was appointed as Lecturer in the School of Physics at the University of New South Wales.

Contents

List of Symbols

Transport coefficients

η	shear viscosity
λ	thermal conductivity
η_v	bulk viscosity
ζ	Brownian friction coefficient
D	self-diffusion coefficient

Thermodynamic fluxes

$\mathbf{P}$	the pressure tensor
$\mathbf{J}_Q$	heat flux
$\mathbf{\Pi}$	viscous pressure tensor

Thermodynamic forces

$\nabla\mathbf{u}$	strain rate tensor
γ	shear rate, $= \partial u_x/\partial y$
∇T	temperature gradient
$\nabla\cdot\mathbf{u}$	dilation rate
$\mathbf{u}$	streaming velocity
$\boldsymbol{\varepsilon}$	elastic deformation
$\nabla\boldsymbol{\varepsilon}$	strain tensor
$\dot{\varepsilon}$	dilation rate, $= 1/3(\nabla\cdot\dot{\varepsilon})$

Thermodynamic state variables

T	temperature
k_B	Boltzmann's constant
β	$1/k_B T$
V	volume
p	hydrostatic pressure, $= 1/3\,\mathrm{tr}(\mathbf{P})$
N	number of particles
ρ	mass density
n	number density

Thermodynamic constants

G	shear modulus
C_V	constant-volume specific heat
C_p	constant-pressure specific heat
c_V	constant-volume specific heat per unit mass
c_p	constant-pressure specific heat per unit mass
D_T	isochoric thermal diffusivity

Thermodynamic potentials

E	internal energy
$U(\mathbf{r}, t)$	internal energy per unit mass
S	entropy
$s(\mathbf{r}, t)$	internal energy per unit volume
σ	entropy source strength = rate of spontaneous entropy production per unit volume
I	enthalpy
Q	heat

Mechanics

L	Lagrangian
H	Hamiltonian
H_0	phase variable whose average is the internal energy
I_0	phase variable whose average is the enthalpy
$J(\mathbf{\Gamma})$	dissipative flux
F_e	external field
α	thermostatting multiplier
iL	p-Liouvillean
iL	f-Liouvillean
$A^\dagger$	Hermitian adjoint of A
Λ	phase-space compression factor
$\exp_R$	right-time-ordered exponential
$\exp_L$	left-time-ordered exponential
$U_R(t_1, t_2)$	incremental p propagator t_1 to t_2
$U_R(t_1, t_2)^{-1}$	inverse of $U_R(t_1, t_2)$, take phase variables from t_2 to t_1, $U_R(t_2, t_1) = U_L(t_1, t_2)$
$U_R(t_1, t_2)^\dagger$	incremental f propagator t_1 to t_2
T_R	right time-ordering operator
T_L	left time-ordering operator
$C_{AB}(t)$	equilibrium time correlation function, $= \langle A(t)B^* \rangle$

$[A, B]$	commutator bracket
$\{A, B\}$	Poisson bracket
$\delta(t)$	Dirac delta function (of time)
$\delta(\mathbf{r} - \mathbf{r}_i)$	Kirkwood delta function $= 0, \quad |\mathbf{r} - \mathbf{r}_i| >$ an 'infinitesimal' macroscopic distance, 1 $= 1/1^3, \quad |\mathbf{r} - \mathbf{r}_i| <$ an 'infinitesimal' macroscopic distance, 1
$f(\mathbf{k}) = \int \mathrm{d}\mathbf{r}\, \mathrm{e}^{i\mathbf{k}\cdot\mathbf{r}} f(\mathbf{r})$	spatial Fourier transform
$f(\omega) = \int_0^\infty \mathrm{d}\mathbf{r}\, \mathrm{e}^{-i\omega t} f(t)$	spatial Fourier–Laplace transform
$\mathrm{d}\mathbf{S}$	infinitesimal vector area element
$\mathbf{J}^\perp$	transverse momentum current
Φ	total intermolecular potential energy
ϕ_{ij}	potential energy of particles i, j
K	total kinetic energy
f_c	canonical distribution function
f_T	isokinetic distribution function
m	particle mass
$\mathbf{r}_i$	position of particle i
$\mathbf{r}_{ij} = \mathbf{r}_j - \mathbf{r}_i$	
$\mathbf{v}_i$	velocity of particle i

1 Introduction

Mechanics provides a complete microscopic description of the state of a system. When the equations of motion are combined with initial conditions and boundary conditions, the subsequent time evolution of a classical system can be predicted. In systems with more than just a few degrees of freedom such an exercise is impossible. There is simply no practical way of measuring the initial microscopic state of, for example, a glass of water, at some instant in time. In any case, even if this was possible we could not then solve the equations of motion for a coupled system of 10^{23} molecules.

In spite of our inability to describe fully the microstate of a glass of water, we are all aware of useful macroscopic descriptions for such systems. Thermodynamics provides a theoretical framework for correlating the equilibrium properties of such systems. If the system is not at equilibrium, fluid mechanics is capable of predicting the macroscopic nonequilibrium behaviour of the system. In order for these macroscopic approaches to be useful, their laws must be supplemented not only with a specification of the appropriate boundary conditions but also with the values of thermophysical constants such as equation-of-state data and transport coefficients. These values cannot be predicted by macroscopic theory. Historically this data has been supplied by experiments. One of the tasks of statistical mechanics is to predict these parameters from knowledge of the interactions of the system's constituent molecules. This then is a major purpose of statistical mechanics. How well have we progressed?

Equilibrium classical statistical mechanics is relatively well developed. The basic ground rules—Gibbsian ensemble theory—have been known for the best part of a century (Gibbs, 1902). The development of electronic computers in the 1950s provided unambiguous tests of the theory of simple liquids, consequently leading to a rapid development of integral equation and perturbation treatments of liquids (Barker and Henderson, 1976). With the possible exceptions of phase equilibria and interfacial phenomena (Rowlinson and Widom, 1982), one could say that the equilibrium statistical mechanics of atomic fluids is a solved problem. Much of the emphasis has moved to molecular and even macromolecular liquids.

The nonequilibrium statistical mechanics of dilute atomic gases—kinetic theory—is, likewise, essentially complete (Ferziger and Kaper, 1972). However, attempts to extend kinetic theory to higher densities have been fraught with severe difficulties. One might have imagined being able to develop a power-series expansion of the transport coefficients in much the same way that one expands the equilibrium equation of state in the virial series. In 1965, Cohen and Dorfman (1965 and 1972) proved that such an expansion does not exist. The Navier–Stokes transport coefficients are nonanalytic functions of density.

It was at about this time that computer simulations began to have an impact on the field. In a celebrated paper, Kubo (1957) showed that linear transport coefficients can be calculated from a knowledge of the equilibrium fluctuations in the flux associated with a particular transport coefficient. For example, the shear viscosity η, is defined as the ratio of the shear stress, $-P_{xy}$, to the strain rate, $\partial u_x/\partial y \equiv \gamma$,

$$P_{xy} \equiv -\eta\gamma \tag{1.1}$$

The Kubo relation predicts that the limiting (small shear rate) viscosity, is given by

$$\eta = \beta V \int_0^\infty ds \langle P_{xy}(0) P_{xy}(s) \rangle \tag{1.2}$$

where β is the reciprocal of the absolute temperature T, multiplied by Boltzmann's constant k_B, V is the system volume and the angle brackets denote an *equilibrium* ensemble average. The viscosity is then the infinite time integral of the equilibrium, autocorrelation function of the shear stress. Similar relations are valid for the other Navier–Stokes transport coefficients such as the self-diffusion coefficient, the thermal conductivity and the bulk viscosity (see Chapter 4).

Alder and Wainwright (1956) were the first to use computer simulations to compute the transport coefficients of atomic fluids. What they found was unexpected. It was believed that at sufficiently long time, equilibrium autocorrelation functions should decay exponentially. Alder and Wainwright discovered that in two-dimensional systems the velocity autocorrelation function which determines the self-diffusion coefficient, only decays as $1/t$. Since the diffusion coefficient is thought to be the integral of this function, we were forced to the reluctant conclusion that the self-diffusion coefficient does not exist for two-dimensional systems. It is presently believed that each of the Navier–Stokes transport coefficients diverge in two dimensions (Pomeau and Resibois, 1975).

This does *not* mean that two-dimensional fluids are infinitely resistant to shear flow. Rather, it means that the Newtonian constitutive relation (1.1) is an inappropriate definition of viscosity in two dimensions. There is no linear regime close to equilibrium where Newton's law (1.1) is valid. It is thought that at small strain rates, $P_{xy} \simeq \gamma \log \gamma$. If this is the case then the limiting value of the shear viscosity ($\lim(\gamma \to 0) - \partial P_{xy}/\partial \gamma$) would be infinite. All this presupposes that steady laminar shear flow is stable in two dimensions. This would be an entirely natural presumption on the basis of the three-dimensional experience. However, there is some evidence that even this assumption may be wrong (Evans and Morriss, 1983). Recent computer simulation data suggests that in two dimensions laminar flow may be unstable at *low* strain rates.

In three dimensions the situation is better. The Navier–Stokes transport coefficients appear to exist. However, the nonlinear Burnett coefficients, the higher order terms in the Taylor series expansion of the shear stress in powers of the strain rate (Sections 2.3 and 9.5), are thought to diverge (Kawasaki and Gunton, 1973). These divergences are sometimes summarized in Dorfman's lemma (Zwanzig, 1982): all relevant fluxes are nonanalytic functions of all relevant variables! The transport coefficients are thought to be nonanalytic functions of density, frequency and the magnitude of the driving thermodynamic force, the strain rate or the temperature gradient, etc.

In this book we will discuss the framework of nonequilibrium statistical mechanics. We will not discuss in detail, the practical results that have been obtained. Rather, we seek to derive a nonequilibrium analogue of the Gibbsian basis for equilibrium statistical mechanics. At equilibrium we have a number of idealizations which serve as standard models for experimental systems. Among these are the well-known microcanonical, canonical and grand-canonical ensembles. The real system of interest will not correspond exactly to any one particular ensemble, but such models furnish useful and reliable information about the experimental system. We have become so accustomed to mapping each real experiment onto its nearest Gibbsian ensemble that we sometimes forget that the canonical ensemble, for example, does not exist in nature—it is an idealization.

A nonequilibrium system can be modelled as a perturbed equilibrium ensemble. We will therefore need to add the perturbing field to the statistical mechanical description. The perturbing field does work on the system—this prevents the system from relaxing to equilibrium. This work is converted to heat, and the heat must be removed in order to obtain a well-defined steady state. Therefore, thermostats will also need to be included in our statistical mechanical models. A major theme of this book is the development of a set of idealized nonequilibrium systems which can play the same role in nonequilibrium statistical mechanics as the Gibbsian ensembles play at equilibrium.

After a brief discussion of linear irreversible thermodynamics in Chapter 2, we address the Liouville equation in Chapter 3. The Liouville equation is the fundamental vehicle of nonequilibrium statistical mechanics. We introduce its formal solution using mathematical operators called propagators (Section 3.3). In Chapter 3, we also outline the procedures by which we identify statistical mechanical expressions for the basic field variables of hydrodynamics.

After this background in both macroscopic and microscopic theory we go on to derive the Green–Kubo relations for linear transport coefficients in Chapter 4 and the basic results of linear response theory in Chapter 5. The Green–Kubo relations derived in Chapter 4 relate *thermal* transport coefficients, such as the Navier–Stokes transport coefficients, to equilibrium fluctuations. Thermal-transport processes are driven by boundary conditions. The expressions derived in Chapter 5 relate *mechanical* transport coefficients to equilibrium fluctuations. A mechanical transport process is one that is driven by a perturbing external field which actually changes the mechanical equations of motion for the system. In Chapter 5 we show how the thermostatted linear mechanical response of many-body systems is related to equilibrium fluctuations.

In Chapter 6 we exploit similarities in the fluctuation formulae for the mechanical and the thermal response, by deriving computer simulation algorithms for calculating the linear Navier–Stokes transport coefficients. Although the algorithms are designed to calculate linear thermal transport coefficients, they employ mechanical methods. The validity of these algorithms is proved using thermostatted linear response theory (Chapter 5) and the knowledge of the Green–Kubo relations provided in Chapter 4.

A diagrammatic summary of some of the common algorithms used to compute shear viscosity, is given in Fig. 1.1. The Green–Kubo method simply consists of simulating an equilibrium field under periodic boundary conditions and making the appropriate analysis of the time dependent stress fluctuations using (1.2). Gosling *et al.* (1973) proposed performing a nonequilibrium simulation of a system subject to a sinusoidal transverse force. The viscosity could be calculated by monitoring the field induced velocity profile and extrapolating the results to infinite wavelength. In 1973, Hoover and Ashurst (1975) used external reservoirs of particles to induce a nearly planar shear in a model fluid. In the reservoir technique the viscosity is calculated by measuring the average ratio of the shear stress to the strain rate, in the bulk of the fluid, away from the reservoir regions. The presence of the reservoir regions gives rise to significant inhomogeneities in the thermodynamic properties of the fluid and in the strain rate in particular. This leads to obvious difficulties in the calculation of the shear viscosity. Lees and Edwards (1972) showed that if one used 'sliding brick' periodic boundary conditions one could induce *planar* Couette flow in a simulation. The so-called Lees–Edwards periodic boundary conditions enable one to perform homogeneous simulations of shear flow in which the low-Reynolds-number velocity profile is linear.

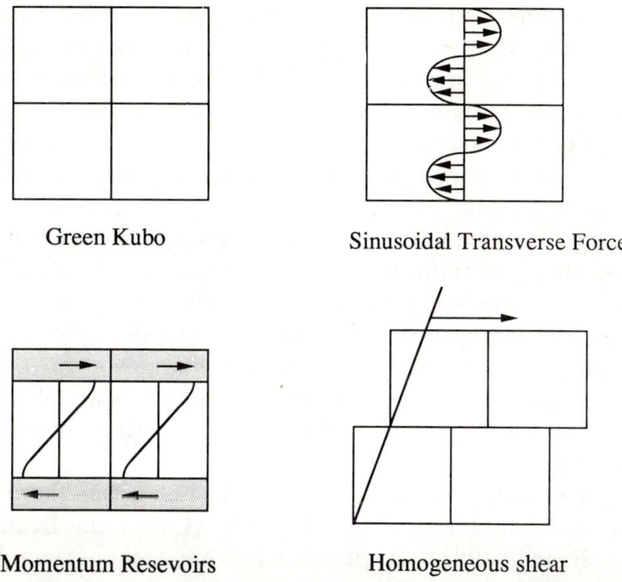

Fig. 1.1 Methods for determining shear viscosity.

With the exception of the Green–Kubo method, these simulation methods all involve nonequilibrium simulations. The Green–Kubo technique is useful in that all linear transport coefficients can, in principle, be calculated from a single simulation. However, the method is restricted to the calculation of linear transport coefficients only, whilst the nonequilibrium methods provide information about the nonlinear as well as the linear response of systems and therefore provide a direct link with rheology.

The use of nonequilibrium computer simulation algorithms, so-called nonequilibrium molecular dynamics (NEMD), leads inevitably to the question of the large field, nonlinear response. Indeed the calculation of linear transport coefficients using NEMD proceeds by calculating the nonlinear response and extrapolating the results to zero field. One of our main aims will be to derive a number of nonlinear generalizations of the Kubo relations which give an exact framework within which one can calculate and characterize transport processes far from equilibrium (Chapters 7 and 8). Because of the divergences alluded to above, the nonlinear theory cannot rely on power-series expansions about the equilibrium state. A major system of interest is the nonequilibrium steady state. Theory enables one to relate the nonlinear transport coefficients and mechanical quantities like the internal energy or the pressure, to transient fluctuations in the thermodynamic flux which generates the nonequilibrium steady state (Chapter 7). We derive the transient time correlation function (TTCF; Section 7.3) and the Kawasaki representations (Section 7.2) of the thermostatted nonlinear response. These results are exact and do not require the nonlinear response to be an analytic function of the perturbing fields. The theory also enables one to calculate specific heats, thermal-expansion coefficients and compressibilities from a knowledge of steady-state fluctuations (Chapter 9). After we have discussed the nonlinear response, we present a resolution of the van Kampen objection to linear response theory and to the Kubo relations in Chapter 7.

An innovation in our theory is the use of reversible equations of motion which incorporate a deterministic thermostat (Section 3.1). This innovation was motivated by the needs imposed by nonequilibrium computer simulation. If one wants to use any of the nonequilibrium methods depicted in Fig. 1.1. to calculate the shear viscosity, one needs a thermostat so that reliable steady-state averages can be accumulated. It is not clear how one could calculate the viscosity of a fluid in which the temperature and pressure are increasing in time.

The first deterministic thermostat, the so-called Gaussian thermostat, was independently and simultaneously developed by Hoover and Evans (Hoover *et al.*, 1982; Evans, 1983). It permitted homogeneous simulations of nonequilibrium steady states using molecular dynamics techniques. Hitherto, molecular dynamics had involved solving Newton's equations for systems of interacting particles. If work was performed on such a system in order to drive it away from equilibrium the system inevitably heated up due to the irreversible conversion of work into heat.

Hoover and Evans showed that if such a system evolved under their thermostatted equations of motion, the so-called Gaussian isokinetic equations of motion, the dissipative heat could be removed by a thermostatting force which is part of the equations of motion themselves. Now, computer simulators had been simulating nonequilibrium steady states for some time but in the past the dissipative heat was removed by simple *ad hoc* rescaling of the second moment of the appropriate velocity. The significance of the Gaussian isokinetic equations of motion was that since the thermostatting was part of the equations of motion it could be analysed theoretically using response theory. Earlier *ad hoc* rescaling or Andersen's stochastic thermostat (Andersen, 1980) could not be so easily analysed. In Chapter 5 we prove that, while the adiabatic (i.e. unthermostatted) linear response of a system can be calculated as the integral of an unthermostatted (i.e. Newtonian) equilibrium time correlation function, the thermostatted linear response is related to the corresponding thermostatted equilibrium time correlation function. These results are quite new and can be proved only because the thermostatting mechanism is reversible and deterministic.

One may ask whether one can talk about the 'thermostatted' response without referring to the details of the thermostatting mechanism. Provided the amount of heat Q, removed by a thermostat within the characteristic microscopic relaxation time of the system τ, is small compared with the enthalpy I, of the fluid (i.e. $(\tau\, \mathrm{d}Q/\mathrm{d}t)/I < 1$), we expect that the microscopic details of the thermostat will be unimportant. In the linear regime close to equilibrium this will always be the case. Even for systems far (but not too far) from equilibrium this condition is often satisfied. In Section 5.4 we give a mathematical proof of the independence of the linear response to the thermostatting mechanism.

Although originally motivated by the needs of nonequilibrium simulations, we have now reached the point where we can simulate equilibrium systems at constant internal energy E, at constant enthalpy I, or at constant temperature T, and pressure p. If we employ the so-called Nosé–Hoover (Hoover, 1985) thermostat, we can allow fluctuations in the state defining variables while controlling their mean values. These methods have had a major impact on computer simulation methodology and practice.

To illustrate the point: in an ergodic system at equilibrium, Newton's equations of motion *generate* the molecular dynamics ensemble in which the number of particles, the total energy, the volume and the total linear momentum are all precisely fixed (N, E, V, $\Sigma\mathbf{p}_i$). Previously this was the only equilibrium ensemble accessible to molecular dynamics simulation. Now, however, we can use Gaussian methods to generate equilibrium ensembles in which the precise value of, say, the enthalpy and pressure are fixed ($N, I, p, \Sigma\mathbf{p}_i$). Alternatively, Nosé–Hoover equations of motion can be used to generate the canonical ensemble ($\mathrm{e}^{-\beta H}$). Gibbs proposed the various ensembles as statistical *distributions* in phase space. In this book we will describe *dynamics* that are capable of generating each of these distributions.

A new element in the theory of nonequilibrium steady states is the abandonment of Hamiltonian dynamics. The Hamiltonian of course plays a central role in Gibbs' equilibrium statistical mechanics. It leads to a compact and elegant description. However, the existence of a Hamiltonian which generates dynamical trajectories is, as we will see, not essential.

In the space of relevant variables, neither the Gaussian thermostatted equations of motion nor the Nosé–Hoover equations of motion can be derived from a Hamiltonian. This is true even in the absence of external perturbing fields. This implies in turn that the usual form of the Liouville equation, $\mathrm{d}f/\mathrm{d}t = 0$, for the N-particle distribution function f, is invalid. Thermostatted equations of motion necessarily imply a compressible phase space.

The abandonment of a Hamiltonian approach to particle dynamics was in fact forced on us somewhat earlier. The Evans–Gillan equations of motion for heat flow (Section 6.5), which predate both the Gaussian and Nosé–Hoover thermostatted dynamics, cannot be derived from a Hamiltonian. The

Evans–Gillan equations provide the most efficient presently known dynamics for describing heat flow in systems close to equilibrium. A synthetic external field was invented so that its interaction with an N-particle system precisely mimics the impact a real temperature gradient would have on the system. Linear response theory is then used to prove that the response of a system to a real temperature gradient is identical to the response to the synthetic Evans–Gillan external field.

We use the term *synthetic* to note the fact that the Evans–Gillan field does not exist in nature. It is a mathematical device used to transform a difficult boundary condition problem, the flow of heat in a system bounded by walls maintained at differing temperatures, into a much simpler mechanical problem. The Evans–Gillan field acts upon the system in a homogeneous way, permitting the use of periodic rather than inhomogeneous boundary conditions. This synthetic field exerts a force on each particle which is proportional to the difference of the particle's enthalpy from the mean enthalpy per particle. The field thereby induces a flow of heat in the absence of either a temperature gradient or of any mass flow. No Hamiltonian is known which can generate the resulting equations of motion.

In a similar way Kawasaki showed that the boundary condition which corresponds to planar Couette shear flow can be incorporated exactly into the equations of motion. These equations are known as the SLLOD equations (Section 6.3); they give an exact description of the shearing motion of systems arbitrarily far from equilibrium. Again no Hamiltonian can be found which is capable of generating these equations.

When external fields or boundary conditions perform work on a system we have at our disposal a very natural set of mechanisms for constructing nonequilibrium ensembles in which different sets of thermodynamic state variables are used to constrain or define the system. Thus nonequilibrium analogues of the canonical, microcanonical or isobaric–isoenthalpic ensembles can be generated on a computer or analysed theoretically.

At equilibrium one is used to the idea of pairs of conjugate thermodynamic variables generating conjugate equilibrium ensembles. In the canonical ensemble particle number N, volume V, and temperature T, are the state variables, whereas in the isothermal–isobaric ensemble the role played by the volume is replaced by the pressure, its thermodynamic conjugate. In the same sense one can generate conjugate pairs of nonequilibrium ensembles. If the driving thermodynamic force is X, it could be a temperature gradient or a strain rate, then one could consider the N, V, T, X ensemble or, alternatively, the conjugate N, p, T, X ensemble.

However, in nonequilibrium steady states one can go much further than this. The dissipation, the heat removed by the thermostat per unit time,

$\mathrm{d}Q/\mathrm{d}t$, can always be written as a product of a thermodynamic force, X, and a thermodynamic flux, $J(\Gamma)$. If, for example, the force is the strain rate, γ, then the conjugate flux is the shear stress, $-P_{xy}$. One can then consider nonequilibrium ensembles in which the thermodynamic flux rather than the thermodynamic force is the independent state variable. For example, we could define the nonequilibrium steady state as an N, V, T, J ensemble. Such an ensemble is, by analogy with electric-circuit theory, called a Norton ensemble, while the case where the force is the state variable, N, V, T, X, is called a Thévenin ensemble. A major postulate in this work is the macroscopic equivalence of corresponding Norton and Thévenin ensembles.

The Kubo relations referred to above only pertain to the Thévenin ensembles. In Section 6.6 we discuss the Norton ensemble analogues to the Kubo relations and show how deep the duality between the two types of ensemble extends. The generalization of Norton ensemble methods to the nonlinear response leads, for the first time, to analytic expressions for the nonlinear Burnett coefficients. The nonlinear Burnett coefficients are simply the coefficients of a Taylor-series expansion, about equilibrium, of a thermodynamic flux in powers of the thermodynamic force. For Navier–Stokes processes, these coefficients are expected to diverge. However, since until recently no explicit expressions were known for the Burnett coefficients, simulation studies of this possible divergence were severely handicapped. In Chapter 9 we discuss Evans and Lynden-Bell's (1988) derivation of, equilibrium time correlation functions for the inverse Burnett coefficients. The inverse Burnett coefficients are so called because they refer to the coefficients of the expansion of the forces in terms of the thermodynamic fluxes rather than vice versa.

In the final chapter we introduce material which is quite recent and perhaps controversial. We attempt to characterize the phase-space distribution of nonequilibrium steady states. This is essential if we are to be able to develop a thermodynamics of nonequilibrium steady states. Presumably such a thermodynamics, a nonlinear generalization of the conventional linear irreversible thermodynamics treated in Chapter 2, will require the calculation of a generalized entropy. The entropy and free energies are functionals of the distribution function and thus are vastly more complex to calculate than nonequilibrium averages.

What we find is surprising. The steady-state nonequilibrium distribution function seen in NEMD simulations, is a fractal object. There is now ample evidence that the *dimension* of the phase space which is accessible to nonequilibrium steady states is *lower* than the dimension of phase space itself. This means that the volume of accessible phase space, as calculated from the ostensible phase space, is zero. This implies that the fine-grained entropy calculated from Gibbs' relation,

$$S = -k_{\mathrm{B}} \int_{\text{all } \Gamma \text{ space}} \mathrm{d}\boldsymbol{\Gamma}\, f(\boldsymbol{\Gamma}, t) \ln(f(\boldsymbol{\Gamma}, t)) \tag{1.3}$$

diverges to negative infinity. (If no thermostat is employed the corresponding nonequilibrium entropy is, as was known to Gibbs (1902), a constant of the motion!) Presumably, the thermodynamic entropy, if it exists, must be computed from *within* the lower dimensional, accessible phase space rather than from the full phase space as in (1.3). We close the book by describing a new method for computing the nonequilibrium entropy.

REFERENCES

Alder, B.J. and Wainwright, T.E. (1958). *Proceedings of the International Symposium on Statistical Mechanical Theory of Transport Processes* (Brussels, 1956).

Andersen, H.C. (1980). *J. Chem. Phys.*, **72**, 2384.

Barker, J.A. and Henderson, D. (1976). *Rev. Mod. Phys.*, **48**, 587.

Cohen, E.G.D. and Dorfman, J.R. (1965). *Phys. Rev. Lett.*, **16**, 124.

Cohen, E.G.D. and Dorfman, J.R. (1972). *Phys. Rev., Ser. A*, **6**, 776.

Evans, D.J. (1983). *J. Chem. Phys.*, **78**, 3297.

Evans, D.J. and Lynden-Bell, R.M. (1988). *Phys. Rev. Ser. A*. **38**, 5249.

Evans, D.J. and Morriss, G.P. (1983). *Phys. Rev. Lett.*, **51**, 1776.

Ferziger, J.H. and Kaper, H.G. (1972). *Mathematical Theory of Transport Processes in Gases*. Amsterdam: North-Holland.

Gibbs, J.W. (1902). *Elementary Principles in Statistical Mechanics*. New Haven, Conn., USA: Yale University Press.

Gosling, E.M., McDonald, I.R. and Singer, K. (1973). *Mol. Phys.*, **26**, 1475.

Hoover, W.G. (1985). *Phys. Rev., Ser. A*, **31**, 1695.

Hoover, W.G. and Ashurst, W.T. (1975). *Adv. Theor. Chem.*, **1**, 1.

Hoover, W.G., Ladd, A.J.C. and Moran, B. (1982). *Phys. Rev. Lett.*, **48**, 1818.

Kawasaki, K. and Gunton, J.D. (1973). *Phys. Rev., Ser. A*, **8**, 2048.

Kubo, R. (1957). *J. Phys. Soc. Jpn.*, **12**, 570.

Lees, A.W. and Edwards, S.F. (1972). *J. Phys., Ser. C*, **5**, 1921.

Pomeau, Y. and Resibois, P. (1975). *Phys. Rep.*, **19**, 63.

Rowlinson, J.S. and Widom, B. (1982). *Molecular Theory of Capillarity*. Oxford: Clarendon Press.

Zwanzig, R. (1982). Remark made at conference on *Nonlinear Fluid Behaviour*, Boulder Colorado, June, 1982.

2 Linear Irreversible Thermodynamics

2.1 THE CONSERVATION EQUATIONS

At the hydrodynamic level we are interested in the macroscopic evolution of densities of conserved extensive variables such as mass, energy and momentum. Because these quantities are conserved, their respective densities can only change by a process of *redistribution*. As we shall see, this means that the relaxation of these densities is slow and, therefore, the relaxation plays a macroscopic role. If this relaxation were fast (i.e. if it occurred on a molecular time-scale for instance) it would be unobservable at a macroscopic level. The macroscopic equations of motion for the densities of conserved quantities are called the Navier–Stokes equations. We will now give a brief description of how these equations are derived. It is important to understand this derivation because one of the objects of statistical mechanics is to provide a microscopic or molecular justification for the Navier–Stokes equations. In the process, statistical mechanics sheds light on the limits of applicability of these equations. Similar treatments can be found in de Groot and Mazur (1962) and Kreuzer (1981).

Let $M(t)$ be the total mass contained in an arbitrary volume V, then

$$M = \int_V \mathrm{d}\mathbf{r}\, \rho(\mathbf{r}, t) \tag{2.1}$$

where $\rho(\mathbf{r}, t)$ is the mass density at position $\mathbf{r}$ and time t. Since mass is conserved, the only way that the mass in the volume V can change is by flowing through the enclosing surface, S (see Fig. 2.1),

$$\begin{aligned} \frac{\mathrm{d}M}{\mathrm{d}t} &= -\int_S \mathrm{d}\mathbf{S}\cdot \rho(\mathbf{r}, t)\mathbf{u}(\mathbf{r}, t) \\ &= -\int_V \mathrm{d}\mathbf{r}\, \nabla\cdot [\rho(\mathbf{r}, t)\mathbf{u}(\mathbf{r}, t)] \end{aligned} \tag{2.2}$$

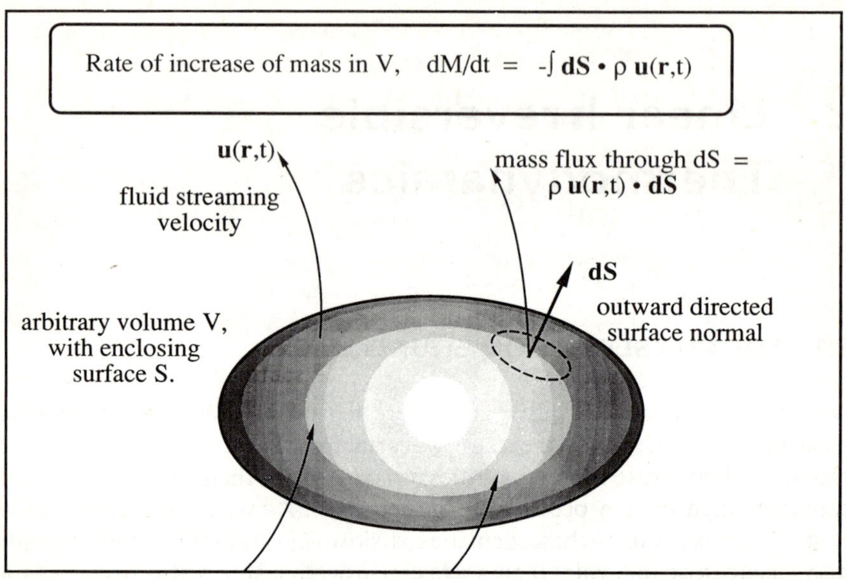

Fig. 2.1 The change in the mass contained in an arbitrary closed volume V can be calculated by integrating the mass flux through the enclosing surface S: $\mathrm{d}M/\mathrm{d}t = -\int \mathrm{d}\mathbf{S}\cdot\rho\mathbf{u}(\mathbf{r},t)$.

Here $\mathbf{u}(\mathbf{r},t)$ is the fluid streaming velocity at position $\mathbf{r}$ and time t. $\mathrm{d}\mathbf{S}$ denotes an area element of the enclosing surface S, and ∇ is the spatial gradient vector operator, $(\partial/\partial x, \partial/\partial y, \partial/\partial z)$. It is clear that the rate of change of the enclosed mass can also be written in terms of the change in mass density $\rho(\mathbf{r},t)$, as

$$\frac{\mathrm{d}M}{\mathrm{d}t} = \int_V \mathrm{d}\mathbf{r}\,\frac{\partial\rho(\mathbf{r},t)}{\partial t} \tag{2.3}$$

If we equate these two expressions for the rate of change of the total mass we find that, since the volume V is arbitrary,

$$\frac{\partial\rho(\mathbf{r},t)}{\partial t} = -\nabla\cdot[\rho(\mathbf{r},t)\mathbf{u}(\mathbf{r},t)] \tag{2.4}$$

This is called the mass-continuity equation and is essentially a statement that mass is conserved. We can write the mass-continuity equation in an alternative form if we use the relation between the total or streaming derivative, and the various partial derivatives. For an arbitrary function of position $\mathbf{r}$ and time t, for example $a(\mathbf{r},t)$, we have

$$\frac{\mathrm{d}}{\mathrm{d}t}a(\mathbf{r},t) = \frac{\partial}{\partial t}a(\mathbf{r},t) + \mathbf{u}\cdot\nabla a(\mathbf{r},t) \tag{2.5}$$

If we let $a(\mathbf{r}, t) \equiv \rho(\mathbf{r}, t)$ in (2.5) and combine this with (2.4), then the mass-continuity equation can be written as

$$\frac{d\rho(\mathbf{r}, t)}{dt} = -\rho(\mathbf{r}, t)\nabla \cdot \mathbf{u}(\mathbf{r}, t) \tag{2.6}$$

In an entirely analogous fashion we can derive an equation of continuity for momentum. Let $\mathbf{G}(t)$ be the total momentum of the arbitrary volume V, then the rate of change of momentum is given by

$$\frac{d\mathbf{G}}{dt} = \int_V d\mathbf{r} \frac{\partial[\rho(\mathbf{r}, t)\mathbf{u}(\mathbf{r}, t)]}{\partial t} \tag{2.7}$$

The total momentum of volume V can change in two ways. Firstly, it can change by convection. Momentum can flow through the enclosing surface. This convective term can be written as,

$$\frac{d\mathbf{G}_c}{dt} = -\int_S d\mathbf{S} \cdot \rho(\mathbf{r}, t)\mathbf{u}(\mathbf{r}, t)\mathbf{u}(\mathbf{r}, t) \tag{2.8}$$

The second way that the momentum could change is by the pressure exerted on V by the surrounding fluid. We call this contribution the stress contribution. The force $d\mathbf{F}$, exerted by the fluid across an elementary area $d\mathbf{S}$, which is moving with the streaming velocity of the fluid, must be proportional to the magnitude of the area $d\mathbf{S}$. The most general such linear relation is,

$$d\mathbf{F} \equiv -d\mathbf{S} \cdot \mathsf{P} \tag{2.9}$$

This is in fact the definition of the pressure tensor P. It is also the negative of the stress tensor. That the pressure tensor is a second-rank tensor rather than a simple scalar, is a reflection of the fact that the force $d\mathbf{F}$, and the area vector $d\mathbf{S}$, need not be parallel. For molecular fluids the pressure tensor is not symmetric in general.

As P is the first tensorial quantity that we have introduced, it is appropriate to define the notational conventions that we will use. P is a second-rank tensor and thus requires two subscripts to specify its elements. In Einstein notation, equation (2.9) reads $dF_\alpha = -dS_\beta P_{\beta\alpha}$, where the first repeated index β implies a summation. Note that the contraction (or dot product) involves the first index of P and that the vector character of the force $d\mathbf{F}$ is determined by the second index of P. We use bold sans serif characters to denote tensors of rank two or greater. Figure 2.2 gives a diagrammatic representation of the tensorial relations in the definition of the pressure tensor.

Using this definition, the stress contribution to the momentum change can be shown to be

$$\frac{d\mathbf{G}_s}{dt} = -\int_S d\mathbf{S} \cdot \mathsf{P} \tag{2.10}$$

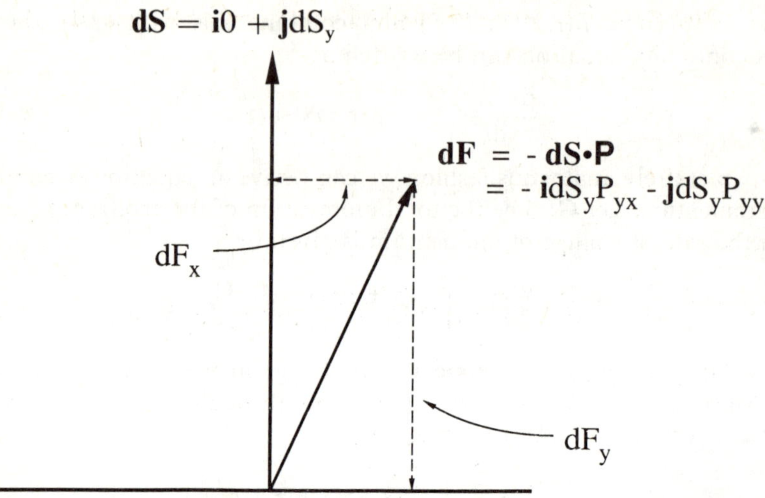

Fig. 2.2 Definition of the pressure tensor. The pressure tensor **P** is defined in terms of the infinitesimal force d**F**, across an infinitesimal area d**S**, for an element which is moving with the streaming fluid velocity.

Combining (2.8) and (2.10) and using the divergence theorem to convert surface integrals to volume integrals gives

$$\frac{d\mathbf{G}}{dt} = \int_V d\mathbf{r} \frac{\partial[\rho(\mathbf{r},t)\mathbf{u}(\mathbf{r},t)]}{\partial t} = -\int_V d\mathbf{r}\, \nabla\cdot[\rho(\mathbf{r},t)\mathbf{u}(\mathbf{r},t)\mathbf{u}(\mathbf{r},t) + \mathbf{P}] \tag{2.11}$$

Since this equation is true for arbitrary V, we conclude that

$$\frac{\partial[\rho(\mathbf{r},t)\mathbf{u}(\mathbf{r},t)]}{\partial t} = -\nabla\cdot[\rho(\mathbf{r},t)\mathbf{u}(\mathbf{r},t)\mathbf{u}(\mathbf{r},t) + \mathbf{P}] \tag{2.12}$$

This is one form of the momentum continuity equation. A simpler form can be obtained by using streaming derivatives of the velocity rather than partial derivatives. Using the chain rule, the left-hand side of (2.12) can be expanded as

$$\rho(\mathbf{r},t)\frac{\partial\mathbf{u}(\mathbf{r},t)}{\partial t} + \mathbf{u}(\mathbf{r},t)\frac{\partial\rho(\mathbf{r},t)}{\partial t} = -\nabla\cdot[\rho(\mathbf{r},t)\mathbf{u}(\mathbf{r},t)\mathbf{u}(\mathbf{r},t) + \mathbf{P}] \tag{2.13}$$

Using the vector identity

$$\nabla\cdot(\rho\mathbf{u}\mathbf{u}) = \mathbf{u}(\nabla\cdot\rho\mathbf{u}) + \rho\mathbf{u}\cdot\nabla\mathbf{u}$$

and the mass continuity equation (2.4), equation (2.13) becomes

$$\begin{aligned}\rho(\mathbf{r},t)\frac{\partial \mathbf{u}(\mathbf{r},t)}{\partial t} &= -\nabla\cdot[\rho(\mathbf{r},t)\mathbf{u}(\mathbf{r},t)\mathbf{u}(\mathbf{r},t)] + \mathbf{u}(\mathbf{r},t)\nabla\cdot[\rho(\mathbf{r},t)\mathbf{u}(\mathbf{r},t)] - \nabla\cdot\mathbf{P} \\ &= -\rho(\mathbf{r},t)\mathbf{u}(\mathbf{r},t)\cdot\nabla\mathbf{u}(\mathbf{r},t) - \nabla\cdot\mathbf{P}\end{aligned} \tag{2.14}$$

Now,

$$\rho(\mathbf{r},t)\frac{\mathrm{d}\mathbf{u}(\mathbf{r},t)}{\mathrm{d}t} = \rho(\mathbf{r},t)\frac{\partial \mathbf{u}(\mathbf{r},t)}{\partial t} + \rho(\mathbf{r},t)\mathbf{u}(\mathbf{r},t)\cdot\nabla\mathbf{u}(\mathbf{r},t) \tag{2.15}$$

so that (2.14) can be written as

$$\rho(\mathbf{r},t)\frac{\mathrm{d}\mathbf{u}(\mathbf{r},t)}{\mathrm{d}t} = -\nabla\cdot\mathbf{P} \tag{2.16}$$

The final conservation equation we will derive is the energy equation. If we denote the total energy per unit mass or the specific total energy as $e(\mathbf{r},t)$, then the total energy density is $\rho(\mathbf{r},t)e(\mathbf{r},t)$. If the fluid is convecting there is obviously a simple convective kinetic-energy component in $e(\mathbf{r},t)$. If this is removed from the energy density then what remains should be a thermodynamic internal energy density, $\rho(\mathbf{r},t)U(\mathbf{r},t)$.

$$\rho(\mathbf{r},t)e(\mathbf{r},t) = \rho(\mathbf{r},t)\frac{\mathbf{u}(\mathbf{r},t)^2}{2} + \rho(\mathbf{r},t)U(\mathbf{r},t) \tag{2.17}$$

Here we have identified the first term on the right-hand side as the convective kinetic energy. Using (2.16) we can show that,

$$\begin{aligned}\rho(\mathbf{r},t)\frac{\mathrm{d}}{\mathrm{d}t}\frac{\mathbf{u}(\mathbf{r},t)^2}{2} &= \rho(\mathbf{r},t)\mathbf{u}(\mathbf{r},t)\cdot\frac{\mathrm{d}\mathbf{u}(\mathbf{r},t)}{\mathrm{d}t} \\ &= -\mathbf{u}(\mathbf{r},t)\cdot[\nabla\cdot\mathbf{P}] = -\mathbf{u}\nabla:\mathbf{P}\end{aligned} \tag{2.18}$$

The second equality is a consequence of the momentum conservation equation (2.16). In this equation we use the dyadic product of two first-rank tensors (or ordinary vectors) $\mathbf{u}$ and ∇ to obtain a second-rank tensor $\mathbf{u}\nabla$. In Einstein notation $(\mathbf{u}\nabla)_{\alpha\beta} \equiv \mathbf{u}_\alpha\nabla_\beta$. In the first form given in equation (2.18) ∇ is contracted into the first index of $\mathbf{P}$, and then $\mathbf{u}$ is contracted into the second remaining index. This defines the meaning of the double contraction notation after the second equals sign in (2.18)—inner indices are contracted first, then outer indices—that is $\mathbf{u}\nabla:\mathbf{P} \equiv (\mathbf{u}\nabla)_{\alpha\beta}\mathbf{P}_{\beta\alpha} \equiv \mathbf{u}_\alpha\nabla_\beta\mathbf{P}_{\beta\alpha}$.

For any variable a, using equation (2.5) we have

$$\rho(\mathbf{r},t)\frac{\mathrm{d}a(\mathbf{r},t)}{\mathrm{d}t} = \rho(\mathbf{r},t)\frac{\partial a(\mathbf{r},t)}{\partial t} + \rho(\mathbf{r},t)\mathbf{u}(\mathbf{r},t)\cdot\nabla a(\mathbf{r},t)$$
$$= \frac{\partial[\rho(\mathbf{r},t)a(\mathbf{r},t)]}{\partial t} + \rho(\mathbf{r},t)\mathbf{u}(\mathbf{r},t)\cdot\nabla a(\mathbf{r},t) - a(\mathbf{r},t)\frac{\partial\rho(\mathbf{r},t)}{\partial t} \tag{2.19}$$

Using the mass-continuity equation (2.4)

$$\rho(\mathbf{r},t)\frac{\mathrm{d}a(\mathbf{r},t)}{\mathrm{d}t} = \frac{\partial[\rho(\mathbf{r},t)a(\mathbf{r},t)]}{\partial t} + \rho(\mathbf{r},t)\mathbf{u}(\mathbf{r},t)\cdot\nabla a(\mathbf{r},t)$$
$$+ a(\mathbf{r},t)\nabla\cdot[\rho(\mathbf{r},t)\mathbf{u}(\mathbf{r},t)]$$
$$= \frac{\partial[\rho(\mathbf{r},t)a(\mathbf{r},t)]}{\partial t} + \nabla\cdot[\rho(\mathbf{r},t)\mathbf{u}(\mathbf{r},t)a(\mathbf{r},t)] \tag{2.20}$$

If we let the total energy inside a volume V be E, then clearly,

$$\frac{\mathrm{d}E}{\mathrm{d}t} = \int_V \mathrm{d}\mathbf{r}\,\frac{\partial[\rho(\mathbf{r},t)e(\mathbf{r},t)]}{\partial t} \tag{2.21}$$

Because the energy is conserved we can make a detailed account of the energy balance in the volume V. The energy can simply convect through the containing surface; it could diffuse through the surface and the surface stresses could do work on the volume V. In order, these terms can be written

$$\frac{\mathrm{d}E}{\mathrm{d}t} = -\int_S \mathrm{d}\mathbf{S}\cdot[\rho(\mathbf{r},t)e(\mathbf{r},t)\mathbf{u}(\mathbf{r},t) + \mathbf{J}_Q] + \int_S (\mathrm{d}\mathbf{S}\cdot\mathbf{P}(\mathbf{r},t))\cdot\mathbf{u}(\mathbf{r},t) \tag{2.22}$$

In (2.22) $\mathbf{J}_Q$ is called the heat-flux vector. It gives the energy flux across a surface which is moving with the local fluid streaming velocity. Using the divergence theorem, (2.22) can be rewritten as,

$$\frac{\mathrm{d}E}{\mathrm{d}t} = -\int_V \mathrm{d}\mathbf{r}\,\nabla\cdot[\rho(\mathbf{r},t)e(\mathbf{r},t)\mathbf{u}(\mathbf{r},t) + \mathbf{J}_Q(\mathbf{r},t) + \mathbf{P}(\mathbf{r},t)\cdot\mathbf{u}(\mathbf{r},t)] \tag{2.23}$$

Comparing equations (2.21) and (2.23) we derive the continuity equation for total energy,

$$\frac{\partial[\rho(\mathbf{r},t)e(\mathbf{r},t)]}{\partial t} = -\nabla\cdot[\rho(\mathbf{r},t)e(\mathbf{r},t)\mathbf{u}(\mathbf{r},t) + \mathbf{J}_Q(\mathbf{r},t) + \mathbf{P}(\mathbf{r},t)\cdot\mathbf{u}(\mathbf{r},t)] \tag{2.24}$$

We can use (2.20) to express this equation in terms of streaming derivatives of the total specific energy

$$\rho(\mathbf{r},t)\frac{\mathrm{d}e(\mathbf{r},t)}{\mathrm{d}t} = -\nabla\cdot[\mathbf{J}_Q(\mathbf{r},t) + \mathbf{P}(\mathbf{r},t)\cdot\mathbf{u}(\mathbf{r},t)] \tag{2.25}$$

Finally, (2.17) and (2.18) can be used to derive a continuity equation for the specific internal energy

$$\rho(\mathbf{r},t)\frac{\mathrm{d}U(\mathbf{r},t)}{\mathrm{d}t} = -\nabla\cdot\mathbf{J}_Q(\mathbf{r},t) - \mathbf{P}(\mathbf{r},t)^{\mathrm{T}}:\nabla\mathbf{u}(\mathbf{r},t) \tag{2.26}$$

where the superscript T denotes transpose. The transpose of the pressure tensor appears as a result of our double-contraction notation, because in (2.25) ∇ is contracted into the first index of $\mathbf{P}$.

The three continuity equations, (2.6), (2.16) and (2.26), are continuum expressions of the fact that mass, momentum and energy are conserved. These equations are exact.

2.2 ENTROPY PRODUCTION

Thus far our description of the equations of hydrodynamics has been exact. We will now derive an equation for the rate at which entropy is produced spontaneously in a nonequilibrium system. The second law of thermodynamics states that entropy is not a conserved quantity. In order to complete this derivation we must assume that we can apply the laws of equilibrium thermodynamics, at least on a local scale, to nonequilibrium systems. This assumption is called the *local thermodynamic equilibrium postulate.* We expect that this postulate should be valid for systems that are sufficiently close to equilibrium (de Groot and Mazur, 1962). This macroscopic theory provides no information on how *small* these deviations from equilibrium should be in order for local thermodynamic equilibrium to hold. It turns out, however, that the local thermodynamic equilibrium postulate is satisfied for a wide variety of systems over a wide range of conditions. One obvious condition that must be met is that the characteristic distances over which inhomogeneities in the nonequilibrium system occur must be large in terms of molecular dimensions. If this is not the case then the thermodynamic state variables will change so rapidly in space that a local thermodynamic state cannot be defined. Similarly, the time-scale for nonequilibrium change in the system must be large compared with the time-scales required for the attainment of local equilibrium.

We let the entropy per unit mass be denoted as, $s(\mathbf{r}, t)$ and the entropy of an arbitrary volume V, be denoted by S. Clearly,

$$\frac{\mathrm{d}S}{\mathrm{d}t} = \int_V \mathrm{d}\mathbf{r} \frac{\partial[\rho(\mathbf{r}, t)s(\mathbf{r}, t)]}{\partial t} \tag{2.27}$$

In contrast to the derivations of the conservation laws we do not expect that by taking account of convection and diffusion, we can totally account for the entropy of the system. The *increase* in the entropy is what we are seeking to calculate. We shall call the entropy produced per unit time per unit volume, the entropy source strength, $\sigma(\mathbf{r}, t)$.

$$\frac{\mathrm{d}S}{\mathrm{d}t} = \int_V \mathrm{d}\mathbf{r}\, \sigma(\mathbf{r}, t) - \int_S \mathrm{d}\mathbf{S} \cdot \mathbf{J}_{ST}(\mathbf{r}, t) \tag{2.28}$$

In this equation $\mathbf{J}_{ST}(\mathbf{r}, t)$ is the total entropy flux. As before, we use the divergence theorem and the arbitrariness of V to calculate

$$\frac{\partial[\rho(\mathbf{r}, t)s(\mathbf{r}, t)]}{\partial t} = \sigma(\mathbf{r}, t) - \nabla \cdot \mathbf{J}_{ST}(\mathbf{r}, t) \tag{2.29}$$

We can decompose $\mathbf{J}_{ST}(\mathbf{r}, t)$ into a streaming or convective term $\rho(\mathbf{r}, t)s(\mathbf{r}, t)\mathbf{u}(\mathbf{r}, t)$ in analogy with equation (2.8), and a diffusive term $\mathbf{J}_S(\mathbf{r}, t)$. Using these terms, (2.29) can be written as

$$\frac{\partial[\rho(\mathbf{r}, t)s(\mathbf{r}, t)]}{\partial t} = \sigma(\mathbf{r}, t) - \nabla \cdot [\mathbf{J}_S(\mathbf{r}, t) + \rho(\mathbf{r}, t)s(\mathbf{r}, t)\mathbf{u}(\mathbf{r}, t)] \tag{2.30}$$

Using (2.5) to convert to total time derivatives, we have

$$\rho(\mathbf{r}, t)\frac{\mathrm{d}s(\mathbf{r}, t)}{\mathrm{d}t} = \sigma(\mathbf{r}, t) - \nabla \cdot \mathbf{J}_S(\mathbf{r}, t) \tag{2.31}$$

At this stage we introduce the assumption of local thermodynamic equilibrium. We *postulate* a local version of the Gibbs relation $T\,\mathrm{d}S = \mathrm{d}U + p\,\mathrm{d}V$. Converting this relation to a local version with extensive quantities replaced by the specific entropy, energy and volume, respectively, and noting that the specific volume V/M is simply $\rho(\mathbf{r}, t)^{-1}$, we find that

$$\begin{aligned} T(\mathbf{r}, t)\frac{\mathrm{d}s(\mathbf{r}, t)}{\mathrm{d}t} &= \frac{\mathrm{d}U(\mathbf{r}, t)}{\mathrm{d}t} + p(\mathbf{r}, t)\frac{\mathrm{d}}{\mathrm{d}t}\rho(\mathbf{r}, t)^{-1} \\ &= \frac{\mathrm{d}U(\mathbf{r}, t)}{\mathrm{d}t} - \frac{p(\mathbf{r}, t)}{\rho(\mathbf{r}, t)^2}\frac{\mathrm{d}\rho(\mathbf{r}, t)}{\mathrm{d}t} \end{aligned} \tag{2.32}$$

We can now use the mass-continuity equation to eliminate the density derivative,

$$T(\mathbf{r},t)\frac{\mathrm{d}s(\mathbf{r},t)}{\mathrm{d}t} = \frac{\mathrm{d}U(\mathbf{r},t)}{\mathrm{d}t} + \frac{p(\mathbf{r},t)}{\rho(\mathbf{r},t)}\nabla\cdot\mathbf{u}(\mathbf{r},t) \tag{2.33}$$

Multiplying (2.33) by $\rho(\mathbf{r},t)$ and dividing by $T(\mathbf{r},t)$ gives

$$\rho(\mathbf{r},t)\frac{\mathrm{d}s(\mathbf{r},t)}{\mathrm{d}t} = \frac{\rho(\mathbf{r},t)}{T(\mathbf{r},t)}\frac{\mathrm{d}U(\mathbf{r},t)}{\mathrm{d}t} + \frac{p(\mathbf{r},t)}{T(\mathbf{r},t)}\nabla\cdot\mathbf{u}(\mathbf{r},t) \tag{2.34}$$

We can substitute the energy continuity expression (2.26) for $\rho\,\mathrm{d}U/\mathrm{d}t$ into (2.34) giving

$$\rho(\mathbf{r},t)\frac{\mathrm{d}s(\mathbf{r},t)}{\mathrm{d}t} = \frac{-1}{T(\mathbf{r},t)}[\nabla\cdot\mathbf{J}_Q(\mathbf{r},t) + \mathbf{P}(\mathbf{r},t)^{\mathrm{T}}:\nabla\mathbf{u}(\mathbf{r},t) - p(\mathbf{r},t)\nabla\cdot\mathbf{u}(\mathbf{r},t)] \tag{2.35}$$

We now have two expressions for the streaming derivative of the specific entropy, $\rho(\mathbf{r},t)\,\mathrm{d}s(\mathbf{r},t)/\mathrm{d}t$, equations (2.31) and (2.35). Using the time derivative of the local equilibrium postulate $\mathrm{d}Q = T\,\mathrm{d}S$, the diffusive entropy flux $\mathbf{J}_S(\mathbf{r},t)$, is equal to the heat flux divided by the absolute temperature and, therefore,

$$\nabla\cdot\mathbf{J}_S(\mathbf{r},t) = \nabla\cdot\left[\frac{\mathbf{J}_Q(\mathbf{r},t)}{T(\mathbf{r},t)}\right] = \frac{\nabla\cdot\mathbf{J}_Q(\mathbf{r},t)}{T(\mathbf{r},t)} - \frac{\mathbf{J}_Q(\mathbf{r},t)\cdot\nabla T(\mathbf{r},t)}{T(\mathbf{r},t)^2} \tag{2.36}$$

Equating (2.31) and (2.35) using (2.36) gives,

$$\begin{aligned}\sigma(\mathbf{r},t) &= \frac{-1}{T(\mathbf{r},t)}\left[\mathbf{P}(\mathbf{r},t)^{\mathrm{T}}:\nabla\mathbf{u}(\mathbf{r},t) - p(\mathbf{r},t)\nabla\cdot\mathbf{u}(\mathbf{r},t) + \frac{\mathbf{J}_Q(\mathbf{r},t)\cdot\nabla T(\mathbf{r},t)}{T(\mathbf{r},t)}\right]\\ &= -\frac{\mathbf{J}_Q(\mathbf{r},t)\cdot\nabla T(\mathbf{r},t)}{T(\mathbf{r},t)^2} - \frac{[\mathbf{P}(\mathbf{r},t)^{\mathrm{T}}:\nabla\mathbf{u}(\mathbf{r},t) - p(\mathbf{r},t)\nabla\cdot\mathbf{u}(\mathbf{r},t)]}{T(\mathbf{r},t)}\end{aligned} \tag{2.37}$$

We define the viscous pressure tensor $\mathbf{\Pi}^{(2)}$, as the nonequilibrium part of the pressure tensor.

$$\mathbf{\Pi}(\mathbf{r},t) = \mathbf{P}(\mathbf{r},t) - p(\mathbf{r},t)\mathbf{I} \tag{2.38}$$

Using this definition the entropy source strength can be written as

$$\sigma(\mathbf{r},t) = -\frac{1}{T(\mathbf{r},t)^2}\mathbf{J}_Q(\mathbf{r},t)\cdot\nabla T(\mathbf{r},t) - \frac{1}{T(\mathbf{r},t)}\mathbf{\Pi}(\mathbf{r},t)^{\mathrm{T}}:\nabla\mathbf{u}(\mathbf{r},t) \tag{2.39}$$

A second postulate of nonlinear irreversible thermodynamics is that the entropy source strength always takes the canonical form (de Groot and Mazur, 1962),

$$\sigma = \sum_i J_i X_i \tag{2.40}$$

This canonical form defines what are known as thermodynamic fluxes, J_i, and their conjugate thermodynamic forces, X_i. We can see immediately that our equation (2.39) takes this canonical form provided we make the identifications that: the thermodynamic fluxes are the various Cartesian elements of the heat-flux vector, $\mathbf{J}_Q(\mathbf{r}, t)$, and the viscous-pressure tensor, $\mathbf{\Pi}(\mathbf{r}, t)$. The thermodynamic forces conjugate to these fluxes are the *corresponding* Cartesian components of the temperature gradient divided by the square of the absolute temperature, $T(\mathbf{r}, t)^{-2}\nabla T(\mathbf{r}, t)$, and the strain-rate tensor divided by the absolute temperature, $T(\mathbf{r}, t)^{-1}\nabla\mathbf{u}(\mathbf{r}, t)$, respectively. We use the term *corresponding* quite deliberately; the αth element of the heat flux is conjugate to the αth element of the temperature gradient. There are no cross-couplings. Similarly, the α, β element of the viscous-pressure tensor is conjugate to the α, β element of the strain-rate tensor.

There is clearly some ambiguity in defining the thermodynamic fluxes and forces. There is no fundamental thermodynamic reason why we included the temperature factors, $T(\mathbf{r}, t)^{-2}$ and $T(\mathbf{r}, t)^{-1}$, into the forces rather than into the fluxes. Either choice is possible. Ours is simply one of convention. More importantly, there is no thermodynamic way of distinguishing between the fluxes and the forces. At a macroscopic level it is simply convention to identify the temperature gradient as a thermodynamic force rather than a flux. The canonical form for the entropy source strength and the associated postulates of irreversible thermodynamics do not permit a distinction to be made between what we should identify as fluxes and what should be identified as a force. Microscopically it is clear that the heat flux is a *flux*: it is the diffusive energy flow across a co-moving surface. At a macroscopic level, however, no such distinction can be made.

Perhaps the simplest example of this macroscopic duality is the Norton constant-current electrical circuit, and the Thevénin constant-voltage equivalent circuit. We can talk of the resistance of a circuit element or of a conductance. At a macroscopic level the choice is simply one of practical convenience or convention.

2.3 CURIE'S THEOREM

Consistent with our use of the local thermodynamic equilibrium postulate, which is assumed to be valid sufficiently close to equilibrium, a linear relation should hold between the conjugate thermodynamic fluxes and forces. We therefore postulate the existence of a set of linear phenomenological transport coefficients $\{L_{ij}\}$ which relate the set forces $\{X_j\}$ to the set of fluxes $\{J_i\}$. We use the term *phenomenological* to indicate that these transport coefficients are to be defined within the framework of linear irreversible thermodynamics and, as we shall see, there may be slight differences between the phenomenological transport coefficients L_{ij} and practical transport coefficients such as the viscosity coefficients or the usual thermal conductivity.

We postulate that *all* the thermodynamic forces appearing in the equation for the entropy source strength (2.40) are related to the various fluxes by a linear equation of the form

$$\mathbf{J}_i = \sum_i \mathbf{L}_{ij}\mathbf{X}_j \tag{2.41}$$

This equation could be thought of as arising from a Taylor-series expansion of the fluxes in terms of the forces. Such a Taylor series will only exist if the flux is an analytic function of the force at $\mathbf{X} = 0$.

$$\mathbf{J}_i(\mathbf{X}) = \mathbf{J}_i(0) + \sum_j \left.\frac{\partial \mathbf{J}_i}{\partial \mathbf{X}_j}\right|_{\mathbf{X}=0} \cdot \mathbf{X}_j + \sum_{j,k} \frac{1}{2!} \left.\frac{\partial^2 \mathbf{J}_i}{\partial \mathbf{X}_j \partial \mathbf{X}_k}\right|_{\mathbf{X}=0} : \mathbf{X}_j\mathbf{X}_k + O(\mathbf{X}^3) \tag{2.42}$$

Clearly the first term is zero as the fluxes vanish when the thermodynamic forces are zero. The term which is linear in the forces is evidently derivable, at least formally, from the equilibrium properties of the system as the functional derivative of the fluxes with respect to the forces computed at equilibrium, $\mathbf{X} = 0$. The quadratic term is related to what are known as the nonlinear Burnett coefficients (see Section 9.5); they represent nonlinear contributions to the linear theory of irreversible thermodynamics.

If we substitute the linear phenomenological relations into the equation for the entropy source strength (2.40), we find that

$$\sigma = \sum_{i,j} \mathbf{X}_i \cdot \mathbf{L}_{ij} \cdot \mathbf{X}_j \tag{2.43}$$

A postulate of linear irreversible thermodynamics is that the entropy source strength is always positive. There is always an increase in the entropy of a system so the transport coefficients are positive. Since this is also true for the mirror image of any system, we conclude that the entropy source strength is a positive polar scalar quantity. (A polar scalar is invariant under a mirror inversion of the coordinate axes. A pseudo-scalar, on the other hand, changes its sign under a mirror inversion. The same distinction between polar and scalar quantities also applies to vectors and tensors.)

Suppose that we are studying the transport processes taking place in a fluid. In the absence of any external non-dissipative fields (such as gravitational or magnetic fields), the fluid is at equilibrium and is assumed to be isotropic. Since the linear transport coefficients can be formally calculated as a zero-field functional derivative they clearly should have the symmetry characteristic of an isotropic system. Furthermore, they should be invariant under a mirror reflection of the coordinate axes.

Suppose that all the fluxes and forces are scalars. The most general linear relation between the forces and fluxes is given by (2.41). Since the transport coefficients must be polar scalars, there cannot be any coupling between a pseudo-scalar flux and a polar force or between a polar flux and a pseudo-scalar force. This is a simple application of the quotient rule in tensor analysis. Only scalars of like parity can be coupled by the transport matrix L_{ij}.

If the forces and fluxes are vectors, the most general linear relation between the forces and fluxes which is consistent with isotropy is

$$\mathbf{J}_i = \sum_j \mathbf{L}_{ij} \cdot \mathbf{X}_j = \sum_j L_{ij} \mathbf{I} \cdot \mathbf{X}_j = \sum_j L_{ij} \mathbf{X}_j \tag{2.44}$$

In this equation $\mathbf{L}_{ij}$ is a second-rank polar tensor because the transport coefficients must be invariant under mirror inversion just like the equilibrium system itself. If the equilibrium system is isotropic, then $\mathbf{L}_{ij}$ must be expressible as a scalar L_{ij} times the only isotropic second-rank tensor $\mathbf{I}$ (the Kronecker delta tensor $\mathbf{I} = \delta_{\alpha\beta}$). The thermodynamic forces and fluxes which couple together must either all be pseudo-vectors or polar vectors; otherwise, since the transport coefficients are polar quantities, the entropy source strength could be pseudo-scalar. By comparing the trace of $\mathbf{L}_{ij}$ with the trace of $L_{ij}\mathbf{I}$, we see that the polar scalar transport coefficients are given as

$$L_{ij} = \tfrac{1}{3}\,\mathrm{tr}(\mathbf{L}_{ij}) = \tfrac{1}{3}\,\mathbf{L}_{ij} : \mathbf{I} \tag{2.45}$$

If the thermodynamic forces and fluxes are all symmetric traceless second-rank tensors $\mathbb{J}_i$, $\mathbb{X}_i$, where $\mathbb{J}_i = \tfrac{1}{2}(\mathbf{J}_i + \mathbf{J}_i^{\mathrm{T}}) - \tfrac{1}{3}\,\mathrm{tr}(\mathbf{J}_i)\mathbf{I}$ (we denote symmetric traceless tensors as outline sans serif characters), then

$$\mathbb{J}_i^{(2)} = \sum_j \mathbf{L}_{ij}^{(4)} : \mathbb{X}_j^{(2)} \tag{2.46}$$

is the most linear general linear relation between the forces and fluxes. $\mathbf{L}_{ij}^{(4)}$ is a symmetric fourth-rank transport tensor. Unlike second-rank tensors there are three linearly independent isotropic fourth-rank polar tensors. (There are no isotropic pseudo-tensors of the fourth rank.) These tensors can be related to the Kronecker delta tensor, and we depict these tensors by the forms

$$\mathsf{UU}_{\alpha\beta\gamma\eta} = \delta_{\alpha\beta}\delta_{\gamma\eta} \tag{2.47a}$$

$$\mathsf{W}_{\alpha\beta\gamma\eta} = \delta_{\alpha\gamma}\delta_{\beta\eta} \tag{2.47b}$$

$$\mathsf{U}_{\alpha\beta\gamma\eta} = \delta_{\alpha\eta}\delta_{\beta\gamma} \tag{2.47c}$$

Since $\mathbf{L}_{ij}^{(4)}$ is an isotropic tensor it must be representable as a linear combination of isotropic fourth-rank tensors. It is convenient to write

$$\mathbf{L}_{ij}^{(4)} = L_{ij}^{\mathrm{s}}\left[\frac{1}{2}\left(\mathsf{U} + \mathsf{W}\right) - \frac{1}{3}\mathbf{UU}\right] + L_{ij}^{\mathrm{a}}\frac{1}{2}\left(\mathsf{U} - \mathsf{W}\right) + L_{ij}^{\mathrm{tr}}\frac{1}{3}\mathbf{UU} \tag{2.48}$$

It is easy to show that for any second-rank tensor $\mathbf{A}$,

$$\mathbf{L}_{ij}:\mathbf{A}^{(2)} = L_{ij}^{\mathrm{s}}\overset{\circ}{\mathbf{A}} + L_{ij}^{\mathrm{a}}\mathbb{A} + L_{ij}^{\mathrm{tr}}A\mathbf{I} \tag{2.49}$$

where $\overset{\circ}{\mathbf{A}}$ is the symmetric traceless part of $\mathbf{A}^{(2)}$, $\mathbb{A} = \frac{1}{2}(\mathbf{A} - \mathbf{A}^{\mathrm{T}})$ is the antisymmetric part of $\mathbf{A}^{(2)}$ (we denote antisymmetric tensors as shadowed sans serif characters), and $A = \frac{1}{3}\operatorname{tr}(\mathbf{A})$. This means that the three isotropic fourth-rank tensors decouple the linear force flux relations into three separate sets of equations which relate respectively, the symmetric second-rank forces and fluxes, the antisymmetric second-rank forces and fluxes, and the traces of the forces and fluxes. These equations can be written as

$$\overset{\circ}{\mathbf{J}}_i = \sum_j L_{ij}^{\mathrm{s}}\overset{\circ}{\mathbf{X}}_j \tag{2.50a}$$

$$\mathbb{J}_i = \sum_j L_{ij}^{\mathrm{a}}\mathbb{X}_j \tag{2.50b}$$

$$J_i = \sum_j L_{ij}^{\mathrm{tr}}X_j \tag{2.50c}$$

where $\mathbb{J}_i$ is the antisymmetric part of $\mathbf{J}$, and $J = \frac{1}{3}\operatorname{tr}(\mathbf{J})$. As $\mathbb{J}_i$ has only three independent elements it turns out that $\mathbb{J}_i$ can be related to a pseudo-vector. This relationship is conveniently expressed in terms of the Levi–Civita isotropic third-rank tensor $\boldsymbol{\varepsilon}^{(3)}$. (Note: $\boldsymbol{\varepsilon}_{\alpha\beta\gamma} = +1$ if $\alpha\beta\gamma$ is an even permutation, -1 if $\alpha\beta\gamma$ is an odd permutation and is zero otherwise.) If we denote the pseudo-vector dual of $\mathbb{J}_i$ as $\mathbf{J}_i^{\mathrm{ps}}$, then

$$\mathbf{J}_i^{\mathrm{ps}} = \frac{1}{2}\boldsymbol{\varepsilon}^{(3)}:\mathbb{J}_i^{(2)}$$

and

$$\mathbb{J}_i^{(2)} = \mathbf{J}_i^{\mathrm{ps}}\cdot\boldsymbol{\varepsilon}^{(3)} \tag{2.51}$$

This means that the second equation in the set (2.50b) can be rewritten as

$$\mathbf{J}_i^{\mathrm{ps}} = \sum_j L_{ij}^{\mathrm{a}}\mathbf{X}_j^{\mathrm{ps}} \tag{2.52}$$

Looking at (2.50) and (2.52) we see that we have decomposed the 81 elements of the (three-dimensional) fourth-rank transport tensor $\mathbf{L}_{ij}^{(4)}$ into three scalar quantities, L_{ij}^{s}, L_{ij}^{a} and L_{ij}^{tr}. Furthermore, we have found that there are three *irreducible* sets of forces and fluxes. Couplings only exist within the sets. There are no couplings of forces of one set with fluxes of another set. The sets naturally represent the symmetric traceless parts, the antisymmetric part and the trace of the second-rank tensors. The three irreducible components can be identified with a symmetric irreducible second-rank polar tensor component an irreducible pseudo-vector and an irreducible polar scalar. Curie's principle states that linear transport couples can only occur between irreducible tensors of the same rank and parity.

If we return to our basic equation for the entropy source strength (2.40) we see that our irreducible decomposition of Cartesian tensors allows us to make the following decomposition for second-rank fields and fluxes,

$$\sigma = \sum_i (J_i \mathbf{I} : X_i \mathbf{I} + \mathbb{J}_i : \mathbb{X}_i + \mathbb{J}_i : \mathbb{X}_i)$$

$$= \sum_i (3 J_i X_i - 2 \mathbf{J}_i^{\mathrm{ps}} \cdot \mathbf{X}_i^{\mathrm{ps}} + \mathbb{J}_i : \mathbb{X}_i) \tag{2.53}$$

The conjugate forces and fluxes appearing in the entropy source equation separate into irreducible sets. This is easily seen when we realize that all cross-couplings between irreducible tensors of different rank vanish; $\mathbf{I} : \mathbb{J}_i = \mathbf{I} : \mathbb{X} = \mathbb{J}_i : \mathbb{X} = 0$, etc. Conjugate thermodynamic forces and fluxes must have the same irreducible rank and parity.

We can now apply Curie's principle to the entropy source equation (2.39),

$$\sigma(\mathbf{r}, t) = \frac{-1}{T(\mathbf{r}, t)} \left[\frac{\mathbf{J}_{\mathrm{Q}}(\mathbf{r}, t) \cdot \nabla T(\mathbf{r}, t)}{T(\mathbf{r}, t)} - \mathbb{P}(\mathbf{r}, t) : \nabla \mathbb{u}(\mathbf{r}, t) \right.$$

$$\left. - \mathbf{\Pi}^{\mathrm{ps}}(\mathbf{r}, t) \cdot \nabla \times \mathbf{u}(\mathbf{r}, t) - \Pi(\mathbf{r}, t) \nabla \cdot \mathbf{u}(\mathbf{r}, t) \right] \tag{2.54}$$

In writing this equation we have used the fact that the transpose of $\mathbb{P}$ is equal to $\mathbb{P}$, and we have used equation (2.51) and the definition of the cross-product $\nabla \times \mathbf{u} = -\boldsymbol{\varepsilon}^{(3)} : \nabla \mathbb{u}$ to transform the antisymmetric part of $\mathbf{P}^{\mathrm{T}}$. Note that the transpose of $\mathbb{P}$ is equal to $-\mathbb{P}$. There is no conjugacy between the vector $\mathbf{J}_{\mathrm{Q}}(\mathbf{r}, t)$ and the pseudo-vector $\nabla \times \mathbf{u}(\mathbf{r}, t)$ because they differ in parity. It can be easily shown that for atomic fluids the antisymmetric part of the pressure tensor is zero so that the terms in (2.54) involving the vorticity $\nabla \times \mathbf{u}(\mathbf{r}, t)$ are identically zero. For molecular fluids, terms involving the vorticity do appear but we also have to consider another conservation equation—the conservation of angular momentum. In our description of the conservation equations we have ignored angular-momentum conservation.

The complete description of the hydrodynamics of molecular fluids must include this additional conservation law.

For single-component atomic fluids we can now use Curie's principle to define the phenomenological transport coefficients.

$$\mathbf{J}_Q = L_Q \mathbf{X}_Q = -L_Q \frac{\nabla T}{T^2} \tag{2.55a}$$

$$\mathbb{\Pi} = L_{\mathbb{\Pi}} \mathbb{X}_{\mathbb{P}} = -L_{\mathbb{\Pi}} \frac{\nabla \mathbb{u}}{T} \tag{2.55b}$$

$$\Pi = L_{\Pi} X_{\Pi} = -L_{\Pi} \frac{3\nabla : \mathbf{u}}{T} \tag{2.55c}$$

The positive sign of the entropy production implies that each of the phenomenological transport coefficients must be positive. As mentioned before, these phenomenological definitions differ slightly from the usual definitions of the Navier–Stokes transport coefficients.

$$\mathbf{J}_Q = -\lambda \nabla T \tag{2.56a}$$

$$\mathbb{\Pi} = -2\eta \nabla \mathbb{u} \tag{2.56b}$$

$$\Pi = -\eta_V \nabla \cdot \mathbf{u} \tag{2.56c}$$

These equations were postulated long before the development of linear irreversible thermodynamics. The first equation is known as Fourier's law of heat conduction. It gives the definition of the thermal conductivity λ. The second equation is known as Newton's law of viscosity (illustrated in Fig. 2.3). It gives a definition of the shear viscosity coefficient η. The third equation is a more recent development. It defines the bulk viscosity coefficient η_V. These equations are known collectively as linear constitutive equations. When they are substituted into the conservation equations they yield the Navier–Stokes equations of hydrodynamics. The conservation equations relate thermodynamic fluxes and forces. They form a system of equations in two unknown fields—the force fields and the flux fields. The constitutive equations relate the forces and the fluxes. By combining the two systems of equations we can derive the Navier–Stokes equations which in their usual form give us a closed system of equations for the thermodynamic forces. Once the boundary conditions are supplied, the Navier–Stokes equations can be solved to give a complete macroscopic description of the nonequilibrium flows expected in a fluid close to equilibrium in the sense required by linear irreversible thermodynamics. It is worth restating the equilibrium in the sense

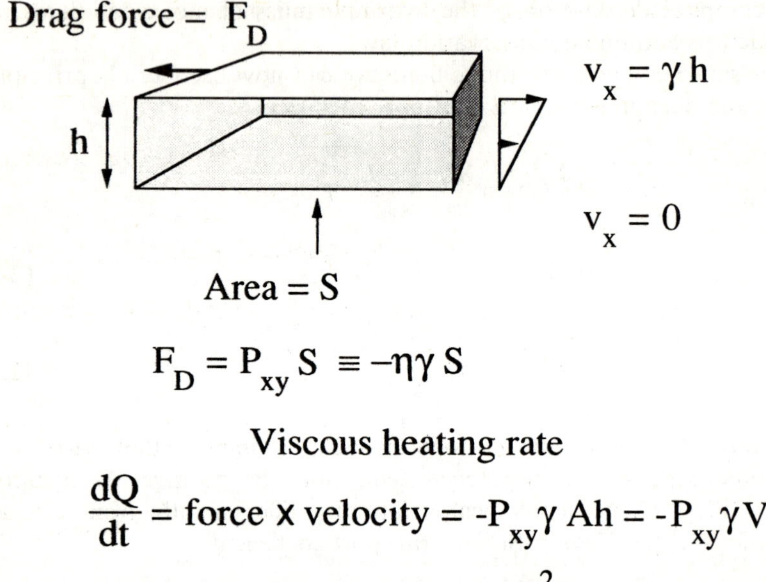

Fig. 2.3 Newton's constitutive relation for shear viscosity.

required by linear irreversible thermodynamics. It is worth restating the expected conditions for the linearity to be observed:

(1) The thermodynamic forces should be sufficiently small so that linear constitutive relations are accurate.
(2) The system should likewise be sufficiently close to equilibrium for the local thermodynamic equilibrium condition to hold. For example, the nonequilibrium equation of state must be the same function of the local position and time-dependent thermodynamic state variables (such as the temperature and density) that it is at equilibrium.
(3) The characteristic distances over which the thermodynamic forces vary should be sufficiently large so that these forces can be viewed as being constant over the microscopic length scale required to define properly a local thermodynamic state.
(4) The characteristic times over which the thermodynamic forces vary should be sufficiently long that these forces can be viewed as being constant over the microscopic times required to define properly a local thermodynamic state.

After some tedious but quite straightforward algebra (de Groot and Mazur, 1962), the Navier–Stokes equations for a single-component atomic fluid are obtained. The first of these is simply the mass conservation equation (2.4).

$$\frac{\partial \rho}{\partial t} = -\nabla \cdot (\rho \mathbf{u}) \tag{2.57}$$

To obtain the second equation we combine equation (2.16) with the definition of the stress tensor from equation (2.12) which gives

$$\rho \frac{\mathrm{d}\mathbf{u}}{\mathrm{d}t} = -\nabla \cdot \mathbf{P} = -\nabla \cdot ((p + \Pi)\mathbf{I} + \overset{\circ}{\Pi}) \tag{2.58}$$

We have assumed that the fluid is atomic and the pressure tensor contains no antisymmetric part. Substituting in the constitutive relations, equations (2.56b) and (2.56c) give

$$\rho \frac{\mathrm{d}\mathbf{u}}{\mathrm{d}t} = -\nabla p + \eta_V \nabla(\nabla \cdot \mathbf{u}) + 2\eta \nabla \cdot (\overset{\circ}{\nabla \mathbf{u}}) \tag{2.59}$$

Here we explicitly assume that the transport coefficients η_V and η are simple constants, independent of position $\mathbf{r}$, time and flow rate $\mathbf{u}$. The $\alpha\beta$ component of the symmetric traceless tensor $\overset{\circ}{\nabla \mathbf{u}}$ is given by

$$(\overset{\circ}{\nabla \mathbf{u}})_{\alpha\beta} = \frac{1}{2}\left(\frac{\partial u_\beta}{\partial x_\alpha} + \frac{\partial u_\alpha}{\partial x_\beta}\right) - \delta_{\alpha\beta} \frac{1}{3} \frac{\partial u_\gamma}{\partial x_\gamma} \tag{2.60}$$

where, as usual, the repeated index γ implies a summation with respect to γ. It is then straightforward to see that

$$\nabla \cdot (\overset{\circ}{\nabla \mathbf{u}}) = \tfrac{1}{2} \nabla^2 \mathbf{u} + \tfrac{1}{6} \nabla(\nabla \cdot \mathbf{u}) \tag{2.61}$$

and it follows that the momentum flow Navier–Stokes equation is

$$\rho \frac{\mathrm{d}\mathbf{u}}{\mathrm{d}t} = -\nabla p + \eta \nabla^2 \mathbf{u} + \left(\frac{\eta}{3} + \eta_V\right) \nabla(\nabla \cdot \mathbf{u}) \tag{2.62}$$

The Navier–Stokes equation for energy flow can be obtained from equation (2.26) and the constitutive relations, equation (2.56). Again we assume that the pressure tensor is symmetric, and the second term on the right-hand side of equation (2.26) becomes

$$\begin{aligned}
\mathbf{P}^{\mathrm{T}} : \nabla \mathbf{u} &= ((p + \Pi)\mathbf{I} + \overset{\circ}{\Pi}) : (\tfrac{1}{3}(\nabla \cdot \mathbf{u})\mathbf{I} + \overset{\circ}{\nabla \mathbf{u}}) \\
&= ((p - \eta_V(\nabla \cdot \mathbf{u}))\mathbf{I} - 2\eta \overset{\circ}{\nabla \mathbf{u}}) : (\tfrac{1}{3}(\nabla \cdot \mathbf{u})\mathbf{I} + \overset{\circ}{\nabla \mathbf{u}}) \\
&= p(\nabla \cdot \mathbf{u}) - \eta_V(\nabla \cdot \mathbf{u})^2 - 2\eta \overset{\circ}{\nabla \mathbf{u}} : \overset{\circ}{\nabla \mathbf{u}}
\end{aligned} \tag{2.63}$$

It is then straightforward to see that

$$\rho \frac{\mathrm{d}U}{\mathrm{d}t} = \lambda \nabla^2 T - p(\nabla \cdot \mathbf{u}) + \eta_V(\nabla \cdot \mathbf{u})^2 + 2\eta \overset{\circ}{\nabla \mathbf{u}} : \overset{\circ}{\nabla \mathbf{u}} \tag{2.64}$$

2.4 NON-MARKOVIAN CONSTITUTIVE RELATIONS: VISCOELASTICITY

Consider a fluid undergoing planar Couette flow. This flow is defined by the streaming velocity,

$$\mathbf{u}(\mathbf{r}, t) = (u_x, u_y, u_z) = (\gamma y, 0, 0) \tag{2.65}$$

According to Curie's principle the only nonequilibrium flux that can couple linearly to the flow is the pressure tensor. According to the constitutive relation equation (2.56) the pressure tensor is

$$\mathbf{P}(\mathbf{r}, t) = \begin{bmatrix} p & -\eta\gamma & 0 \\ -\eta\gamma & p & 0 \\ 0 & 0 & 0 \end{bmatrix} \tag{2.66}$$

where η is the shear viscosity and γ is the strain rate. If the strain rate is time dependent then the shear stress, $-\mathbf{P}_{xy} = -\mathbf{P}_{yx} = \eta\gamma(t)$. It is known that many fluids do not satisfy this relation regardless of how small the strain rate is. There must, therefore, be a *linear* but time-dependent constitutive relation for shear flow which is more general than the Navier–Stokes constitutive relation.

Poisson (1829) pointed out that there is a deep correspondence between the shear stress induced by a strain rate in a fluid, and the shear stress induced by a strain in an elastic solid. The strain tensor is, $\nabla\boldsymbol{\varepsilon}$ where $\boldsymbol{\varepsilon}(\mathbf{r}, t)$ gives the displacement of atoms at $\mathbf{r}$ from their equilibrium lattice sites. It is clear that

$$\frac{\mathrm{d}\nabla\boldsymbol{\varepsilon}}{\mathrm{d}t} = \nabla\mathbf{u} \tag{2.67}$$

Maxwell (1873) realized that if a displacement were applied to a liquid then for a short time the liquid must behave as if it were an elastic solid. After a *Maxwell relaxation time* the liquid would relax to equilibrium since *by definition* a liquid cannot support a strain (Frenkel, 1955).

It is easier to analyse this matter by transforming to the frequency domain. Maxwell said that at low frequencies the shear stress of a liquid is generated by the Navier–Stokes constitutive relation for a Newtonian fluid (2.66). In the frequency domain this states that

$$\tilde{P}_{xy}(\omega) = -\eta\tilde{\gamma}(\omega) \tag{2.68}$$

where

$$\tilde{A}(\omega) = \int_0^\infty \mathrm{d}t\, \mathrm{e}^{-i\omega t} A(t) \tag{2.69}$$

denotes the Fourier–Laplace transform of $A(t)$.

At very high frequencies we should have,

$$\tilde{P}_{xy}(\omega) = -G\frac{\partial\tilde{\varepsilon}_x}{\partial y} \tag{2.70}$$

where G is the infinite frequency shear modulus. From equation (2.67) we can transform the terms involving the strain into terms involving the strain rate (we assume that at $t = 0$, the strain $\varepsilon(0) = 0$). At high frequencies, therefore,

$$\tilde{P}_{xy}(\omega) = -\frac{G}{i\omega}\frac{\partial\tilde{u}_x}{\partial y} = -\frac{G}{i\omega}\tilde{\gamma}(\omega) \tag{2.71}$$

The Maxwell model of viscoelasticity is obtained by simply summing the high- and low-frequency expressions for the compliances $i\omega/G$ and η^{-1},

$$\tilde{\gamma}(\omega) = -\left(\frac{i\omega}{G} + \frac{1}{\eta}\right)\tilde{P}_{xy}(\omega) = -\frac{\tilde{P}_{xy}(\omega)}{\tilde{\eta}_{\mathrm{M}}(\omega)} \tag{2.72}$$

The expression for the frequency dependent Maxwell viscosity is

$$\tilde{\eta}_{\mathrm{M}}(\omega) = \frac{\eta}{1 + i\omega\tau_{\mathrm{M}}} \tag{2.73}$$

It is easily seen that this expression smoothly interpolates between the high- and low-frequency limits. The Maxwell relaxation time $\tau_{\mathrm{M}} = \eta/G$ controls the transition frequency between low-frequency viscous behaviour and high-frequency elastic behaviour (Fig. 2.4). The Maxwell model provides

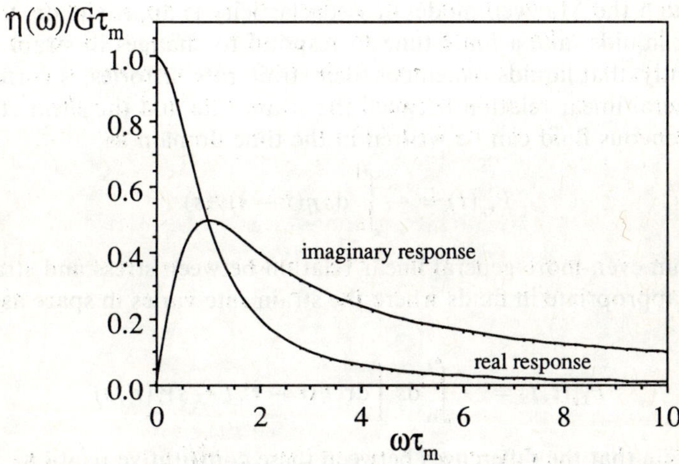

Fig. 2.4 Frequency-dependent viscosity of the Maxwell model.

a rough approximation to the viscoelastic behaviour of so-called viscoelastic fluids such as polymer melts or colloidal suspensions. It is important to remember that viscoelasticity is a linear phenomenon. The resulting shear stress is a linear function of the strain rate. It is also important to point out that Maxwell believed that all fluids are viscoelastic. The reason why polymer melts are observed to exhibit viscoelasticity is that their Maxwell relaxation times are macroscopic, of the order of seconds. However, the Maxwell relaxation time for argon at its triple point is approximately 10^{-12} seconds! Using standard viscometric techniques elastic effects are completely unobservable in argon.

If we rewrite the Maxwell constitutive relation in the time domain using an inverse Fourier–Laplace transform, we see that

$$P_{xy}(t) = -\int_0^t \mathrm{d}s\, \eta_{\mathrm{M}}(t-s)\gamma(s) \tag{2.74}$$

In this equation $\eta_{\mathrm{M}}(t)$ is called the Maxwell memory function (Figs 2.5 and 2.6). It is called a memory function because the shear stress at time t is not simply linearly proportional to the strain rate at the current time t, but to the entire strain rate *history*, over times s where $0 \leq s \leq t$. Constitutive relations which are history dependent are called non-Markovian. A Markovian process is one in which the present state of the system is all that is required to determine its future. The Maxwell model of viscoelasticity describes non-Markovian behaviour. The Maxwell memory function is easily identified as an exponential,

$$\eta_{\mathrm{M}}(t) = G\,\mathrm{e}^{-t/\tau_{\mathrm{M}}} \tag{2.75}$$

Although the Maxwell model of viscoelasticity is approximate, the basic idea that liquids take a finite time to respond to changes in strain rate or, equivalently, that liquids remember their strain rate histories, is correct. The most general linear relation between the strain rate and the shear stress for a homogeneous fluid can be written in the time domain as

$$P_{xy}(t) = -\int_0^t \mathrm{d}s\, \eta(t-s)\gamma(s) \tag{2.76}$$

There is an even more general linear relation between stress and strain rate which is appropriate in fluids where the strain rate varies in space as well as in time,

$$P_{xy}(\mathbf{r}, t) = -\int_0^t \mathrm{d}s \int \mathrm{d}\mathbf{r}'\, \eta(\mathbf{r}-\mathbf{r}', t-s)\gamma(\mathbf{r}', s) \tag{2.77}$$

We reiterate that the differences between these constitutive relations and the Newtonian constitutive relation, equations (2.56b), are only observable if the

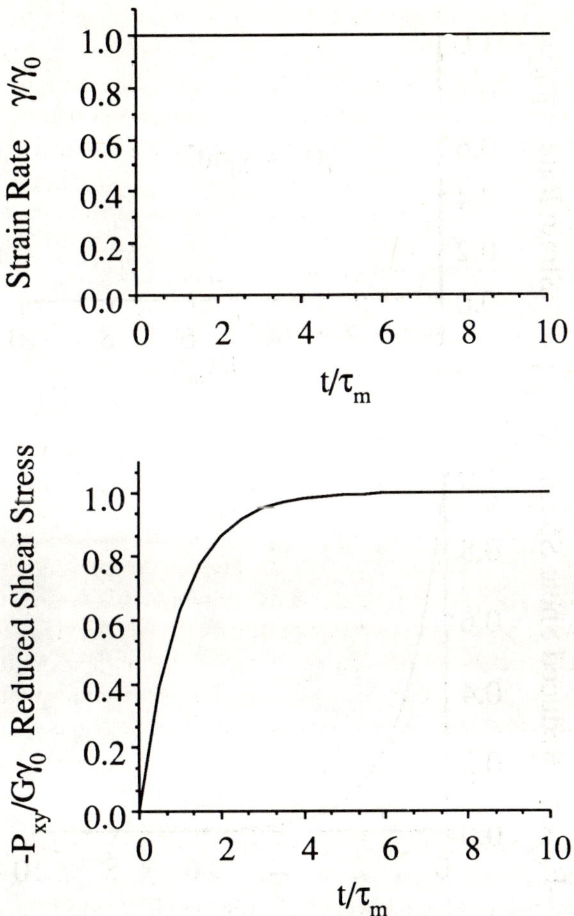

Fig. 2.5 The transient response of the Maxwell fluid to a step-function strain rate is the integral of the memory function for the model, $\eta_M(t)$.

strain rate varies significantly over either the time- or length-scales characteristic of the molecular relaxation for the fluid. The surprise is not so much that the validity of the Newtonian constitutive relation is limited, but that, for example in argon, the strain rates can vary in time from essentially zero frequency to 10^{12} Hz, or in space from zero wavevector of 10^{-9} m^{-1}, before non-Newtonian effects are observable. It is clear from this discussion that analogous corrections will be needed for all the other Navier–Stokes transport coefficients if their corresponding thermodynamic fluxes vary on molecular time or distance scales.

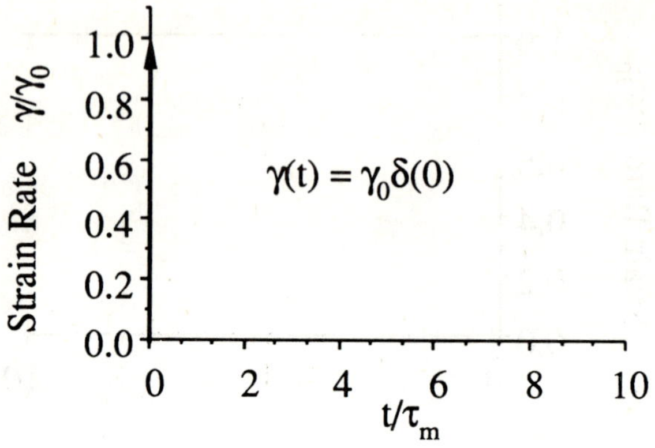

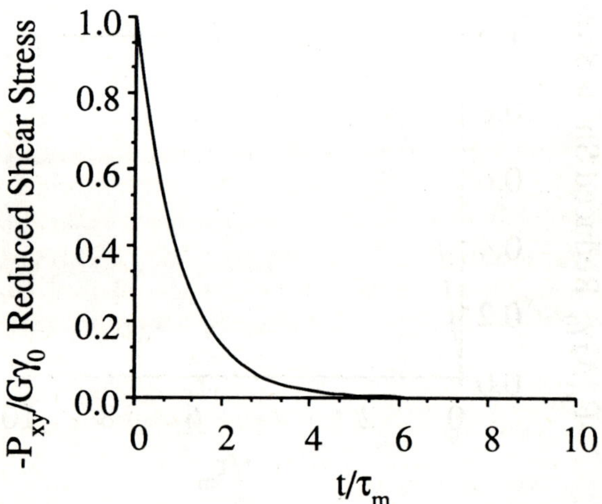

Fig. 2.6 The transient response of the Maxwell model to a zero time delta function in the strain rate is the memory function itself, $\eta_M(t)$.

REFERENCES

de Groot, S.R. and Mazur, P. (1962). *Non-equilibrium Thermodynamics.* Amsterdam: North-Holland.

Frenkel, J. (1955). *Kinetic Theory of Fluids.* New York: Dover.

Kreuzer, H.J. (1981). *Nonequilibrium Thermodynamics and its Statistical Foundations.* Oxford: Oxford University Press.

Maxwell, J.C. (1873). *Proc. R. Soc.*, **148**, 48.

Poisson, (1829). *Journal de l'Ecole*, Polytechnique tome **xiii**, cah. xx.

3 The Microscopic Connection

3.1 CLASSICAL MECHANICS

In nonequilibrium statistical mechanics we seek to model transport processes beginning with an understanding of the motion and interactions of individual atoms or molecules. The laws of classical mechanics govern the motion of atoms and molecules so in this chapter we begin with a brief description of the mechanics of Newton, Lagrange and Hamilton. It is often useful to be able to treat *constrained* mechanical systems. We will use a principle due to Gauss to treat many different types of constraint—from simple bond-length constraints, to constraints on kinetic energy. As we shall see, kinetic-energy constraints are useful for constructing various constant-temperature ensembles. We will then discuss the Liouville equation and its formal solution. This equation is the central vehicle of nonequilibrium statistical mechanics. We will then need to establish the link between the microscopic dynamics of individual atoms and molecules and the macroscopic hydrodynamic description discussed in the last chapter. We will discuss two procedures for making this connection. The Irving and Kirkwood procedure relates hydrodynamic variables to nonequilibrium *ensemble averages* of microscopic quantities. A more direct procedure which we will describe succeeds in deriving *instantaneous* expressions for the hydrodynamic field variables.

Newtonian mechanics

Classical mechanics (Goldstein, 1980) is based on Newton's three laws of motion. This theory introduced the concepts of a force and an acceleration. Prior to Newton's work, the connection had been made between forces and velocities. Newton's laws of motion were supplemented by the notion of a force acting at a distance. With the identification of the force of gravity and an appropriate initial condition—initial coordinates and velocities—trajectories could be computed. Philosophers of science have debated the content of Newton's laws, but when augmented with a force which is expressible as a function of time, position or possibly of velocity, those laws lead to the equation,

$$m\ddot{\mathbf{r}} = \mathbf{F}(\mathbf{r}, \dot{\mathbf{r}}, t) \tag{3.1}$$

which is well posed and possesses a unique solution.

Lagrange's equations

After Newton, scientists discovered different sets of equivalent laws or axioms upon which classical mechanics could be based. More *elegant* formulations are due to Lagrange and Hamilton. Newton's laws are less general than they might seem. For instance, the position **r**, that appears in Newton's equation must be a Cartesian vector in a Euclidian space. One does not have the freedom of, say, using angles as measures of position. Lagrange solved the problem of formulating the laws of mechanics in a form which is valid for *generalized* coordinates.

Let us consider a system with generalized coordinates q. These coordinates may be Cartesian positions, angles or any other convenient parameters that can be found to specify the configuration of the system uniquely. The kinetic energy T will, in general, be a function of the coordinates and their time derivatives $\mathrm{d}q/\mathrm{d}t$. If $V(q)$ is the potential energy, we define the Lagrangian to be $L \equiv T(q, \mathrm{d}q/\mathrm{d}t) - V(q)$. The fundamental dynamical postulate states that the motion of a system is such that the *action*, S, is an extremum

$$\delta S = \delta \int_{t_0}^{t_1} L(q, \dot{q}, t) = 0 \tag{3.2}$$

Let $q(t)$ be the coordinate trajectory that satisfies this condition and let $q(t) + \delta q(t)$ where $\delta q(t)$ is an arbitrary variation in $q(t)$, be an arbitrary trajectory. The varied motion must be consistent with the initial and final positions. So that, $\delta q(t_1) = \delta q(t_0) = 0$. We consider the change in the action due to this variation.

$$\delta S = \int_{t_0}^{t_1} \mathrm{d}t\, L(q + \delta q, \dot{q} + \delta \dot{q}, t) - \int_{t_0}^{t_1} \mathrm{d}t\, L(q, \dot{q}, t) = \int_{t_0}^{t_1} \mathrm{d}t \left(\frac{\partial L}{\partial q} \delta q + \frac{\partial L}{\partial \dot{q}} \delta \dot{q} \right) \tag{3.3}$$

Integrating the second term by parts gives

$$\delta S = \left[\frac{\partial L}{\partial \dot{q}} \delta q \right]_{t_0}^{t_1} + \int_{t_0}^{t_1} \left[\frac{\partial L}{\partial q} - \frac{\mathrm{d}}{\mathrm{d}t} \left(\frac{\partial L}{\partial \dot{q}} \right) \right] \delta q \,\mathrm{d}t \tag{3.4}$$

The first term vanishes because δq is zero at both limits. Since for $0 < t < t_1$, $\delta q(t)$ is arbitrary, the only way that the variation in the action δS, can vanish is if the equation,

$$\frac{\mathrm{d}}{\mathrm{d}t}\left(\frac{\partial L}{\partial \dot{q}}\right) - \frac{\partial L}{\partial q} = 0 \tag{3.5}$$

holds for all time. This is Lagrange's equation of motion. If the coordinates are taken to be Cartesian, it is easy to see that Lagrange's equation reduces to that of Newton.

Hamiltonian mechanics

Although Lagrange's equation has removed the special status attached to Cartesian coordinates, it has introduced a new difficulty. The Lagrangian is a function of generalized coordinates, their time derivatives and possibly of time. The equation is not symmetric with respect to the interchange of coordinates and velocities. Hamilton derived an equivalent set of equations in which the roles played by coordinates and *velocities* can be interchanged. Hamilton defined the *canonical momentum*, p, as

$$p \equiv \frac{\partial L(q, \dot{q}, t)}{\partial \dot{q}} \tag{3.6}$$

and introduced the function

$$H(q, p, t) \equiv \frac{\partial L}{\partial \dot{q}}\dot{q} - L = p\dot{q} - L \tag{3.7}$$

This function is of course now known as the Hamiltonian. Consider a change in the Hamiltonian which can be written as

$$\mathrm{d}H = \dot{q}\,\mathrm{d}p + q\,\mathrm{d}\dot{q} - \mathrm{d}L \tag{3.8}$$

The Lagrangian is a function of q, $\mathrm{d}q/\mathrm{d}t$ and t so that the change $\mathrm{d}L$, can be written as

$$\mathrm{d}L = \frac{\partial L}{\partial q}\mathrm{d}q + \frac{\partial L}{\partial \dot{q}}\mathrm{d}\dot{q} + \frac{\partial L}{\partial t}\mathrm{d}t \tag{3.9}$$

Using the definition of the canonical momentum, p, and substituting for $\mathrm{d}L$, the expression for $\mathrm{d}H$ becomes

$$\mathrm{d}H = \dot{q}\,\mathrm{d}p - \frac{\partial L}{\partial q}\mathrm{d}q - \frac{\partial L}{\partial t}\mathrm{d}t \tag{3.10}$$

Lagrange's equation of motion (3.5), rewritten in terms of the canonical momenta is

$$\dot{p} = \frac{\partial L}{\partial q} \tag{3.11}$$

so that the change in H is

$$\mathrm{d}H = \dot{q}\,\mathrm{d}p - \dot{p}\,\mathrm{d}q - \frac{\partial L}{\partial t}\,\mathrm{d}t \tag{3.12}$$

Since the Hamiltonian is a function of q, p and t, it is easy to see that Hamiltonian equations of motion are

$$\dot{q} = \frac{\partial H}{\partial p} \qquad \text{and} \qquad \dot{p} = -\frac{\partial H}{\partial q} \tag{3.13}$$

As mentioned above, these equations are symmetric with respect to coordinates and momenta. Each has equal status in Hamilton's equations of motion. If H has no explicit time dependence, its value is a constant of the motion. Other formulations of classical mechanics such as the Hamilton–Jacobi equations will not concern us in this book.

Gauss' principle of least constraint

Apart from relativistic or quantum corrections, classical mechanics is thought to give an exact description of motion. In this section our point of view will change somewhat. Newtonian or Hamiltonian mechanics imply a certain set of constants of the motion: energy, and linear and angular momentum. In thermodynamically interesting systems the natural fixed quantities are the thermodynamic state variables: the number of molecules N, the volume V and the temperature T. Often the pressure rather than the volume may be preferred. Thermodynamically interesting systems usually exchange energy, momentum and mass with their surroundings. This means that within thermodynamic systems none of the classical constants of the motion are actually constant.

Typical thermodynamic systems are characterized by fixed values of thermodynamic variables: temperature, pressure, chemical potential, density, enthalpy or internal energy. The system is maintained at a fixed thermodynamic state (say temperature) by placing it in contact with a reservoir, with which it exchanges energy (heat) in such a manner as to keep the temperature of the system of interest fixed. The heat capacity of the reservoir must be much larger than that of the system, so that the heat exchanged from the reservoir does not affect the reservoir temperature.

Classical mechanics is an awkward vehicle for describing this type of system. The only way that thermodynamic systems can be treated in Newtonian or Hamiltonian mechanics is by explicitly modelling the system, the reservoir and the exchange processes. This is complex and tedious and, as we will see below, it is also unnecessary. We will now describe a little-known principle of classical mechanics which is extremely useful for *designing* equations of motion which are more useful from a macroscopic or thermodynamic viewpoint. This principle does indeed allow us to modify classical mechanics so that thermodynamic variables may be made constants of the motion.

Just over 150 years ago, Gauss formulated a mechanics more general than Newton's. This mechanics has as its foundation Gauss' principle of least constraint. Gauss (1829) referred to this as *the most fundamental dynamical principle* (Whittacker, 1904; Pars, 1979). Suppose that the *Cartesian* coordinates and velocities of a system are given at time t. Consider the function C, referred to by Hertz as the square of the curvature, where

$$C = \tfrac{1}{2} \sum_{i=1}^{N} m_i \left(\ddot{\mathbf{r}}_i - \frac{\mathbf{F}_i}{m_i} \right)^2 \tag{3.14}$$

C is a function of the set of accelerations $\{d^2\mathbf{r}_i/dt^2\}$. Gauss' principle states that the actual physical acceleration corresponds to the minimum value of C. Clearly, if the system is not subject to a constraint, then $C = 0$ and the system evolves under Newton's equations of motion. For a constrained system it is convenient to change variables from $\mathbf{r}_i$ to $\mathbf{w}_i$ where

$$\begin{aligned} \mathbf{w}_i &= m_i^{\frac{1}{2}} \mathbf{r}_i \\ \zeta_i &= m_i^{-\frac{1}{2}} \mathbf{F}_i \end{aligned} \tag{3.15}$$

Because the $\{\mathbf{w}_i\}$ are related to the Jacobi metric we will refer to this coordinate system as the Jacobi frame.

The types of constraints which might be applied to a system fall naturally into two types: *holonomic* and *nonholonomic*. A holonomic constraint is one which can be integrated out of the equations of motion. For instance, if a certain generalized coordinate is fixed, its conjugate momentum is zero for all time, so we can simply consider the problem in the reduced set of unconstrained variables. We need not be conscious of the fact that a force of constraint is acting upon the system to fix the coordinate and the momentum. An analysis of the two-dimensional motion of an ice skater need not refer to the fact that the gravitational force is exactly resisted by the stress on the ice surface fixing the vertical coordinate and velocity of the ice skater. We can ignore these degrees of freedom.

Nonholonomic constraints usually involve velocities. These constraints are not integrable. In general a nonholonomic constraint will do work on a system. Thermodynamic constraints are invariably nonholonomic. It is known that the 'action principle' cannot be used to describe motion under nonholonomic constraints (Evans and Morriss, 1984b).

We can write a general constraint in the Jacobi frame in the form

$$g(\mathbf{w}, \dot{\mathbf{w}}, t) = 0 \tag{3.16}$$

where g is a function of Jacobi positions, velocities and, possibly, time. Either type of constraint function, holonomic or nonholonomic, can be written in this form. If this equation is differentiated with respect to time, once for nonholonomic constraints and twice for holonomic constraints, we see that

$$\mathbf{n}(\mathbf{w}, \dot{\mathbf{w}}, t) \cdot \ddot{\mathbf{w}} = s(\mathbf{w}, \dot{\mathbf{w}}, t) \tag{3.17}$$

We refer to this equation as the *differential constraint equation* and it plays a fundamental role in Gauss' 'principle of least constraint'. It is the equation for a plane which we refer to as the *constraint plane*; $\mathbf{n}$ is the vector normal to the constraint plane.

Our problem is to solve Newton's equation subject to the constraint. Newton's equation gives us the acceleration in terms of the unconstrained forces. The differential constraint equation places a condition on the acceleration vector for the system. The differential constraint equation says that the constrained acceleration vector must terminate on a hyper-plane in the $3N$-dimensional Jacobi acceleration space (3.17).

Imagine for the moment that at some initial time the system satisfies the constraint equation $g = 0$. In the absence of the constraint the system would evolve according to Newton's equations of motion where the acceleration is given by

$$\ddot{\mathbf{w}}_i^u = \zeta_i \tag{3.18}$$

This trajectory would, in general, not satisfy the constraint. Furthermore, the constraint function g tells us that the only accelerations which do continuously satisfy the constraint, are those which terminate on the constraint plane. To obtain the constrained acceleration we must project the unconstrained acceleration back into the constraint plane.

Gauss' principle of least constraint gives us a prescription for constructing this projection. *Gauss' principle states that the trajectories actually followed are those which deviate as little as possible, in a least-squares sense, from the unconstrained Newtonian trajectories.* The projection which the system actually follows is the one which minimizes the magnitude of the Jacobi frame constraint force. This means that the force of constraint must be parallel to the normal of the constraint surface. The Gaussian equations of motion are then

$$\ddot{\mathbf{w}}_i = \zeta_i - \lambda \mathbf{n} \tag{3.19}$$

where λ is a Gaussian multiplier which is a function of position, velocity and time (Fig. 3.1).

To calculate the multiplier we use the differential form of the constraint function. Substituting for the acceleration we obtain

$$\lambda = \frac{\mathbf{n}\cdot\boldsymbol{\zeta} - s}{\mathbf{n}\cdot\mathbf{n}} \tag{3.20}$$

It is worthwhile at this stage to make a few comments about the procedure outlined above. First, note that the original constraint equation is never used explicitly. Gauss' principle only refers to the *differential* form of the constraint equation. This means that the precise *value* of the constrained quantity is undetermined. The constraint acts only to stop its value changing. In the holonomic case, Gauss' principle and the principle of least action are of course completely equivalent. In the nonholonomic case the equations resulting from the application of Gauss' principle cannot be derived from a Hamiltonian and the principle of least action *cannot* be used to derive constraint satisfying equations. In the nonholonomic case, Gauss' principle does *not* yield equations of motion for which the work done by the constraint forces is a minimum.

The derivation of constrained equations of motion given above is geometric, and is done in the transformed coordinates which we have termed the Jacobi frame. It is not always convenient to write a constraint function in the Jacobi frame and, from an operational point of view, a much simpler

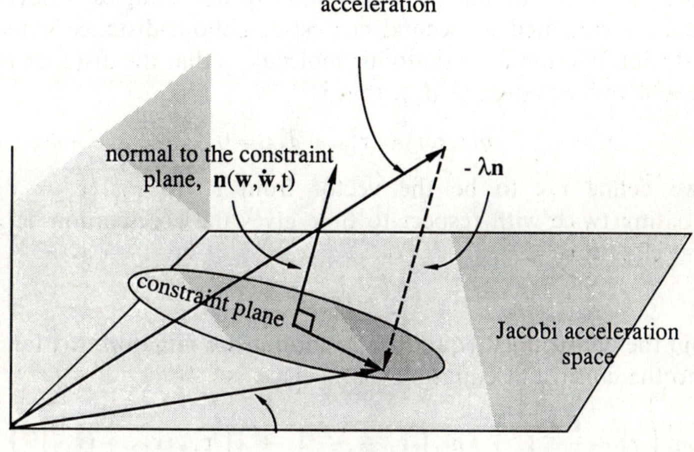

Fig. 3.1 Gauss' principle of least constraint.

derivation of constrained equations of motion is possible using Lagrange multipliers. The square of the curvature C is a function of accelerations only (the Cartesian coordinates and velocities are considered to be given parameters). Gauss' principle reduces to finding the minimum of C, subject to the constraint. The constraint function must also be written as a function of accelerations, but this is easily achieved by differentiating with respect to time. If G is the acceleration dependent form of the constraint, then the constrained equations of motion are obtained from

$$\frac{\partial}{\partial \ddot{\mathbf{r}}}(C - \lambda G) = 0 \tag{3.21}$$

It is easy to see that the Lagrange multiplier, λ, is (apart from the sign) equal to the Gaussian multiplier. We will illustrate Gauss' principle by considering some useful examples.

Gauss' principle for holonomic constraints

The most common type of holonomic constraint in statistical mechanics is probably that of fixing bond lengths and bond angles in molecular systems. The vibrational degrees of freedom typically have a relaxation time-scale which is an order of magnitude faster than translational degrees of freedom and are, therefore, often irrelevant to the processes under study. As an example of the application of Gauss' principle of least constraint for holonomic constraints we consider a diatomic molecule with a fixed bond length. The generalization of this method to more than one bond length is straightforward (see Edberg *et al.*, 1986) and the application to bond angles is trivial since they can be formulated as second-nearest-neighbour-distance constraints. The constraint function for a diatomic molecule is that the distance between sites one and two be equal to d_{12}, that is

$$g(\mathbf{r}, \dot{\mathbf{r}}, t) = \mathbf{r}_{12}^2 - d_{12}^2 = 0 \tag{3.22}$$

where we define $\mathbf{r}_{12}$ to be the vector from $\mathbf{r}_1$ to $\mathbf{r}_2$, $(\mathbf{r}_{12} \equiv \mathbf{r}_2 - \mathbf{r}_1)$. Differentiating twice with respect to time gives the acceleration-dependent constraint equation,

$$\mathbf{r}_{12} \cdot \ddot{\mathbf{r}}_{12} + (\dot{\mathbf{r}}_{12})^2 = 0 \tag{3.23}$$

To obtain the constrained equations of motion we minimize the function C subject to the constraint equation (3.23). That is

$$\frac{\partial}{\partial \ddot{\mathbf{r}}_i}\left(\tfrac{1}{2}m_1\left(\ddot{\mathbf{r}}_1 - \frac{\mathbf{F}_1}{m_1}\right)^2 + \tfrac{1}{2}m_2\left(\ddot{\mathbf{r}}_2 - \frac{\mathbf{F}_2}{m_2}\right)^2 - \lambda\left(\mathbf{r}_{12} \cdot \ddot{\mathbf{r}}_{12} + (\dot{\mathbf{r}}_{12})^2\right)\right) = 0 \tag{3.24}$$

For $i = 1$ or 2 this gives

$$\begin{aligned} m_1\ddot{\mathbf{r}}_1 &= \mathbf{F}_1 - \lambda\mathbf{r}_{12} \\ m_2\ddot{\mathbf{r}}_2 &= \mathbf{F}_2 + \lambda\mathbf{r}_{12} \end{aligned} \tag{3.25}$$

Notice that the extra terms in these equations of motion have opposite signs. This is because the coefficients of the $\mathbf{r}_1$ and $\mathbf{r}_2$ accelerations have opposite signs. The total constraint force on the molecule is zero so there is no change in the total momentum of the molecule. To obtain an expression for the multiplier λ we combine these two equations to give an equation of motion for the bond vector $\mathbf{r}_{12}$,

$$\ddot{\mathbf{r}}_{12} = \left(\frac{\mathbf{F}_2}{m_2} - \frac{\mathbf{F}_1}{m_1}\right) + \lambda\left(\frac{1}{m_2} + \frac{1}{m_1}\right)\mathbf{r}_{12} \tag{3.26}$$

Substituting this into the differential form of the constraint function (3.23), gives

$$\lambda = -\frac{\mathbf{r}_{12}\cdot(m_1\mathbf{F}_2 - m_2\mathbf{F}_1) + m_1 m_2 \dot{\mathbf{r}}_{12}^2}{(m_1 + m_2)\mathbf{r}_{12}^2} \tag{3.27}$$

It is very easy to implement these constrained equations of motion as the multiplier is a simple explicit function of the positions, velocities and Newtonian forces. For more complicated systems with multiple bond-length and bond-angle constraints (all written as distance constraints) we obtain a set of coupled linear equations to solve for the multipliers.

Gauss' principle for nonholonomic constraints

One of the simplest and most useful applications of Gauss' principle is to derive equations of motion for which the ideal gas temperature (i.e. the kinetic energy) is a constant of the motion (Evans *et al.*, 1983). Here the constraint function is

$$g(\mathbf{r}, \dot{\mathbf{r}}, t) = \sum_{i=1}^{N} \frac{m_i \dot{\mathbf{r}}_i^2}{2} - \frac{3Nk_B T}{2} = 0 \tag{3.28}$$

Differentiating once with respect to time gives the equation for the constraint plane

$$\sum_{i=1}^{N} m_i \dot{\mathbf{r}}_i \cdot \ddot{\mathbf{r}}_i = 0 \tag{3.29}$$

Therefore to obtain the constrained Gaussian equations we minimize C subject to the constraint equation (3.29). That is

$$\frac{\partial}{\partial \ddot{\mathbf{r}}_i}\left(\tfrac{1}{2}\sum_{j=1}^{N} m_j\left(\ddot{\mathbf{r}}_j - \frac{\mathbf{F}_j}{m_j}\right)^2 + \lambda \sum_{j=1}^{N} m_j \dot{\mathbf{r}}_j \cdot \ddot{\mathbf{r}}_j\right) = 0 \tag{3.30}$$

This gives

$$m_i\ddot{\mathbf{r}}_i = \mathbf{F}_i - \lambda m_i\dot{\mathbf{r}}_i \tag{3.31}$$

Substituting the equations of motion into the differential form of the constraint equation, we find that the multiplier is given by

$$\lambda = \frac{\sum_{i=1}^{N} \mathbf{F}_i \cdot \dot{\mathbf{r}}_i}{\sum_{i=1}^{N} m_i\dot{\mathbf{r}}_i^2} \tag{3.32}$$

As before, λ is a simple function of the forces and velocities so that the implementation of the constant-kinetic-energy constraint in a molecular dynamics computer program only requires a trivial modification of the equations of motion in a standard program. Equations (3.31) and (3.32) constitute what have become known as the *Gaussian isokinetic equations of motion.* These equations were first proposed simultaneously and independently by Hoover *et al.* (1982) and Evans (1983). However, in these papers original Gauss' principle was not referred to. It was a year before the connection with Gauss' principle was made.

With regard to the general application of Gauss' principle of least constraint one should always examine the statistical mechanical properties of the resulting dynamics. If one applies Gauss' principle to the problem of maintaining a constant heat flow, then a comparison with linear-response theory shows that the Gaussian equations of motion *cannot* be used to calculate thermal conductivity (Hoover, 1986). The correct application of Gauss' principle is limited to arbitrary holonomic constraints and, apparently, to nonholonomic constraint functions which are *homogeneous* functions of the momenta.

3.2 PHASE SPACE

To give a complete description of the state of a three-dimensional N-particle system at any given time it is necessary to specify the $3N$ coordinates and $3N$ momenta. The $6N$-dimensional space of coordinates and momenta is called *phase space* (or $\mathbf{\Gamma}$ space). As time progresses the phase point $\mathbf{\Gamma}$ traces out a path which we call the phase-space trajectory of the system (Fig. 3.2). As the equations of motion for $\mathbf{\Gamma}$ are $6N$ first-order differential equations, there are $6N$ constants of integration (they may be, for example, the $6N$ initial conditions $\mathbf{\Gamma}(0)$). Rewriting the equations of motion in terms of these constants shows that the trajectory of $\mathbf{\Gamma}$ is completely determined by specifying these $6N$ constants. An alternate description of the time evolution

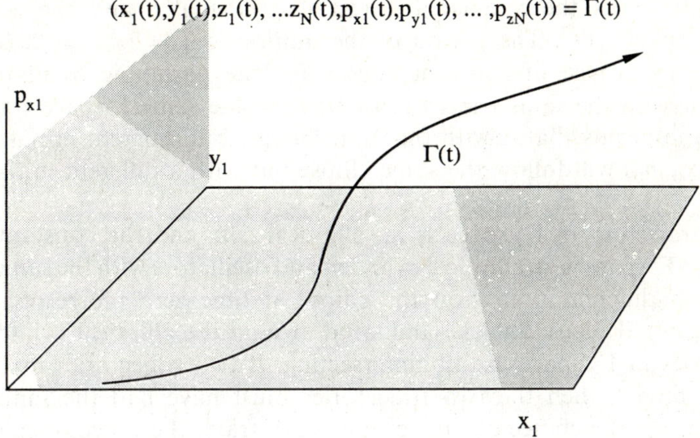

Fig. 3.2 Phase-space trajectory: $6N$-dimensional $\mathbf{\Gamma}$ space. As time evolves the system traces out a trajectory.

of the system is given by the trajectory in the extended $\mathbf{\Gamma}'$ space, where $\mathbf{\Gamma}' = (\mathbf{\Gamma}, t)$. As the $6N$ initial conditions uniquely determine the trajectory, two points in phase space with different initial conditions form distinct noninteresecting trajectories in $\mathbf{\Gamma}'$ space.

To illustrate the basic ideas of $\mathbf{\Gamma}$ space and $\mathbf{\Gamma}'$ space it is useful to consider one of the simplest mechanical systems, the harmonic oscillator. The Hamiltonian for the harmonic oscillator is $H = \frac{1}{2}(kx^2 + p^2/m)$ where m is the mass of the oscillator and k is the spring constant. The equations of motion are

$$\begin{aligned} \dot{x} &= \frac{\partial H}{\partial p} = \frac{p}{m} \\ \dot{p} &= -\frac{\partial H}{\partial x} = -kx \end{aligned} \tag{3.33}$$

and the energy (or the Hamiltonian) is a constant of the motion. The $\mathbf{\Gamma}$ space for this system is two-dimensional (x, p) and the $\mathbf{\Gamma}$-space trajectory is given by

$$(x(t), p(t)) = \left(x_0 \cos \omega t + \frac{p_0}{m\omega} \sin \omega t, p_0 \cos \omega t - m\omega x_0 \sin \omega t \right) \tag{3.34}$$

The constants x_0 and p_0 are the two integration constants, written in this case as an initial condition. The frequency ω is related to the spring constant and mass by $\omega^2 = k/m$. The $\mathbf{\Gamma}$-space trajectory is an ellipse,

$$m^2\omega^2 x(t)^2 + p(t)^2 = m^2\omega^2 x_0^2 + p_0^2 \tag{3.35}$$

which intercepts the x axis at $\pm(x_0^2 + p_0^2/m^2\omega^2)^{1/2}$, and the p axis at $\pm(p_0^2 + m^2\omega^2 x_0^2)^{1/2}$. The period of the motion is $T = 2\pi/\omega = 2\pi(m/k)^{1/2}$. This is the surface of constant energy for the harmonic oscillator. Any oscillator with the same energy must traverse the same $\mathbf{\Gamma}$-space trajectory; that is another oscillator with the same energy, but different initial starting points (x_0, p_0) will follow the same ellipse but with a different initial phase angle.

The trajectory in $\mathbf{\Gamma}'$ space is an elliptical coil, and the constant-energy surface in $\mathbf{\Gamma}'$ space is an elliptical cylinder, and oscillators with the same energy start from different points on the ellipse at time zero (corresponding to different initial phase angles), and wind around the elliptical cylinder. The trajectories in $\mathbf{\Gamma}'$ space are nonintersecting. If two trajectories in $\mathbf{\Gamma}'$ space meet at time t, then the two trajectories must have had the same initial condition. As the choice of time origin is arbitrary, the trajectories must be the same for all time.

In $\mathbf{\Gamma}$ space the situation is somewhat different. The trajectory for the harmonic oscillator winds around the ellipse, returning to its initial phase point (x_0, p_0) after a time T. The period of time taken for a system to return to (or to within an ε neighbourhood of) its initial starting phase is called the Poincaré recurrence time. For a simple system such as harmonic oscillator the recurrence time is trivial to calculate, but for higher dimensional systems the recurrence time quickly exceeds the estimated age of the universe.

3.3 DISTRIBUTION FUNCTIONS AND THE LIOUVILLE EQUATION

In the first few sections of this chapter we have given a description of the mechanics of individual N-particle systems. The development which follows describes an *ensemble* of such systems; that is an essentially infinite number of systems characterized by identical dynamics and identical state variables (N, V, E or T, etc.) but different intitial conditions, ($\mathbf{\Gamma}(0)$). We wish to consider the average behaviour of a collection of macroscopically identical systems distributed over a range of initial states (microstates). In generating the ensemble we make the usual assumptions of classical mechanics. We assume that it is possible to know all the positions and momenta of an N-particle system to arbitrary precision at some initial time, and that the motion can be calculated exactly from the equations of motion.

The ensemble contains an infinite number of individual systems so that the number of systems in a particular state may be considered to change continuously as we pass to neighbouring states. This assumption allows us to define a density function $f(\mathbf{\Gamma}, t)$, which assigns a probability to points in phase space. Implicit in this assumption is the requirement that $f(\mathbf{\Gamma}, t)$, has continuous partial derivatives with respect to all its variables, otherwise the phase density will not change *continuously* as we move to neighbouring states. If the system is Hamiltonian and all trajectories are confined to the energy surface then $f(\mathbf{\Gamma}, t)$ will not have a continuous partial derivative with respect to energy. Problems associated with this particular source of discontinuity can obviously be avoided by eliminating the energy as a variable, and by considering $f(\mathbf{\Gamma}, t)$ to be a density function defined on a surface of constant energy (effectively reducing the dimensionality of the system). However, it is worth pointing out that other sources of discontinuity in the phase-space density, may not be so easily removed.

To define a distribution function for a particular system we consider an ensemble of identical systems whose initial conditions *span* the phase space specified by the macroscopic constraints. We consider an infinitesimal element of phase space located at $\mathbf{q}, \mathbf{p} \equiv \mathbf{\Gamma}$. The fraction of systems δN, which at time t have coordinates and momenta within $\delta\mathbf{q}, \delta\mathbf{p}$ of $\mathbf{q}, \mathbf{p}$ is used to define the phase-space distribution function $f(\mathbf{q}, \mathbf{p}, t)$ by

$$\delta N = f(\mathbf{q}, \mathbf{p}, t)\delta\mathbf{q}\, \delta\mathbf{p} \tag{3.36}$$

The total number of systems in the ensemble is fixed, so by integrating over the whole phase space we can normalize the distribution function,

$$1 = \int f(\mathbf{q}, \mathbf{p}, t)\, \mathrm{d}\mathbf{q}\, \mathrm{d}\mathbf{p} \tag{3.37}$$

If we consider a small volume element of phase space, the number of trajectories entering the volume element $\delta\mathbf{q}\delta\mathbf{p}$ through some face will, in general, be different from the number which leave through an opposite face. For the faces normal to the q_1 axis, located at q_1, and $q_1 + \delta q_1$, the fraction of ensemble members entering the first face is

$$f(q_1, \ldots)\dot{q}_1(q_1, \ldots)\delta q_2 \ldots \delta q_{3N}\delta\mathbf{p}$$

Similarly the fraction of points leaving through the second face is

$$f(q_1 + \delta q_1, \ldots)\dot{q}_1(q_1 + \delta q_1, \ldots)\delta q_2 \ldots \delta q_{3N}\delta\mathbf{p}$$
$$\approx \left(f(q_1, \ldots) + \frac{\partial f}{\partial q_1}\delta q_1\right)\left(\dot{q}_1(q_1, \ldots) + \frac{\partial \dot{q}_1}{\partial q_1}\delta q_1\right)\delta q_2 \ldots \partial q_{3N}\delta\mathbf{p}$$

Combining these expressions gives the change in δN due to fluxes in the q_1 direction

$$\frac{\mathrm{d}}{\mathrm{d}t}\delta N_{q_1} = -\left(\dot{q}_1\frac{\partial f}{\partial q_1} + f\frac{\partial \dot{q}_1}{\partial q_1}\right)\delta\mathbf{q}\delta\mathbf{p} \tag{3.38}$$

Summing over all coordinate (and momentum) directions gives the total fractional change δN as

$$\frac{\mathrm{d}}{\mathrm{d}t}\delta N = -\sum_{i=1}^{N}\left[f\left(\frac{\partial}{\partial \mathbf{q}_i}\cdot\dot{\mathbf{q}}_i + \frac{\partial}{\partial \mathbf{p}_i}\cdot\dot{\mathbf{p}}_i\right) + \dot{\mathbf{q}}_i\cdot\frac{\partial f}{\partial \mathbf{q}_i} + \dot{\mathbf{p}}_i\cdot\frac{\partial f}{\partial \mathbf{p}_i}\right]\delta\mathbf{q}\delta\mathbf{p} \quad (3.39)$$

Dividing through by the phase-space volume element $\delta\mathbf{q}\delta\mathbf{p}$ we obtain the rate of change in density $f(\mathbf{q}, \mathbf{p})$, at the point $(\mathbf{q}, \mathbf{p})$,

$$\frac{1}{\delta\mathbf{q}\delta\mathbf{p}}\frac{\mathrm{d}}{\mathrm{d}t}\delta N = \frac{\partial}{\partial t}\left(\frac{\delta N}{\delta\mathbf{q}\delta\mathbf{p}}\right) = \left.\frac{\partial f}{\partial t}\right|_{\mathbf{q},\mathbf{p}} \quad (3.40)$$

Using the notation, $\mathbf{\Gamma} = (\mathbf{q}, \mathbf{p}) = (q_1, q_2, \ldots q_{3N}, p_1, p_2, \ldots p_{3N})$ for the $6N$-dimensional phase point, this may be written as

$$\left.\frac{\partial f}{\partial t}\right|_{\mathbf{\Gamma}} = -f\frac{\partial}{\partial \mathbf{\Gamma}}\cdot\dot{\mathbf{\Gamma}} - \dot{\mathbf{\Gamma}}\cdot\frac{\partial f}{\partial \mathbf{\Gamma}} = -\frac{\partial}{\partial \mathbf{\Gamma}}\cdot(\dot{\mathbf{\Gamma}}f) \quad (3.41)$$

This is the Liouville equation for the phase-space distribution function. Using the streaming or total time derivative of the distribution function, we can rewrite the Liouville equation in an equivalent form as,

$$\frac{\mathrm{d}f}{\mathrm{d}t} = \frac{\partial f}{\partial t} + \dot{\mathbf{\Gamma}}\cdot\frac{\partial f}{\partial \mathbf{\Gamma}} = -f\frac{\partial}{\partial \mathbf{\Gamma}}\cdot\dot{\mathbf{\Gamma}} \equiv -f\Lambda(\mathbf{\Gamma}) \quad (3.42)$$

This equation has been obtained without reference to the equations of motion. Its correctness does not require the existence of a Hamiltonian to generate the equations of motion. The equation rests on two conditions: that ensemble members cannot be created or destroyed and that the distribution function is sufficiently smooth that the appropriate derivatives exist. The term $\Lambda(\mathbf{\Gamma})$ is called the *phase-space compression factor* since it is equal to the negative time derivative of the logarithm of the phase-space distribution function.

$$\frac{\mathrm{d}}{\mathrm{d}t}\ln[f(\mathbf{\Gamma}, t)] = -\Lambda(\mathbf{\Gamma}) \quad (3.43)$$

The Liouville equation is usually written in a slightly simpler form. If the equations of motion can be generated from a Hamiltonian, then it is a simple matter to show that $\Lambda(\mathbf{\Gamma}) = 0$. This is so even in the presence of external fields which may be driving the system away from equilibrium by performing work on the system.

$$\Lambda(\mathbf{\Gamma}) = \sum_{i=1}^{N}\left(\frac{\partial}{\partial \mathbf{q}_i}\cdot\dot{\mathbf{q}}_i + \frac{\partial}{\partial \mathbf{p}_i}\cdot\dot{\mathbf{p}}_i\right) = \sum_{i=1}^{N}\left(\frac{\partial}{\partial \mathbf{q}_i}\cdot\frac{\partial H}{\partial \mathbf{p}_i} - \frac{\partial}{\partial \mathbf{p}_i}\cdot\frac{\partial H}{\partial \mathbf{q}_i}\right) = 0 \quad (3.44)$$

The existence of a Hamiltonian is a sufficient, but not necessary, condition for the phase-space compression factor to vanish. If phase space is incompressible, then the Liouville equation takes on its simplest form,

$$\frac{\mathrm{d}f}{\mathrm{d}t} = 0 \tag{3.45}$$

Time evolution of the distribution function

The following sections will be devoted to developing a formal operator algebra for manipulating the distribution function and averages of mechanical phase variables. This development is an extension of the treatment given by Berne (1977) which is applicable to Hamiltonian systems only. We will use the compact operator notation

$$\frac{\partial f}{\partial t} = -iLf = -\left(\left(\frac{\partial}{\partial \boldsymbol{\Gamma}}\cdot\dot{\boldsymbol{\Gamma}}\right) + \dot{\boldsymbol{\Gamma}}\cdot\frac{\partial}{\partial \boldsymbol{\Gamma}}\right)f \tag{3.46}$$

for the Liouville equation (3.41). The operator iL is called the distribution function (or f-) Liouvillean. Both the distribution function f, and the f-Liouvillean are functions of the initial phase $\boldsymbol{\Gamma}$. We assume that there is no explicit time dependence in the equations of motion (time-varying external fields will be treated in Chapter 8). Using this notation we can write the formal solution of the Liouville equation for the time-dependent N-particle distribution function $f(t)$ as

$$f(\boldsymbol{\Gamma}, t) = \exp(-iLt) f(\boldsymbol{\Gamma}, 0) \tag{3.47}$$

where $f(0)$, is the initial distribution function. This representation for the distribution function contains the exponential of an operator, which is a symbolic representation for the infinite series of operators. The *f-propagator* is defined as

$$\exp(-iLt) = \sum_{n=0}^{\infty} \frac{(-t)^n}{n!}(iL)^n \tag{3.48}$$

The formal solution given above can therefore be written as

$$f(t) = \sum_{n=0}^{\infty} \frac{(-t)^n}{n!}(iL)^n f(0) = \sum_{n=0}^{\infty} \frac{t^n}{n!}\frac{\partial^n}{\partial t^n} f(0) \tag{3.49}$$

This form makes it clear that the formal solution derived above is the Taylor-series expansion of the *explicit* time dependence of $f(\boldsymbol{\Gamma}, t)$, about $f(\boldsymbol{\Gamma}, 0)$.

Time evolution of phase variables

We will need to consider the time evolution of functions of the phase of the system. Such functions are called phase variables. An example would be the phase variable for the internal energy of a system, $H_0 = \Sigma_i p_i^2/2m + \Phi(q) = H_0(\mathbf{\Gamma})$. Phase variables, by definition, do not depend on time explicitly, their time dependence comes solely from the time dependence of the phase $\mathbf{\Gamma}$. Using the chain rule, the equation of motion for an arbitrary phase variable $B(\mathbf{\Gamma})$ can be written as

$$\dot{B}(\mathbf{\Gamma}) = \dot{\mathbf{\Gamma}} \cdot \frac{\partial}{\partial \mathbf{\Gamma}} B = \sum_{i=1}^{N} \left[\dot{\mathbf{q}}_i \cdot \frac{\partial}{\partial \mathbf{q}_i} + \dot{\mathbf{p}}_i \cdot \frac{\partial}{\partial \mathbf{p}_i} \right] B(\mathbf{\Gamma}) \equiv \mathrm{iL}(\mathbf{\Gamma}) B(\mathbf{\Gamma}) \quad (3.50)$$

The operator associated with the time derivative of a phase variable $\mathrm{iL}(\mathbf{\Gamma})$ is referred to as the phase variable (or p-) Liouvillean. The formal solution of this equation can be written in terms of the *p-propagator*, $\mathrm{e}^{\mathrm{iL}t}$. This gives the value of the phase variable as a function of time

$$B(t) = \exp(\mathrm{iL}t) B(0) \quad (3.51)$$

This expression is very similar in form to that for the distribution function. It is the Taylor-series expansion of the total time dependence of $B(t)$, expanded about $B(0)$. If the phase-space compression factor $\Lambda(\mathbf{\Gamma})$ is identically zero, then the p-Liouvillean is equal to the f-Liouvillean. In general this is not the case.

Properties of Liouville operators

In this section we will derive some of the more important properties of the Liouville operators. These will lead us naturally to a discussion of various *representations* of the properties of classical systems. The first property we shall discuss relates the p-Liouvillean to the f-Liouvillean as follows,

$$\int \mathrm{d}\mathbf{\Gamma}\, f(\mathbf{\Gamma}) \mathrm{iL} B(\mathbf{\Gamma}) = -\int \mathrm{d}\mathbf{\Gamma}\, B(\mathbf{\Gamma}) iLf(\mathbf{\Gamma}) \quad (3.52)$$

This is true for an arbitrary distribution function $f(0)$. To prove this identity the left-hand side can be written as

$$\begin{aligned} \int \mathrm{d}\mathbf{\Gamma}\, f(\mathbf{\Gamma}) \dot{\mathbf{\Gamma}} \cdot \frac{\partial}{\partial \mathbf{\Gamma}} B(\mathbf{\Gamma}) &= [f(\mathbf{\Gamma}) \dot{\mathbf{\Gamma}} B(\mathbf{\Gamma})]_S - \int \mathrm{d}\mathbf{\Gamma}\, B(\mathbf{\Gamma}) \frac{\partial}{\partial \mathbf{\Gamma}} \cdot (f(\mathbf{\Gamma}) \dot{\mathbf{\Gamma}}) \\ &= -\int \mathrm{d}\mathbf{\Gamma}\, B(\mathbf{\Gamma}) \left(\dot{\mathbf{\Gamma}} \cdot \frac{\partial}{\partial \mathbf{\Gamma}} + \frac{\partial}{\partial \mathbf{\Gamma}} \cdot \dot{\mathbf{\Gamma}} \right) f(\mathbf{\Gamma}) \\ &= -\int \mathrm{d}\mathbf{\Gamma}\, B(\mathbf{\Gamma}) iLf(\mathbf{\Gamma}) \end{aligned} \quad (3.53)$$

The boundary term (or surface integral) is zero because $f \to 0$ as any component of the momentum goes to infinity, and f can be taken to be periodic in all coordinates. If the coordinate space for the system is bounded then the surface S is the system boundary, and the surface integral is again zero as there can be no flow through the boundary.

Equations (3.52) and (3.53) show that L, L are adjoint operators. If the equations of motion are such that the phase-space compression factor, (3.43), is identically zero, then obviously $\mathrm{L} = L$ and the Liouville operator is self-adjoint, or *Hermitian.*

Schrödinger and Heisenberg representations

We can calculate the value of a phase variable $B(t)$ at time t by following B as it changes along a single trajectory in phase space. The average $\langle B(\boldsymbol{\Gamma}(t))\rangle$ can then be calculated by summing the values of $B(t)$ with a weighting factor determined by the probability of starting from each initial phase $\boldsymbol{\Gamma}$. These probabilities are chosen from an initial distribution function $f(\boldsymbol{\Gamma},0)$. This is the Heisenberg picture of phase space averages.

$$\langle B(t)\rangle = \int \mathrm{d}\boldsymbol{\Gamma}\, B(t) f(\boldsymbol{\Gamma}) = \int \mathrm{d}\boldsymbol{\Gamma}\, f(\boldsymbol{\Gamma}) \exp(\mathrm{iL}t) B(\boldsymbol{\Gamma}) \qquad (3.54)$$

The Heisenberg picture is exactly analogous to the Lagrangian formulation of fluid mechanics; we can imagine that the phase space *mass point* has a differential box $\mathrm{d}\boldsymbol{\Gamma}$ surrounding it which changes shape (and volume for a compressible fluid) with time as the phase point follows its trajectory. The probability of the differential element, or mass $f(\boldsymbol{\Gamma})\,\mathrm{d}\boldsymbol{\Gamma}$ remains constant, but the value of the observable changes implicitly in time.

The second view is the Schrödinger, or distribution based picture (Fig. 3.3). In this case we note that $\langle B(t)\rangle$ can be calculated by siting at a particular point in phase space and calculating the density of ensemble points as a function of time. This will give us the time-dependent N-particle distribution function $f(\boldsymbol{\Gamma},t)$. The average of B can now be calculated by summing the values of $B(\boldsymbol{\Gamma})$ but weighting these values by the current value of the distribution function at that place in phase space. Just as in the Eulerian formulation of fluid mechanics, the observable takes on a fixed value $B(\boldsymbol{\Gamma})$ for all time, while mass points with different probability flow through the box.

$$\langle B(t)\rangle = \int \mathrm{d}\boldsymbol{\Gamma}\, B(\boldsymbol{\Gamma}) f(\boldsymbol{\Gamma},t) = \int \mathrm{d}\boldsymbol{\Gamma}\, B(\boldsymbol{\Gamma}) \exp(-iLt) f(\boldsymbol{\Gamma},0) \qquad (3.55)$$

The average value of B changes with time as the distribution function changes. The average of B is computed by multiplying the value of $B(\boldsymbol{\Gamma})$, by the probability of finding the phase point $\boldsymbol{\Gamma}$ at time t, that is $f(\boldsymbol{\Gamma},t)$.

As we have just seen, these two pictures are of course equivalent. One can also prove their equivalence using the Liouville equation. This proof is

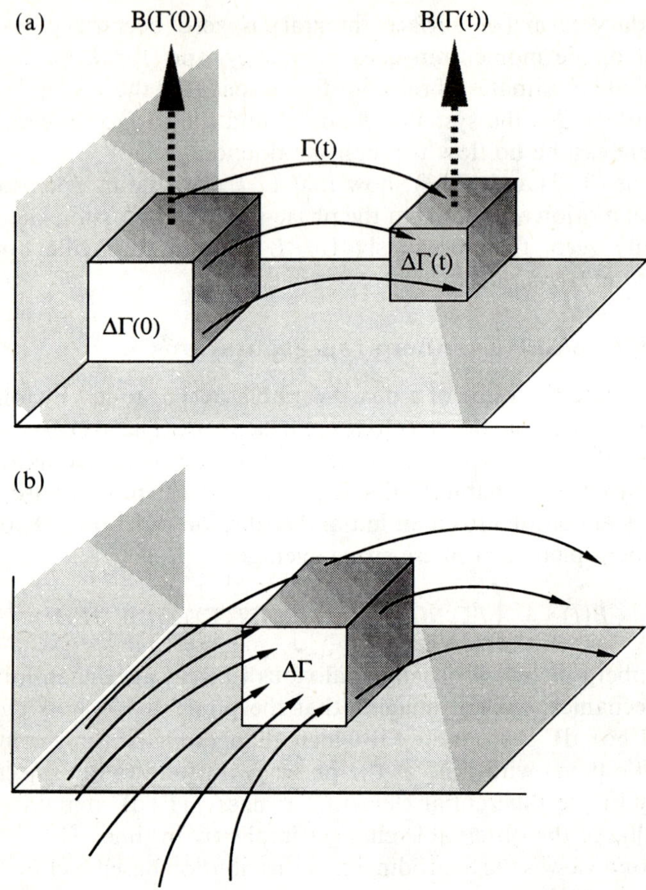

Fig. 3.3 The Schrödinger–Heisenberg equivalence. (a) The Heisenberg picture: $\dot{B} = \dot{\mathbf{\Gamma}} \cdot \partial/\partial\mathbf{\Gamma} B \equiv \mathrm{iL}(\mathbf{\Gamma})B(\mathbf{\Gamma})$. (b) The Schrödinger picture: $\partial f(\mathbf{\Gamma}, t)/\partial t = -\partial/\partial\mathbf{\Gamma} \cdot \dot{\mathbf{\Gamma}} f(\mathbf{\Gamma}, t) \equiv iLf(\mathbf{\Gamma}, t)$; $f(\mathbf{\Gamma}, t) \simeq \Delta N(t)/\Delta\mathbf{\Gamma}$ = the local density of ensemble representatives.

obtained by successive integrations by parts, or equivalently by repeated applications of (3.52). Consider

$$
\begin{aligned}
\int \mathrm{d}\mathbf{\Gamma}\, f(\mathbf{\Gamma})B(t) &= \int \mathrm{d}\mathbf{\Gamma}\, f(\mathbf{\Gamma})\exp(\mathrm{iL}t)B(\mathbf{\Gamma}) \\
&= \sum_{n=0}^{\infty} \frac{1}{n!} \int \mathrm{d}\mathbf{\Gamma}\, f(\mathbf{\Gamma})(\mathrm{iL}t)^n B(\mathbf{\Gamma}) \\
&= \sum_{n=0}^{\infty} \frac{1}{n!} \int \mathrm{d}\mathbf{\Gamma}\, f(\mathbf{\Gamma})\left(t\dot{\mathbf{\Gamma}} \cdot \frac{\partial}{\partial \mathbf{\Gamma}} \right)^n B(\mathbf{\Gamma}) \qquad (3.56)
\end{aligned}
$$

One can *unroll* each *p*-Liouvillean in turn from the phase variable to the distribution function (for the first transfer we consider $(\mathrm{iL})^{n-1}B$ to be a composite phase variable) so that (3.56) becomes

$$\int \mathrm{d}\boldsymbol{\Gamma}\, f(\boldsymbol{\Gamma})B(t) = \sum_{n=0}^{\infty} \frac{1}{n!} \int \mathrm{d}\boldsymbol{\Gamma} \left\{ -t \frac{\partial}{\partial \boldsymbol{\Gamma}} \cdot (\dot{\boldsymbol{\Gamma}} f(\boldsymbol{\Gamma})) \right\} \left(t\dot{\boldsymbol{\Gamma}} \cdot \frac{\partial}{\partial \boldsymbol{\Gamma}} \right)^{n-1} B(\boldsymbol{\Gamma})$$

This is essentially the property of phase- and distribution-function Liouvilleans which, as we have already proved, apply to the *n*th Liouvillean. Repeated application of this result leads to

$$\int \mathrm{d}\boldsymbol{\Gamma}\, f(\boldsymbol{\Gamma})B(t) = \sum_{n=0}^{\infty} \frac{(-t)^n}{n!} \int \mathrm{d}\boldsymbol{\Gamma} \left[\left(\frac{\partial}{\partial \boldsymbol{\Gamma}} \cdot \dot{\boldsymbol{\Gamma}} \right)^n f(\boldsymbol{\Gamma}) \right] B(\boldsymbol{\Gamma})$$

$$= \int \mathrm{d}\boldsymbol{\Gamma}\, B(\boldsymbol{\Gamma}) \exp(-iLt) f(\boldsymbol{\Gamma})$$

So, finally, we have the result

$$\int \mathrm{d}\boldsymbol{\Gamma}\, f(\boldsymbol{\Gamma})B(t) = \int \mathrm{d}\boldsymbol{\Gamma}\, B(\boldsymbol{\Gamma}) f(\boldsymbol{\Gamma}, t) \tag{3.57}$$

The derivation we have used assumes that the Liouvillean for the system has no explicit time dependence. (In Chapter 8 we will extend the derivation of these and other results to the time-dependent case.) Our present derivation makes no other references to the details of either the initial distribution function, or the equations of motion for the system. This means that these results are valid for systems subject to time-independent external fields, whether or not those equations are derivable from a Hamiltonian. These results are also independent of whether or not the phase-space compression factor vanishes identically.

A final point that can be made concerning the Schrödinger and Heisenberg pictures is that these two ways of computing phase averages by no means exhaust the range of possibilities. The Schrödinger and Heisenberg pictures differ in terms of the time chosen to calculate the distribution function, $f(\boldsymbol{\Gamma}, t)$. In the Heisenberg picture this time is zero, while in the Schrödinger picture it is t. One can of course develop intermediate representations corresponding any time between 0 and t (e.g. the interaction representation).

3.4 ERGODICITY, MIXING AND LYAPUNOV EXPONENTS

For many systems it is apparent that after possible initial transients lasting a time t_0, the N-particle distribution function $f(\boldsymbol{\Gamma}, t)$, becomes essentially time independent. This is evidenced by the fact that the macroscopic properties of the system relax to fixed average values. This obviously happens for equilibrium systems. It also occurs in some nonequilibrium systems, so-called nonequilibrium steady states. We will call all such systems stationary.

For a stationary system, we may define the ensemble average of a phase variable $B(\mathbf{\Gamma})$, using the stationary distribution function $f(\mathbf{\Gamma})$, so that

$$\langle B \rangle = \int \mathrm{d}\mathbf{\Gamma}\, f(\mathbf{\Gamma}) B(\mathbf{\Gamma}) \tag{3.58}$$

However, we may define a time average of the same phase variable as

$$\langle B \rangle_t \equiv \lim_{T \to \infty} \frac{1}{T} \int_{t_0}^{t_0 + T} \mathrm{d}t\, B(t) \tag{3.59}$$

where t_0 is the relaxation time required for the establishment of the stationary state. An ergodic system is a stationary system for which the ensemble and time averages of *usual* phase variables exist and are equal. By *usual* we mean phase-variable representations of the common macroscopic thermodynamic variables (see Section 3.7).

It is commonly believed that all realistic nonlinear many body systems are ergodic.

Example

We can give a simple example of ergodic flow if we take the energy surface to be the two-dimensional unit square $0 < p < 1$ and $0 < q < 1$. We shall assume that the equations of motion are given by

$$\begin{aligned} \dot{p} &= \alpha \\ \dot{q} &= 1 \end{aligned} \tag{3.60}$$

and we impose periodic boundary conditions on the system. These equations of motion can be solved to give

$$\begin{aligned} p(t) &= p_0 + \alpha t \\ q(t) &= q_0 + t \end{aligned} \tag{3.61}$$

The phase-space trajectory on the energy surface is given by eliminating t from these two equations

$$p = p_0 + \alpha(q - q_0) \tag{3.62}$$

If α is a rational number, $\alpha = m/n$, then the trajectory will be periodic and will repeat after a period $T = n$. If α is irrational, then the trajectory will be dense on the unit square but will not fill it. When α is irrational the system is ergodic. To show this explicitly consider the Fourier series expansion of an arbitrary phase function $A(q, p)$,

$$A(q, p) = \sum_{j,k=-\infty}^{\infty} A_{jk} \exp(2\pi i(jq + kp)) \tag{3.63}$$

We wish to show that the time average and phase average of $A(q, p)$ are equal for α irrational. The time average is given by

$$\begin{aligned} \langle A \rangle_t &= \lim_{T\to\infty} \int_{t_0}^{t_0+T} \mathrm{d}t \sum_{j,k=-\infty}^{\infty} A_{jk} \exp(2\pi i[j(q_0 + t) + k(p_0 + \alpha t)]) \\ &= A_{00} + \lim_{T\to\infty} \frac{1}{T} \sum_{j,k\neq 0} A_{jk} \mathrm{e}^{2\pi i[j(q_0+t_0)+k(p_0+\alpha t_0)]} \frac{\mathrm{e}^{2\pi i(j+\alpha k)T} - 1}{2\pi i(j + \alpha k)} \end{aligned} \tag{3.64}$$

For irrational α, the denominator can never be equal to zero, therefore

$$\langle A \rangle_t = A_{00} \tag{3.65}$$

Similarly, we can show that the phase-space average of A is

$$\langle A \rangle_{qp} = \int_0^1 \mathrm{d}q \int_0^1 \mathrm{d}p \, A(q, p) = A_{00} \tag{3.66}$$

and hence the system is ergodic. For rational α the denominator in (3.64) does become singular for a particular jk mode. The system is in the pure state labelled by jk. There is no mixing.

Ergodicity does *not* guarantee the relaxation of a system toward a stationary state. Consider a probability density which is not constant over the unit square; for example, let $f(q, p, t = 0)$ be given by

$$f(q, p, 0) = \sin(\pi p_0) \sin(\pi q_0) \tag{3.67}$$

then, at time t, under the above dynamics (with irrational α), it will be

$$f(q, p, t) = \sin(\pi(p_0 - \alpha t)) \sin(\pi(q_0 - t)) \tag{3.68}$$

The probability distribution is not changed in shape, it is only displaced. It has also *not* relaxed to a time-independent equilibrium distribution function. However, after an infinite length of time it will have wandered uniformly over the entire energy surface. It is therefore ergodic but it is termed *nonmixing*.

It is often easier to show that a system is not ergodic, rather than to show that it is ergodic. For example, the phase space of a system must be metrically transitive for it to be ergodic. That is, all of phase space, except possibly a set of measure zero, must be accessible to *almost all* the trajectories of the system. The reference to *almost all*, is because of the possibility that a set of initial starting states of measure zero, may remain forever within a subspace of phase space which is itself of measure zero. Ignoring the more pathological cases, if it is possible to divide phase space into two (or more) finite regions of nonzero measure, so that trajectories initially in a particular region remain there forever, then the system is not ergodic. A typical example would be a system in which a particle is trapped in a certain region of configuration space. Later we will give examples of this specific type.

Lyapunov exponents

If we consider two harmonic oscillators (see Section 3.2) which have the same frequency ω but different initial conditions (x_1, p_1) and (x_2, p_2), we can define the *distance* between the two phase points as

$$d = \|\mathbf{\Gamma}\| = (\mathbf{\Gamma}\cdot\mathbf{\Gamma})^{\frac{1}{2}} = \sqrt{(x_2 - x_1)^2 + \frac{(p_2 - p_1)^2}{m^2\omega^2}} \tag{3.69}$$

Using the equation for the trajectory of the harmonic oscillator (3.34), we see that as a function of time this distance is given by

$$d(t) = \sqrt{(x_2(t) - x_1(t))^2 + \frac{(p_2(t) - p_1(t))^2}{m^2\omega^2}} = d(0) \tag{3.70}$$

where $x_i(t)$ and $p_i(t)$ are the position and momenta of oscillator i at time t. This means that the trajectories of two independent harmonic oscillators always remain the same distance apart in $\mathbf{\Gamma}$ space.

This is not the typical behaviour of *nonlinear* systems. The neighbouring trajectories of most N-body nonlinear systems tend to drift apart with time. Indeed it is clear that if a system is to be mixing then the separation of neighbouring trajectories is a precondition. Weakly coupled harmonic oscillators are exceptions to the generally observed trajectory separation. This was a cause of some concern in the earliest dynamical simulations (Fermi *et al.*, 1955).

As the separation between neighbouring trajectories can be easily calculated in a classical mechanical simulation, this has been used to obtain quantitative measures of the mixing properties of nonlinear many-body systems. If we consider two N-body systems composed of particles which interact via identical sets of interparticle forces, but whose initial conditions differ by a small amount, then the phase-space separation is observed to change exponentially as

$$d(t) \equiv \sqrt{(\mathbf{\Gamma}_1(t) - \mathbf{\Gamma}_2(t))^2} \cong c \exp(\lambda t) \tag{3.71}$$

At intermediate times the exponential growth of $d(t)$ will be dominated by the fastest growing direction in phase space (this direction will change continuously with time). This equation defines the largest *Lyapunov exponent*,

λ, for the system (by convention λ is defined to be real, so any oscillating part of the trajectory separation is ignored). For the harmonic oscillator the phase separation is a constant of the motion and, therefore, the Lyapunov exponent, λ, is zero. In practical applications this exponential separation for an N-particle system continues until it approaches a limit imposed by the externally imposed boundary conditions—the container walls, or the energy, or other thermodynamic constraints on the system (Section 7.8). If the system has energy as a constant of the motion then the maximum separation is the maximum distance between two points on the energy hypersphere. This depends upon the value of the energy and the dimension of the phase space.

The largest Lyapunov exponent indicates the rate of growth of trajectory separation in phase space. If we consider a third phase point $\mathbf{\Gamma}_3(t)$, which is constrained such that the vector between $\mathbf{\Gamma}_1$ and $\mathbf{\Gamma}_3$ is always orthogonal to the vector between $\mathbf{\Gamma}_1$ and $\mathbf{\Gamma}_2$, then we can follow the rate of change of a two-dimensional area in phase space. We can use these two directions to define an area element $V_2(t)$, and rate of change of the volume element is given by

$$V_2(t) = V_2(0)\exp([\lambda_1 + \lambda_2]t) \tag{3.72}$$

As we already know the value of λ_1, this defines the second-largest Lyapunov exponent λ_2. In a similar way, if we construct a third phase-space vector $\mathbf{\Gamma}_{14}(t)$ which is constrained to be orthogonal to both $\mathbf{\Gamma}_{12}(t)$ and $\mathbf{\Gamma}_{13}(t)$, then we can follow the rate of change of a three-dimensional volume element $V_3(t)$ and calculate the third-largest exponent λ_3

$$V_3(t) = V_3(0)\exp([\lambda_1 + \lambda_2 + \lambda_3]t) \tag{3.73}$$

This construction can be generalized to calculate the full spectrum of Lyapunov exponents for an N-particle system. We consider the trajectory $\mathbf{\Gamma}(t)$ of a dynamical system in phase space and study the convergence or divergence of neighbouring trajectories by taking a set of basis vectors (tangent vectors) in phase space $\{\boldsymbol{\delta}_1, \boldsymbol{\delta}_2, \boldsymbol{\delta}_3, \ldots\}$, where $\boldsymbol{\delta}_i = \mathbf{\Gamma}_i - \mathbf{\Gamma}_0$. Some care must be exercised in forming the set of basis vectors to ensure that the full dimension of phase space is spanned by the basis set, and that the basis set is minimal. This simply means that the constants of motion must be considered when calculating the dimension of accessible phase space. If the equation of motion for a trajectory is of the form

$$\dot{\mathbf{\Gamma}} = \mathbf{G}(\mathbf{\Gamma}) \tag{3.74}$$

then the equation of motion for the tangent vector $\boldsymbol{\delta}_i$ is

$$\dot{\boldsymbol{\delta}}_i = \mathbf{F}_i(\mathbf{\Gamma}) = \mathbf{T}(\mathbf{\Gamma})\cdot\boldsymbol{\delta}_i + \mathrm{O}(\boldsymbol{\delta}_i^2) \tag{3.75}$$

Here $\mathbf{T}(\boldsymbol{\Gamma})$ is the Jacobian matrix (or stability matrix $\partial\mathbf{G}/\partial\boldsymbol{\Gamma}$) for the system. If the magnitude of the tangent vector is small enough the nonlinear terms in equation (3.75) can be neglected. The formal solution of this equation is

$$\boldsymbol{\delta}_i(t) = \exp\left[\int_0^t \mathrm{d}s\, \mathbf{T}(s)\right]\boldsymbol{\delta}_i(0) \tag{3.76}$$

The mean exponential rate of growth of the ith tangent vector, gives the ith Lyapunov exponent

$$\lambda_i(\boldsymbol{\Gamma}(0), \boldsymbol{\delta}_i(0)) = \lim_{t\to\infty} \frac{1}{t} \ln \frac{\|\boldsymbol{\delta}_i(t)\|}{\|\boldsymbol{\delta}_i(0)\|} \tag{3.77}$$

The existence of the limit is ensured by the multiplicative ergodic theorem of Oseledec (1968) (see also Eckmann and Ruelle (1985)). The Lyapunov exponents can be ordered $\lambda_1 > \lambda_2 > \ldots > \lambda_M$ and if the system is ergodic, the exponents are independent of the initial phase $\boldsymbol{\Gamma}(0)$ and the initial phase-space separation $\boldsymbol{\delta}_i(0)$.

If we consider the volume element V_N where N is the dimension of phase space then we can show that the phase-space compression factor gives the rate of change of phase-space volume, and that this is simply related to the sum of the Lyapunov exponents by

$$\dot{V}_N = \left\langle \frac{\partial}{\partial\boldsymbol{\Gamma}} \cdot \dot{\boldsymbol{\Gamma}} \right\rangle V_N = \left(\sum_{i=1}^{N} \lambda_i \right) V_N \tag{3.78}$$

For a Hamiltonian system, the phase-space compression factor is identically zero, so the phase-space volume is conserved. This is a simple consequence of Liouville's theorem. From (3.78) it follows that the sum of the Lyapunov exponents is also equal to zero. If the system is time reversible then the Lyapunov exponents occur in pairs $(-\lambda_i, \lambda_i)$. This ensures that $\mathrm{d}(t)$, $V_2(t)$, $V_3(t)$, etc., change at the same rate with both forward and backward time evolution. It is generally believed that it is necessary to have at least one positive Lyapunov exponent for the system to be mixing. In Chapters 7 and 10 we will return to consider Lyapunov exponents in both equilibrium and nonequilibrium systems.

3.5 EQUILIBRIUM TIME-CORRELATION FUNCTIONS

We shall often refer to averages over equilibrium distribution functions f_0 (we use the subscript zero to denote equilibrium, which should not be confused with $f(0)$, a distribution function at $t = 0$). Distribution functions are called equilibrium if they pertain to steady, unperturbed equations of motion and they have no explicit time dependence. An equilibrium distribution function satisfies a Liouville equation of the form

$$\frac{\partial}{\partial t} f_0 = -iLf_0 = 0 \tag{3.79}$$

This implies that the equilibrium average of any phase variable is a stationary quantity. That is, for an arbitrary phase variable B,

$$\begin{aligned}
\frac{\mathrm{d}}{\mathrm{d}t} \langle B(t) \rangle_0 &= \frac{\mathrm{d}}{\mathrm{d}t} \int \mathrm{d}\boldsymbol{\Gamma}\, f_0(\boldsymbol{\Gamma}) \exp(\mathrm{i}\mathrm{L}t) B(\boldsymbol{\Gamma}) \\
&= \int \mathrm{d}\boldsymbol{\Gamma}\, f_0 \frac{\partial}{\partial t} \exp(\mathrm{i}\mathrm{L}t) B(\boldsymbol{\Gamma}) \\
&= \int \mathrm{d}\boldsymbol{\Gamma}\, f_0 \mathrm{iL} \exp(\mathrm{i}\mathrm{L}t) B(\boldsymbol{\Gamma}) \\
&= -\int \mathrm{d}\boldsymbol{\Gamma}\, [iLf_0(\boldsymbol{\Gamma})] \exp(\mathrm{i}\mathrm{L}t) B(\boldsymbol{\Gamma}) \\
&= 0
\end{aligned} \tag{3.80}$$

We will often need to calculate the equilibrium time-correlation function of a phase variable A with another phase variable B at some other time. We define the equilibrium time-correlation function of A and B by

$$C_{AB}(t) \equiv \int \mathrm{d}\boldsymbol{\Gamma}\, f_0 B^* \mathrm{e}^{\mathrm{i}\mathrm{L}t} A = \langle A(t) B^* \rangle_0 \tag{3.81}$$

where B^* denotes the complex conjugate of the phase variable B. Sometimes we will refer to the autocorrelation function of a phase variable A. If this variable is real, one can form a simple graphical representation of how such functions are calculated (see Fig. 3.4).

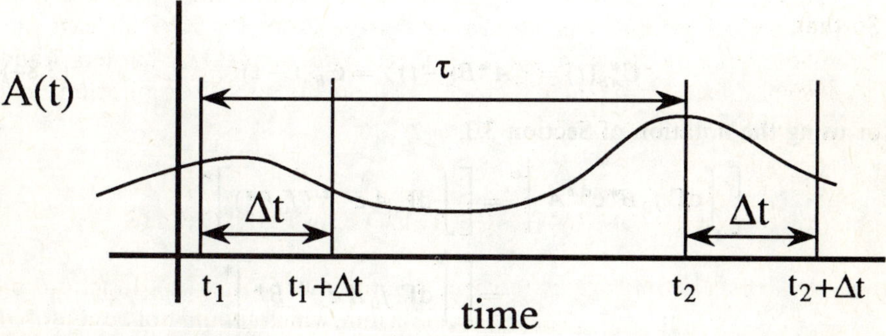

Fig. 3.4 Equilibrium time autocorrelation function of real variable A. $\langle A(0)A(\Delta t)\rangle = 1/N_\tau \Sigma_{i=1}^{N_\tau} A(t_i)A(t_i + \Delta t)$; $t_i = 0, \tau, 2\tau, \ldots$. For samples in the sum to be independent, τ should be chosen so that $\langle A(0)A(\tau)\rangle \ll \langle A(0)^2\rangle$.

Because the averages are to be taken over a stationary equilibrium distribution function, time-correlation functions are only sensitive to time difference between which A and B are evaluated. $C_{AB}(t)$ is independent of the particular choice of the time origin. If iL *generates* the distribution function f_0, then the propagator $\exp(-iLt)$ preserves f_0. (The converse is not necessarily true.) To be more explicit $f_0(t_1) = \exp(-iLt_1)f_0 = f_0$, so that $C_{AB}(t)$ becomes

$$C_{AB}(t) = \int \mathrm{d}\boldsymbol{\Gamma}\, f_0 B^* \mathrm{e}^{\mathrm{i}Lt} A = \int \mathrm{d}\boldsymbol{\Gamma}\, f_0(t_1) B^* \mathrm{e}^{\mathrm{i}Lt} A$$

$$= \int \mathrm{d}\boldsymbol{\Gamma} (\exp[-iLt_1] f_0) B^* \exp[\mathrm{i}Lt] A$$

$$= \int \mathrm{d}\boldsymbol{\Gamma}\, f_0 (\exp[\mathrm{i}Lt_1] B^*)(\exp[\mathrm{i}L(t + t_1)] A$$

$$= \int \mathrm{d}\boldsymbol{\Gamma}\, f_0 A(t_1 + t) B^*(t_1) \qquad (3.82)$$

In deriving the last form of (3.82) we have used the important fact that since $\mathrm{iL} = \mathrm{d}\boldsymbol{\Gamma}/\mathrm{d}t \cdot \partial/\partial\boldsymbol{\Gamma}$ and the equations of motion are real it follows that L is purely imaginary. Thus, $(\mathrm{iL})^* = \mathrm{iL}$ and $(\mathrm{e}^{\mathrm{i}Lt})^* = \mathrm{e}^{\mathrm{i}Lt}$. Comparing (3.82) with the definition of $C_{AB}(t)$ above, we see that the equilibrium time-correlation function is independent of the choice of time origin. It is solely a function of the difference in time of the two arguments, A and B. A further identity follows from this result if we choose t_1 to be $-t$. We find that

$$C_{AB}(t) = \langle A(t) B^*(0) \rangle_0 = \langle A(0) B^*(-t) \rangle_0 \qquad (3.83)$$

So that,

$$C^*_{AB}(t) = \langle A^* B(-t) \rangle = C_{BA}(-t) \qquad (3.84)$$

or using the notation of Section 3.3,

$$\left[\int \mathrm{d}\boldsymbol{\Gamma}\, f_0 B^* \mathrm{e}^{\mathrm{i}Lt} A \right]^* = \left[\int \mathrm{d}\boldsymbol{\Gamma}\, A\, \mathrm{e}^{-\mathrm{i}Lt} (f_0 B^*) \right]^*$$

$$= \left[\int \mathrm{d}\boldsymbol{\Gamma}\, f_0 A\, \mathrm{e}^{-\mathrm{i}Lt} B^* \right]^*$$

$$= \int \mathrm{d}\boldsymbol{\Gamma}\, f_0 A^* \mathrm{e}^{-\mathrm{i}Lt} B \qquad (3.85)$$

The second equality in equation (3.85) follows by expanding the operator $\exp(-iLt)$ and repeatedly applying the identity

$$iLt[f_0 B^*) = \frac{\partial}{\partial \mathbf{\Gamma}} \cdot (\dot{\mathbf{\Gamma}} f_0 B^*) = B^* \frac{\partial}{\partial \mathbf{\Gamma}} \cdot (\dot{\mathbf{\Gamma}} f_0) + f_0 \dot{\mathbf{\Gamma}} \cdot \frac{\partial}{\partial \mathbf{\Gamma}} B^*$$

$$= B^* iLf_0 + f_0 \mathrm{iL} B^*$$

$$= f_0 \mathrm{iL}$$

The term iLf_0 is zero from equation (3.79).

Over the scalar product defined by equation (3.81), L is an Hermitian operator. The Hermitian adjoint of L denoted, $\mathrm{L}^\dagger$ can be defined by the equation,

$$\left[\int \mathrm{d}\mathbf{\Gamma}\, f_0 B^* \mathrm{e}^{\mathrm{iL}t} A\right]^* \equiv \int \mathrm{d}\mathbf{\Gamma}\, f_0 A^* \exp[-\mathrm{iL}^\dagger t] B \qquad (3.86)$$

Comparing (3.86) with (3.85) we see two things: we see that the Liouville operator L is self-adjoint or *Hermitian* ($\mathrm{L} = \mathrm{L}^\dagger$); and, therefore, the propagator $\mathrm{e}^{\mathrm{iL}t}$, is *unitary*. This result stands in contrast to those of Section 3.3, for arbitrary distribution functions.

We can use the autocorrelation function of A to define a norm in Liouville space. This length, or norm, of a phase variable $\| A \|$, is defined by the equation

$$\| A \|^2 = \int \mathrm{d}\mathbf{\Gamma}\, f_0 A(\mathbf{\Gamma}) A^*(\mathbf{\Gamma}) = \int \mathrm{d}\mathbf{\Gamma}\, f_0 |A(\mathbf{\Gamma})|^2$$

$$= \langle |A(\mathbf{\Gamma})|^2 \rangle_0 \geq 0 \qquad (3.87)$$

We can see immediately that the norm of any phase variable is time independent because

$$\| A(t) \|^2 = \int \mathrm{d}\mathbf{\Gamma}\, f_0 A(t) A^*(t)$$

$$= \int \mathrm{d}\mathbf{\Gamma}\, f_0 (\exp(\mathrm{iL}t) A(\mathbf{\Gamma}))(\exp(\mathrm{iL}t) A^*(\mathbf{\Gamma}))$$

$$= \int \mathrm{d}\mathbf{\Gamma}\, f_0 \exp(\mathrm{iL}t)(A(\mathbf{\Gamma}) A^*(\mathbf{\Gamma}))$$

$$= \int \mathrm{d}\mathbf{\Gamma}\, (\exp(-iLt) f_0) |A|^2 = \| A(0) \|^2 \qquad (3.88)$$

The propagator is said to be norm preserving (Fig. 3.5). This is a direct result of the fact that the propagator is a unitary operator. The propagator can be thought of as a rotation operator in Liouville space.

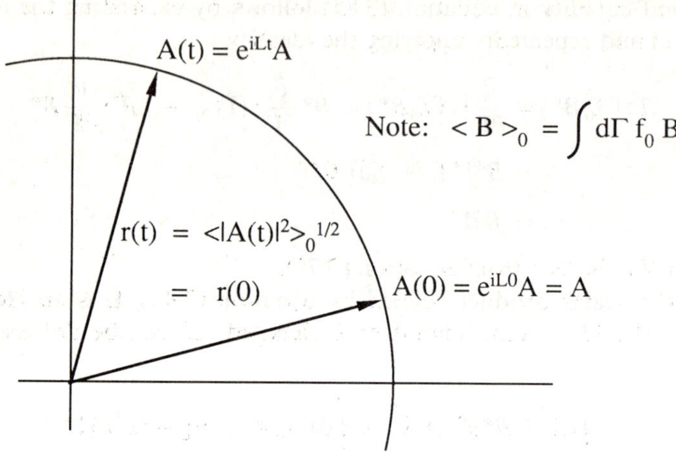

Fig. 3.5 The propagator is norm preserving.

A phase variable whose norm is unity is said to be normalized. The scalar product, (A, B^*) of two phase variables A and B is simply the equilibrium average of A and B^*, namely $\langle AB^* \rangle_0$. The norm of a phase variable is simply the scalar product of the variable with itself. The autocorrelation function $C_{AA}(t)$ has a zero time value which is equal to the norm of A. The propagator increases the angle between A^* and $A(t)$, and the scalar product which is the projection of $A(t)$ along A^* therefore decreases. The autocorrelation function of a given phase variable therefore measures the rate at which the $6N$-dimensional rotation occurs.

We will now derive some relations for the time derivatives of time-correlation functions. It is easy to see that

$$\frac{d}{dt} C_{AB}(t) = \frac{d}{dt} \int d\mathbf{\Gamma}\, f_0(\exp(iLt)A(\mathbf{\Gamma}))B^*(\mathbf{\Gamma})$$

$$= \int d\mathbf{\Gamma}\, f_0(iL\exp(iLt)A(\mathbf{\Gamma}))B^*(\mathbf{\Gamma}) = C_{\dot{A}B}(t) \tag{3.89}$$

$$= -\int d\mathbf{\Gamma} (e^{iLt}A) iL(f_0 B^*) = -\int d\mathbf{\Gamma}\, A(t) \frac{\partial}{\partial \mathbf{\Gamma}} (\dot{\mathbf{\Gamma}} f_0 B^*)$$

$$= -\int d\mathbf{\Gamma}\, f_0(\exp(iLt)A(\mathbf{\Gamma}))iLB^*(\mathbf{\Gamma})$$

$$= -\int d\mathbf{\Gamma}\, f_0(\exp(iLt)A(\mathbf{\Gamma}))\dot{B}^*(\mathbf{\Gamma})$$

$$= -C_{A\dot{B}}(t) \tag{3.90}$$

3.6 OPERATOR IDENTITIES

In this section we develop the operator algebra that we will need to manipulate expressions containing Liouvilleans and their associated propagators. Most of the identities which we obtain are valid for arbitrary time-independent operators. Thus far we have dealt with propagators in the time domain. For many problems it is more useful to consider the frequency-dependent Laplace, or Fourier–Laplace, transforms. A useful mathematical object is the Laplace transform of the propagator. This is called the resolvent. The resolvent is an operator in the domain of the Laplace transform variable s,

$$G(s) = \int_0^\infty dt\, e^{-st} e^{-iLt} \tag{3.91}$$

Our first operator identity is obtained by considering two arbitrary operators A and B,

$$(A + B)^{-1} = A^{-1} - A^{-1}B(A + B)^{-1} \tag{3.92}$$

This identity is easily verified by operating from the right-hand side of this equation with $(A + B)$, so that

$$\begin{aligned}(A + B)^{-1}(A + B) &= [A^{-1} - A^{-1}B(A + B)^{-1}](A + B)\\ &= A^{-1}(A + B) - A^{-1}B\\ &= A^{-1}A + A^{-1}B - A^{-1}B = I\end{aligned} \tag{3.93}$$

The operator expression $(A + B)^{-1}$ is the inverse of the operator $(A + B)$. To interpret an operator inverse of $(A + B)^{-1}$, we use the series expansion

$$(I + A)^{-1} = \sum_{n=0}^{\infty} (-A)^n \tag{3.94}$$

First we prove that the right-hand side of this expression is indeed the inverse of the operator $(I + A)$. To do this consider

$$\sum_{n=0}^{\infty} (-A)^n(I + A) = \sum_{n=0}^{\infty} (-A)^n - \sum_{n=1}^{\infty} (-A)^n = I \tag{3.95}$$

so that this series expansion allows us to represent the inverse of $(I + A)$ in terms of an infinite series of products of the operator A.

The Dyson decomposition of propagators

We can now investigate the Laplace transform (or resolvent) of the exponential of an operator in more detail. We use the expansion of the exponential to show that

$$\int_0^\infty dt\, e^{-st} e^{-At} = \int_0^\infty dt\, e^{-st} \sum_{n=0}^\infty \frac{(-)^n}{n!} A^n t^n = \sum_{n=0}^\infty \frac{(-)^n}{n!} A^n \int_0^\infty dt\, t^n e^{-st}$$

$$= \sum_{n=0}^\infty (-)^n \frac{A^n}{s^{n+1}} = \frac{1}{s}\left(I + \frac{A}{s}\right)^{-1} = (s + A)^{-1} \quad (3.96)$$

This means that the resolvent of the operator, e^{-At}, is simply $(s + A)^{-1}$. We can now consider the resolvent derived from the operator $A + B$, and using the first identity above, relate this resolvent to the resolvent of A. We can write

$$(s + A + B)^{-1} = (s + A)^{-1} - (s + A)^{-1} B (s + A + B)^{-1} \quad (3.97)$$

Substituting the Laplace integrals for the operators $(s + A)^{-1}$ and $(s + A + B)^{-1}$ into this equation gives

$$\int_0^\infty dt\, e^{-st} e^{-(A+B)t} = \int_0^\infty dt\, e^{-st} e^{-At} - \int_0^\infty dt_1\, e^{-st_1} e^{-At_1} B \int_0^\infty dt_2\, e^{-st_2} e^{-(A+B)t_2}$$

$$= \int_0^\infty dt\, e^{-st} \left\{ e^{-At} - \int_0^t dt_1\, e^{-At_1} B\, e^{-(A+B)(t-t_1)} \right\} \quad (3.98)$$

As the equality holds for all values of s, the integrands must be equal, so

$$e^{-(A+B)t} = e^{-At} - \int_0^t dt_1\, e^{-At_1} B\, e^{-(A+B)(t-t_1)} \quad (3.99)$$

This result is a very important step towards understanding the relationship between different propagators and is referred to as the *Dyson decomposition* when applied to propagators (Dyson, 1949). The derivation that we have used here is only valid if both operators A and B have no explicit time dependence. (We consider the time-dependent case in Chapter 8.) If we consider the propagators $\exp((A + B)t)$ and $\exp(At)$, then a second Dyson decomposition can be obtained:

$$e^{(A+B)t} = e^{At} + \int_0^t dt_1\, e^{At_1} B\, e^{(A+B)(t-t_1)} \quad (3.100)$$

It is handy to use a graphical shorthand notation for the Dyson equation. Using this shorthand notation these two respective equations become

$$\Leftarrow\, = \,\leftarrow - \Leftarrow(\blacklozenge - \bigcirc)\leftarrow \quad (3.101)$$

and

$$\Rightarrow\, = \,\rightarrow + \Rightarrow(\blacklozenge - \bigcirc)\rightarrow \quad (3.102)$$

The diamond (◆) denotes the $(A + B)$ Liouvillean and the circle (○) denotes the A Liouvillean; the arrows (⇐) and (⇒) denote the propagators $\exp(-i(A+B)t)$ and $\exp(i(A+B)t)$, respectively, while (←) and (→) denote $\exp(-At)$ and $\exp(At)$, respectively. An $n-1$ fold convolution is implied by a chain of n arrows.

As an example of the application of this result, consider the case where B is a small perturbation to the operator A. In this case the Dyson decomposition gives the full $(A+B)$ propagator as the sum of the unperturbed A propagator plus a correction term. One often needs to compare the operation of different propagators on either a phase variable or a distribution function. For example, one might like to know the difference between the value of a phase variable $A(\boldsymbol{\Gamma})$ propagated under the combined influence of the N-particle interactions *and* an applied external field F_e, with the value which the phase variable might take at the same time in the absence of the external field. In that case (Evans and Morriss, 1984a), recursive substitution in (3.102) gives,

$$\begin{aligned} A(t, F_e) &= \Rightarrow A(\boldsymbol{\Gamma}) \\ &= [\rightarrow + \rightarrow(\blacklozenge - \bigcirc)\rightarrow + \rightarrow(\blacklozenge - \bigcirc)\rightarrow(\blacklozenge - \bigcirc)\rightarrow \\ &\quad + \rightarrow(\blacklozenge - \bigcirc)\rightarrow(\blacklozenge - \bigcirc)\rightarrow(\blacklozenge - \bigcirc)\rightarrow \\ &\quad + \rightarrow(\blacklozenge - \bigcirc)\rightarrow(\blacklozenge - \bigcirc)\rightarrow(\blacklozenge - \bigcirc)\rightarrow(\blacklozenge - \bigcirc)\rightarrow \\ &\quad + \ldots]A(\boldsymbol{\Gamma}) \end{aligned}$$

Therefore, we can write,

$$\Rightarrow A(\boldsymbol{\Gamma}) = \sum_{n=0}^{\infty} \rightarrow[(\blacklozenge - \bigcirc)\rightarrow]^n A(\boldsymbol{\Gamma}) \tag{3.103}$$

This equation is of limited usefulness because, in general, ◆ and → do not commute. This means that the Liouvillean frequently represented by ◆, is locked inside a convolution chain of propagators with which it does not commute. A more useful expression can be derived from (3.102) by realizing that ◆ commutes with its own propagator, namely ⇒. Similarly, ○ commutes with its own propagator →. We can 'unlock' the respectively Liouvilleans from the chain in (3.102) by writing

$$\Rightarrow\; = \;\rightarrow + \blacklozenge \Rightarrow\rightarrow - \Rightarrow\rightarrow\bigcirc \tag{3.104}$$

We can recursively substitute for ⇒, yielding

$$\begin{aligned} \Rightarrow\; &= \;\rightarrow + \blacklozenge\rightarrow\rightarrow - \rightarrow\rightarrow\bigcirc \\ &\quad + \blacklozenge\blacklozenge\rightarrow\rightarrow\rightarrow - 2\blacklozenge\rightarrow\rightarrow\rightarrow\bigcirc + \rightarrow\rightarrow\rightarrow\bigcirc\bigcirc \\ &\quad + \ldots \end{aligned} \tag{3.105}$$

It is easy to show that

$$(\rightarrow)^n = (t^n/n!)\rightarrow \tag{3.106}$$

Thus (3.105) can be written as

$$\Rightarrow = \{1 + t(\blacklozenge - \bigcirc) + (t^2/2!)(\blacklozenge\blacklozenge - 2\blacklozenge\bigcirc + \bigcirc\bigcirc) + (t^3/3!)(\blacklozenge\blacklozenge\blacklozenge - 3\blacklozenge\blacklozenge\bigcirc + 3\blacklozenge\bigcirc\bigcirc - \bigcirc\bigcirc\bigcirc) + \ldots\}\rightarrow \tag{3.107}$$

This equation was first derived by Evans and Morriss (1984a). Its utility arises from the fact that by 'unrolling' the Liouville operators to the left and the propagator to the right, *explicit* formulae for the expansion can usually be derived. A limitation of the formula is that successive terms on the right-hand side do *not* constitute a power-series expansion of the difference in the two propagators in powers of the difference between the respective Liouvilleans. To be more explicit, the term, $(t^3/3!)(\blacklozenge\blacklozenge\blacklozenge - 3\blacklozenge\blacklozenge\bigcirc + 3\blacklozenge\bigcirc\bigcirc - \bigcirc\bigcirc\bigcirc)$ is *not* in general of the order $(\blacklozenge - \bigcirc)^3$.

Campbell–Barker–Hausdorff theorem

If A and B are noncommuting operators then the operator expression $\exp(A)\exp(B)$ can be written in the form $\exp(C)$ where C is given by

$$C = A + B + \tfrac{1}{2}[A,B] + \tfrac{1}{12}\{[[A,B],B] + [[B,A],A]\} + \ldots \tag{3.108}$$

The notation [,] is the usual quantum-mechanical commutator. A rearrangement of this expansion, known as the Magnus expansion is well known to quantum theorists (Magnus, 1954). Any finite truncation of the Magnus expansion for the time-displacement operator, gives a unitary time-displacement operator approximation (Pechukas and Light, 1966). This result has not proved as useful for nonequilibrium statistical mechanics as it is for quantum theory.

3.7 THE IRVING–KIRKWOOD PROCEDURE

In Chapter 2 we gave a brief outline of the structure of macroscopic hydrodynamics. We saw that, given appropriate boundary conditions, it is possible to use the Navier–Stokes equations to describe the resulting macroscopic flow patterns. In this chapter we began the microscopic description of nonequilibrium systems using the Liouville equation. We will now follow a procedure first outlined by Irving and Kirkwood (1950) to derive microscopic expressions for the thermodynamic forces and fluxes appearing in the phenomenological equations of hydrodynamics.

In our treatment of the macroscopic equations we stressed the role played by the densities of conserved quantities. Our first task here will be to define microscopic expressions for the local densities of mass, momentum and energy. If the mass of the individual atoms in our system is m then the mass per unit volume at a position $\mathbf{r}$ and time t can be obtained by taking an appropriate average over the normalized N-particle distribution function $f(\mathbf{\Gamma}, t)$. To specify that the particles should be at the macroscopic position $\mathbf{r}$, we will use a *macroscopic* delta function, $\delta(\mathbf{r} - \mathbf{r}_i)$. This *macroscopic* delta function is zero if atom i is outside some microscopic volume δV; it is a constant if atom i is inside this volume (δ is a smeared-out version of the usual point delta function). We will assume that particle dynamics are given by field-free Newtonian equations of motion. The value of the constant is determined from the normalization condition,

$$\int_V \mathrm{d}\mathbf{r}\, \delta(\mathbf{r}) = 1 \tag{3.109}$$

The volume V is envisioned to be infinitesimal on a macroscopic scale.

The mass density $\rho(\mathbf{r}, t)$ can be calculated from the following average,

$$\begin{aligned} \rho(\mathbf{r}, t) &= \int \mathrm{d}\mathbf{\Gamma}\, f(\mathbf{\Gamma}, t) \sum_i m\delta(\mathbf{r} - \mathbf{r}_i) \\ &= \int \mathrm{d}\mathbf{\Gamma}\, f(\mathbf{\Gamma}, 0) \sum_i m\delta(\mathbf{r} - \mathbf{r}_i(t)) \\ &= \left\langle \sum_i m | \mathbf{r}_i(t) = \mathbf{r} \right\rangle \end{aligned} \tag{3.110}$$

The first line of this equation is a Schrödinger representation of the density while the second and third lines are written in the Heisenberg representation. The equivalence of these two representations is easily seen by 'unrolling' the propagator from the distribution function onto the phase variable. Since $\mathbf{r}$ is a constant, a nominated position, it is unchanged by this 'unrolling' procedure.

The momentum density, $\rho\mathbf{u}$, and total energy density, ρe, are defined in an analogous manner.

$$\begin{aligned} \rho(\mathbf{r}, t)\mathbf{u}(\mathbf{r}, t) &= \int \mathrm{d}\mathbf{\Gamma}\, f(\mathbf{\Gamma}, t) \sum_i m\mathbf{v}_i \delta(\mathbf{r} - \mathbf{r}_i) \\ &= \int \mathrm{d}\mathbf{\Gamma}\, f(\mathbf{\Gamma}, t) \sum_i \mathbf{p}_i \delta(\mathbf{r} - \mathbf{r}_i) \\ &= \left\langle \sum_i \mathbf{p}_i(t) | \mathbf{r}_i(t) = \mathbf{r} \right\rangle \end{aligned} \tag{3.111}$$

$$\rho(\mathbf{r},t)e(\mathbf{r},t) = \int d\mathbf{\Gamma}\, f(\mathbf{\Gamma},t)\left[\tfrac{1}{2}\sum_i m\mathbf{v}_i^2 + \tfrac{1}{2}\sum_{i,j}\phi_{ij}\right]\delta(\mathbf{r}-\mathbf{r}_i)$$
$$= \left\langle \tfrac{1}{2}\sum_i m\mathbf{v}_i^2(t) + \tfrac{1}{2}\sum_{i,j}\phi_{ij}(t)|\mathbf{r}_i(t) = \mathbf{r}\right\rangle \tag{3.112}$$

In these equations $\mathbf{v}_i$ is the velocity of particle i, $\mathbf{p}_i$ is its momentum, $\mathbf{r}_{ij} \equiv \mathbf{r}_j - \mathbf{r}_i$, and we assume that the total potential energy of the system, Φ, is pair-wise additive and can be written as

$$\Phi = \tfrac{1}{2}\sum_{i,j}\phi_{ij} \tag{3.113}$$

We arbitrarily assign one half of the potential energy to each of the two particles which contribute ϕ_{ij} to the total potential energy of the system.

The conservation equations involve time derivatives of the averages of the densities of conserved quantities. To begin, we will calculate the time derivative of the mass density.

$$\frac{\partial\rho(\mathbf{r},t)}{\partial t} = \int d\mathbf{\Gamma}\,\frac{\partial f(\mathbf{\Gamma},t)}{\partial t}\sum_i m\delta(\mathbf{r}-\mathbf{r}_i)$$
$$= -\int d\mathbf{\Gamma}\sum_i m\delta(\mathbf{r}-\mathbf{r}_i)iLf(\mathbf{\Gamma},t)$$
$$= \int d\mathbf{\Gamma}\, f(\mathbf{\Gamma},t)\mathrm{iL}\sum_i m\delta(\mathbf{r}-\mathbf{r}_i)$$
$$= \int d\mathbf{\Gamma}\, f(\mathbf{\Gamma},t)\sum_i m\mathbf{v}_i\cdot\frac{\partial\delta(\mathbf{r}-\mathbf{r}_i)}{\partial\mathbf{r}_i}$$
$$= -\int d\mathbf{\Gamma}\, f(\mathbf{\Gamma},t)\sum_i m\mathbf{v}_i\cdot\frac{\partial\delta(\mathbf{r}-\mathbf{r}_i)}{\partial\mathbf{r}}$$
$$= -\nabla\cdot\int d\mathbf{\Gamma}\, f(\mathbf{\Gamma},t)\sum_i m\mathbf{v}_i\delta(\mathbf{r}-\mathbf{r}_i)$$
$$= -\nabla\cdot[\rho(\mathbf{r},t)\mathbf{u}(\mathbf{r},t)] \tag{3.114}$$

The fifth equality follows using the delta function identity

$$\frac{\partial\delta(\mathbf{x}-\mathbf{y})}{\partial\mathbf{x}} = -\frac{\partial\delta(\mathbf{x}-\mathbf{y})}{\partial\mathbf{y}}$$

We have shown that the time derivative of the mass density yields the mass continuity equation (2.4) as expected. Strictly speaking, therefore, we did not really need to define the momentum density in equation (3.111), as given the mass-density definition, the mass-continuity equation yields the momentum density expression. We will now use exactly the same procedure to differentiate the momentum density.

$$\frac{\partial[\rho(\mathbf{r},t)\mathbf{u}(\mathbf{r},t)]}{\partial t} = \int d\boldsymbol{\Gamma}\,\frac{\partial f(\boldsymbol{\Gamma},t)}{\partial t}\sum_i m\mathbf{v}_i\delta(\mathbf{r}-\mathbf{r}_i)$$
$$= -\int d\boldsymbol{\Gamma}\sum_i m\mathbf{v}_i\delta(\mathbf{r}-\mathbf{r}_i)iLf(\boldsymbol{\Gamma},t)$$
$$= \int d\boldsymbol{\Gamma}\,f(\boldsymbol{\Gamma},t)\mathrm{iL}\sum_i m\mathbf{v}_i\delta(\mathbf{r}-\mathbf{r}_i)$$
$$= \int d\boldsymbol{\Gamma}\,f(\boldsymbol{\Gamma},t)\sum_i [\delta(\mathbf{r}-\mathbf{r}_i)\mathrm{iL}m\mathbf{v}_i + m\mathbf{v}_i\mathrm{iL}\delta(\mathbf{r}-\mathbf{r}_i)]$$
$$= \int d\boldsymbol{\Gamma}\,f(\boldsymbol{\Gamma},t)\sum_i \left[\delta(\mathbf{r}-\mathbf{r}_i)\mathbf{F}_i + m\mathbf{v}_i\mathbf{v}_i\cdot\frac{\partial\delta(\mathbf{r}-\mathbf{r}_i)}{\partial\mathbf{r}_i}\right] \tag{3.115}$$

We have used Newtonian equations of motion for the Liouvillean iL.

$$\frac{\partial[\rho(\mathbf{r},t)\mathbf{u}(\mathbf{r},t)]}{\partial t} = \int d\boldsymbol{\Gamma}\,f(\boldsymbol{\Gamma},t)\sum_i \left[\delta(\mathbf{r}-\mathbf{r}_i)\mathbf{F}_i - \frac{\partial}{\partial\mathbf{r}}\cdot m\mathbf{v}_i\mathbf{v}_i\delta(\mathbf{r}-\mathbf{r}_i)\right]$$
$$= \int d\boldsymbol{\Gamma}\,f(\boldsymbol{\Gamma},t)\sum_i \delta(\mathbf{r}-\mathbf{r}_i)\mathbf{F}_i - \frac{\partial}{\partial\mathbf{r}}\cdot\int d\boldsymbol{\Gamma}\,f(\boldsymbol{\Gamma},t)\sum_i m\mathbf{v}_i\mathbf{v}_i\delta(\mathbf{r}-\mathbf{r}_i)$$

If we consider the second term on the right-hand side then

$$\frac{\partial}{\partial\mathbf{r}}\cdot\int d\boldsymbol{\Gamma}\,f(\boldsymbol{\Gamma},t)\sum_i m\mathbf{v}_i\mathbf{v}_i\delta(\mathbf{r}-\mathbf{r}_i) = \frac{\partial}{\partial\mathbf{r}}\cdot\left\langle \sum_i m(\mathbf{v}_i(t)-\mathbf{u}(\mathbf{r},t))(\mathbf{v}_i(t)-\mathbf{u}(\mathbf{r},t))|\mathbf{r}_i(t)=\mathbf{r}\right\rangle$$
$$+\frac{\partial}{\partial\mathbf{r}}\cdot\left\langle \sum_i m\mathbf{u}(\mathbf{r},t)\mathbf{u}(\mathbf{r},t)|\mathbf{r}_i(t)=\mathbf{r}\right\rangle \tag{3.116}$$

In the final term in equation (3.116), $\mathbf{u}(\mathbf{r},t)$ is independent of the particle index and can be factored outside the summation. The remaining summation is, using (3.110), simply equal to the mass density $\rho(\mathbf{r},t)$. Combining these results it follows that

$$\frac{\partial[\rho(\mathbf{r},t)\mathbf{u}(\mathbf{r},t)]}{\partial t} + \frac{\partial}{\partial\mathbf{r}}\cdot[\rho(\mathbf{r},t)\mathbf{u}(\mathbf{r},t)\mathbf{u}(\mathbf{r},t)] = \int d\boldsymbol{\Gamma}\,f(\boldsymbol{\Gamma},t)\sum_i\delta(\mathbf{r}-\mathbf{r}_i)\mathbf{F}_i$$
$$-\frac{\partial}{\partial\mathbf{r}}\cdot\left\langle\sum_i m(\mathbf{v}_i(t)-\mathbf{u}(\mathbf{r},t))(\mathbf{v}_i(t)-\mathbf{u}(\mathbf{r},t))|\mathbf{r}_i(t)=\mathbf{r}\right\rangle \tag{3.117}$$

We will now consider the first term on the right-hand side of this equation in some detail.

$$\int d\boldsymbol{\Gamma}\, f(\boldsymbol{\Gamma},t)\sum_i \delta(\mathbf{r}-\mathbf{r}_i)\mathbf{F}_i$$

$$= -\int d\boldsymbol{\Gamma}\, f(\boldsymbol{\Gamma},t)\sum_{i,j} \delta(\mathbf{r}-\mathbf{r}_i)\frac{\partial \phi_{ij}}{\partial \mathbf{r}_i}$$

$$= -\tfrac{1}{2}\int d\boldsymbol{\Gamma}\, f(\boldsymbol{\Gamma},t)\sum_{i,j}\left[\delta(\mathbf{r}-\mathbf{r}_i)\frac{\partial \phi_{ij}}{\partial \mathbf{r}_i} + \delta(\mathbf{r}-\mathbf{r}_j)\frac{\partial \phi_{ji}}{\partial \mathbf{r}_j}\right]$$

$$= -\tfrac{1}{2}\int d\boldsymbol{\Gamma}\, f(\boldsymbol{\Gamma},t)\sum_{i,j}\left[\delta(\mathbf{r}-\mathbf{r}_i)\frac{\partial \phi_{ij}}{\partial \mathbf{r}_i} - \delta(\mathbf{r}-\mathbf{r}_j)\frac{\partial \phi_{ij}}{\partial \mathbf{r}_i}\right]$$

$$= -\tfrac{1}{2}\int d\boldsymbol{\Gamma}\, f(\boldsymbol{\Gamma},t)\sum_{i,j}[\delta(\mathbf{r}-\mathbf{r}_i) - \delta(\mathbf{r}-\mathbf{r}_j)]\frac{\partial \phi_{ij}}{\partial \mathbf{r}_i} \tag{3.118}$$

Treating the macroscopic delta function as an analytic function, we may expand $\delta(\mathbf{r}-\mathbf{r}_j)$ as a Taylor series about $\delta(\mathbf{r}-\mathbf{r}_i)$. This gives

$$\delta(\mathbf{r}-\mathbf{r}_j) = \delta(\mathbf{r}-\mathbf{r}_i) + \mathbf{r}_{ij}\cdot\frac{\partial \delta(\mathbf{r}-\mathbf{r}_i)}{\partial \mathbf{r}_i} + \frac{1}{2!}\mathbf{r}_{ij}\mathbf{r}_{ij}:\frac{\partial^2 \delta(\mathbf{r}-\mathbf{r}_i)}{\partial \mathbf{r}_i^2} + \ldots$$

$$= \delta(\mathbf{r}-\mathbf{r}_i) - \mathbf{r}_{ij}\cdot\frac{\partial \delta(\mathbf{r}-\mathbf{r}_i)}{\partial \mathbf{r}} + \frac{1}{2!}\mathbf{r}_{ij}\mathbf{r}_{ij}:\frac{\partial^2 \delta(\mathbf{r}-\mathbf{r}_i)}{\partial \mathbf{r}^2} - \ldots \tag{3.119}$$

Thus the difference between the two delta functions is

$$\delta(\mathbf{r}-\mathbf{r}_i) - \delta(\mathbf{r}-\mathbf{r}_j) = \mathbf{r}_{ij}\cdot\frac{\partial \delta(\mathbf{r}-\mathbf{r}_i)}{\partial \mathbf{r}} - \frac{1}{2!}\mathbf{r}_{ij}\mathbf{r}_{ij}:\frac{\partial^2 \delta(\mathbf{r}-\mathbf{r}_i)}{\partial \mathbf{r}^2} + \ldots$$

$$= \frac{\partial}{\partial \mathbf{r}}\cdot\mathbf{r}_{ij}O_{ij}\delta(\mathbf{r}-\mathbf{r}_i) \tag{3.120}$$

where the operator O_{ij} is given by,

$$O_{ij} \equiv 1 - \frac{1}{2!}\mathbf{r}_{ij}\cdot\frac{\partial}{\partial \mathbf{r}} + \ldots + \frac{1}{n!}\left[-\mathbf{r}_{ij}\cdot\frac{\partial}{\partial \mathbf{r}}\right]^{n-1} + \ldots \tag{3.121}$$

Using this equation for the difference of the two delta functions $\delta(\mathbf{r}-\mathbf{r}_i)$ and $\delta(\mathbf{r}-\mathbf{r}_j)$ leads to

$$\frac{\partial[\rho(\mathbf{r},t)\mathbf{u}(\mathbf{r},t)]}{\partial t} + \frac{\partial}{\partial \mathbf{r}}\cdot[\rho(\mathbf{r},t)\mathbf{u}(\mathbf{r},t)\mathbf{u}(\mathbf{r},t)]$$

$$= -\frac{\partial}{\partial \mathbf{r}}\cdot\left[\tfrac{1}{2}\int d\boldsymbol{\Gamma}\, f(\boldsymbol{\Gamma},t)\sum_{i,j}\mathbf{r}_{ij}O_{ij}\delta(\mathbf{r}-\mathbf{r}_i)\frac{\partial \phi_{ij}}{\partial \mathbf{r}_i}\right.$$

$$\left. + \left\langle \sum_i m(\mathbf{v}_i(t)-\mathbf{u}(\mathbf{r},t))(\mathbf{v}_i(t)-\mathbf{u}(\mathbf{r},t))|\mathbf{r}_i(t)=\mathbf{r}\right\rangle\right] \tag{3.122}$$

Comparing this equation with the momentum conservation equation (2.12) we see that the pressure tensor is

$$\mathbf{P}(\mathbf{r},t) = \left\langle \sum_i m(\mathbf{v}_i(t) - \mathbf{u}(\mathbf{r},t))(\mathbf{v}_i(t) - \mathbf{u}(\mathbf{r},t)) - \tfrac{1}{2}\sum_{i,j} \mathbf{r}_{ij}(t)O_{ij}(t)\mathbf{F}_{ij}(t) | \mathbf{r}_i(t) = \mathbf{r} \right\rangle \tag{3.123}$$

where $\mathbf{F}_{ij} = -\partial\phi_{ij}/\partial\mathbf{r}_i$ is the force on particle i due to particle j.

We will now use the same technique to calculate the microscopic expression for the heat flux vector. The partial time derivative of the energy density is (from (3.112))

$$\begin{aligned}
\frac{\partial[\rho(\mathbf{r},t)e(\mathbf{r},t)]}{\partial t} &= \int d\mathbf{\Gamma} \frac{\partial f(\mathbf{\Gamma},t)}{\partial t}\left[\tfrac{1}{2}\sum_i m\mathbf{v}_i^2 + \tfrac{1}{2}\sum_{i,j}\phi_{ij}\right]\delta(\mathbf{r}-\mathbf{r}_i) \\
&= \int d\mathbf{\Gamma}\, f(\mathbf{\Gamma},t)\left\{ \mathrm{i}L\left[\tfrac{1}{2}\sum_i m\mathbf{v}_i^2 + \tfrac{1}{2}\sum_{i,j}\phi_{ij}\right]\right\}\delta(\mathbf{r}-\mathbf{r}_i) \\
&\quad + \int d\mathbf{\Gamma}\, f(\mathbf{\Gamma},t)\left[\tfrac{1}{2}\sum_i m\mathbf{v}_i^2 + \tfrac{1}{2}\sum_{i,j}\phi_{ij}\right]\mathrm{i}L\delta(\mathbf{r}-\mathbf{r}_i) \\
&= \int d\mathbf{\Gamma}\, f(\mathbf{\Gamma},t)\left[\sum_i \mathbf{v}_i\cdot\mathbf{F}_i - \tfrac{1}{2}\sum_{i,j}[\mathbf{v}_i\cdot\mathbf{F}_{ij} + \mathbf{v}_j\cdot\mathbf{F}_{ji}]\right]\delta(\mathbf{r}-\mathbf{r}_i) \\
&\quad + \int d\mathbf{\Gamma}\, f(\mathbf{\Gamma},t)\left[\tfrac{1}{2}\sum_i m\mathbf{v}_i^2 + \tfrac{1}{2}\sum_{i,j}\phi_{ij}\right]\mathbf{v}_i\cdot\frac{\partial\delta(\mathbf{r}-\mathbf{r}_i)}{\partial\mathbf{r}_i} \\
&= \int d\mathbf{\Gamma}\, f(\mathbf{\Gamma},t)\left[\tfrac{1}{2}\sum_{i,j}\mathbf{v}_i\cdot\mathbf{F}_{ij}[\delta(\mathbf{r}-\mathbf{r}_i) - \delta(\mathbf{r}-\mathbf{r}_j)]\right. \\
&\quad \left. - \frac{\partial}{\partial\mathbf{r}}\cdot\left[\tfrac{1}{2}\sum_i m\mathbf{v}_i^2 + \tfrac{1}{2}\sum_{i,j}\phi_{ij}\right]\mathbf{v}_i\delta(\mathbf{r}-\mathbf{r}_i)\right]
\end{aligned} \tag{3.124}$$

In the second term, the gradient operator $\partial/\partial\mathbf{r}$ is contracted into $\mathbf{v}_i$. Using our previous result for the difference of two delta functions, (3.120), gives

$$\begin{aligned}
\frac{\partial[\rho(\mathbf{r},t)e(\mathbf{r},t)]}{\partial t} &= \frac{\partial}{\partial\mathbf{r}}\cdot\int d\mathbf{\Gamma}\, f(\mathbf{\Gamma},t)\left[\tfrac{1}{2}\sum_{i,j}\mathbf{r}_{ij}O_{ij}\mathbf{v}_i\cdot\mathbf{F}_{ij}\right. \\
&\quad \left. - \left(\tfrac{1}{2}\sum_i m\mathbf{v}_i^2 + \tfrac{1}{2}\sum_{i,j}\phi_{ij}\right)\mathbf{v}_i\right]\delta(\mathbf{r}-\mathbf{r}_i)
\end{aligned} \tag{3.125}$$

From equation (2.24) we conclude that,

$$\begin{aligned}
&\mathbf{J}_Q(\mathbf{r},t) + \mathbf{P}(\mathbf{r},t)\cdot\mathbf{u}(\mathbf{r},t) + \rho(\mathbf{r},t)e(\mathbf{r},t)\mathbf{u}(\mathbf{r},t) \\
&\quad = \int d\mathbf{\Gamma}\, f(\mathbf{\Gamma},t)\left[-\tfrac{1}{2}\sum_{i,j}\mathbf{r}_{ij}O_{ij}\mathbf{v}_i\cdot\mathbf{F}_{ij} + \left(\tfrac{1}{2}\sum_i m\mathbf{v}_i^2 + \tfrac{1}{2}\sum_{i,j}\phi_{ij}\right)\mathbf{v}_i\right]\delta(\mathbf{r}-\mathbf{r}_i)
\end{aligned} \tag{3.126}$$

Now the definition of the energy density, equation (3.112) gives

$$\rho(\mathbf{r},t)e(\mathbf{r},t)\mathbf{u}(\mathbf{r},t) = \int d\Gamma\, f(\Gamma,t)\Big[\tfrac{1}{2}\sum_i m\mathbf{v}_i^2 + \tfrac{1}{2}\sum_{i,j}\phi_{ij}\Big]\delta(\mathbf{r}-\mathbf{r}_i)\mathbf{u}(\mathbf{r},t) \tag{3.127}$$

so that,

$$\begin{aligned}\mathbf{J}_Q(\mathbf{r},t) + \mathbf{P}(\mathbf{r},t)\cdot\mathbf{u}(\mathbf{r},t) = \int d\Gamma\, f(\Gamma,t)\Big[&-\tfrac{1}{2}\sum_{i,j}\mathbf{r}_{ij}O_{ij}\mathbf{v}_i\cdot\mathbf{F}_{ij}\\ &+\Big(\tfrac{1}{2}\sum_i m\mathbf{v}_i^2 + \tfrac{1}{2}\sum_{i,j}\phi_{ij}\Big)(\mathbf{v}_i - \mathbf{u}(\mathbf{r},t))\Big]\delta(\mathbf{r}-\mathbf{r}_i)\end{aligned} \tag{3.128}$$

Similarly, from the definition of the pressure tensor $\mathbf{P}(\mathbf{r},t)$ (see (3.123)), we know that

$$\mathbf{P}(\mathbf{r},t)\cdot\mathbf{u}(\mathbf{r},t) = \int d\Gamma\, f(\Gamma,t)\Big[\sum_i m(\mathbf{v}_i-\mathbf{u})(\mathbf{v}_i-\mathbf{u}) - \tfrac{1}{2}\sum_{i,j}\mathbf{r}_{ij}O_{ij}\mathbf{F}_{ij}\Big]\cdot\mathbf{u}\,\delta(\mathbf{r}-\mathbf{r}_i) \tag{3.129}$$

thus we identify the heat flux vector as

$$\begin{aligned}\mathbf{J}_Q(\mathbf{r},t) &= \int d\Gamma\, f(\Gamma,t)\Big[\Big(\tfrac{1}{2}\sum_i m\mathbf{v}_i^2 + \tfrac{1}{2}\sum_{i,j}\phi_{ij}\Big)(\mathbf{v}_i-\mathbf{u}) - \sum_i m(\mathbf{v}_i-\mathbf{u})(\mathbf{v}_i-\mathbf{u})\cdot\mathbf{u}\\ &\qquad - \tfrac{1}{2}\sum_{i,j}\mathbf{r}_{ij}(\mathbf{v}_i-\mathbf{u})\cdot\mathbf{F}_{ij}O_{ij}\Big]\delta(\mathbf{r}-\mathbf{r}_i)\\ &= \int d\Gamma\, f(\Gamma,t)\Big[\sum_i \tfrac{1}{2}m\{(\mathbf{v}_i-\mathbf{u})^2 + \mathbf{u}^2\}(\mathbf{v}_i-\mathbf{u}) + \tfrac{1}{2}\sum_{i,j}\phi_{ij}(\mathbf{v}_i-\mathbf{u})\\ &\qquad - \tfrac{1}{2}\sum_{i,j}\mathbf{r}_{ij}(\mathbf{v}_i-\mathbf{u})\cdot\mathbf{F}_{ij}O_{ij}\Big]\delta(\mathbf{r}-\mathbf{r}_i)\end{aligned} \tag{3.130}$$

From the definitions of the mass density and momentum density ((3.110) and (3.111)) we find that

$$\int d\Gamma\, f(\Gamma,t)\sum_i(\mathbf{v}_i-\mathbf{u})m\mathbf{u}^2\delta(\mathbf{r}-\mathbf{r}_i) = 0 \tag{3.131}$$

so there is no contribution from the $\mathbf{u}^2$ term. Furthermore, if we define the peculiar energy of particle i to be

$$e_i = \frac{m}{2}(\mathbf{v}_i-\mathbf{u})^2 + \tfrac{1}{2}\sum_j\phi_{ij} \tag{3.132}$$

then the heat-flux vector can be written as

$$\mathbf{J}_Q(\mathbf{r},t) = \int d\mathbf{\Gamma}\, f(\mathbf{\Gamma},t)\left[\sum_i (\mathbf{v}_i - \mathbf{u})e_i - \tfrac{1}{2}\sum_{i,j} \mathbf{r}_{ij}(\mathbf{v}_i - \mathbf{u})\cdot \mathbf{F}_{ij} O_{ij}\right]\delta(\mathbf{r} - \mathbf{r}_i)$$

or

$$\mathbf{J}_Q(\mathbf{r},t) = \Big\langle \sum_i (\mathbf{v}_i(t) - \mathbf{u}(t))e_i(t) - \tfrac{1}{2}\sum_{i,j} \mathbf{r}_{ij}(t)(\mathbf{v}_i(t) - \mathbf{u}(t))\cdot \mathbf{F}_{ij}(t) O_{ij}(t) | \mathbf{r}_i(t) = \mathbf{r} \Big\rangle \tag{3.133}$$

3.8 INSTANTANEOUS MICROSCOPIC REPRESENTATION OF FLUXES

The Irving–Kirkwood procedure has given us microscopic expressions for the thermodynamic fluxes in terms of ensemble averages. At equilibrium in a uniform fluid, the Irving–Kirkwood expression for the pressure tensor is the same expression as that derived using Gibbs' ensemble theory for equilibrium statistical mechanics. If the fluid density is uniform in space, the O_{ij} operator appearing in the above expressions reduces to unity. This is easier to see if we calculate microscopic expressions for the fluxes in **k** space rather than real space. In the process we will better understand the nature of the Irving–Kirkwood expressions.

In this section we derive *instantaneous* expressions for the fluxes rather than the ensemble-based Irving–Kirkwood expressions. The reason for considering instantaneous expressions is two-fold. The fluxes are based upon conservation laws and these laws are valid instantaneously for every member of the ensemble. They do not require ensemble averaging to be true. Secondly, most computer simulation involves calculating system properties from a single system trajectory. Ensemble averaging is rarely used because it is relatively expensive in computer time. The ergodic hypothesis, that the result obtained by ensemble averaging is equal to that obtained by time averaging the same property along a single phase-space trajectory, implies that one should be able to develop expressions for the fluxes which do not require ensemble averaging. For this to be practically realizable it is clear that the mass, momentum and energy densities must be definable at each instant along the trajectory.

We define the Fourier transform pair by

$$f(\mathbf{k}) = \int d\mathbf{r}\, e^{i\mathbf{k}\cdot\mathbf{r}} f(\mathbf{r})$$
$$f(\mathbf{r}) = \frac{1}{(2\pi)^3}\int d\mathbf{k}\, e^{-i\mathbf{k}\cdot\mathbf{r}} f(\mathbf{k}) \tag{4.134}$$

In the spirit of the Irving–Kirkwood procedure, we define the instantaneous mass density to be,

$$\rho(\mathbf{r},t) = \sum_{i=1}^{N} m\delta(\mathbf{r} - \mathbf{r}_i(t)) \tag{3.135}$$

where the explicit time dependence of $\rho(\mathbf{r},t)$ (that is the time dependence differentiated by the hydrodynamic derivative $\partial/\partial t$, with $\mathbf{r}$ fixed) is through the time dependence of $\mathbf{r}_i(t)$. The $\mathbf{k}$ space instantaneous mass density is then

$$\begin{aligned}\rho(\mathbf{k},t) &= \int \mathrm{d}\mathbf{r} \sum_{i=1}^{N} m\delta(\mathbf{r} - \mathbf{r}_i(t))\, \mathrm{e}^{i\mathbf{k}\cdot\mathbf{r}} \\ &= \sum_{i=1}^{N} m\, \mathrm{e}^{i\mathbf{k}\cdot\mathbf{r}_i(t)} \end{aligned}\tag{3.136}$$

We will usually allow the context to distinguish whether we are using ensemble averages or instantaneous expressions. The time dependence of the mass density is solely through the time dependence of $\mathbf{r}_i$, so that

$$\frac{\partial\rho(\mathbf{k},t)}{\partial t} = i\mathbf{k}\cdot \sum_{i=1}^{N} m\mathbf{v}_i\, \mathrm{e}^{i\mathbf{k}\cdot\mathbf{r}_i(t)} \tag{3.137}$$

Comparing this with the Fourier transform of (2.4) (noting that $\partial/\partial t|_{\mathbf{k}}$ in (3.137) corresponds to $\partial/\partial t|_{\mathbf{r}}$ in (2.4)) we see that if we define $\mathbf{J}(\mathbf{r},t) \equiv \rho(\mathbf{r},t)\mathbf{u}(\mathbf{r},t)$, then

$$\mathbf{J}(\mathbf{k},t) = \sum_{i=1}^{N} m\mathbf{v}_i\, \mathrm{e}^{i\mathbf{k}\cdot\mathbf{r}_i(t)} \tag{3.138}$$

This equation is clearly the instantaneous analogue of the Fourier transform of the Irving–Kirkwood expression for the momentum density. There is no ensemble average required in (3.137). To look at the *instantaneous pressure tensor* we only need to differentiate equation (3.138) over time.

$$\frac{\partial\mathbf{J}(\mathbf{k},t)}{\partial t} = \sum_{i=1}^{N} (i\mathbf{k}\cdot m\mathbf{v}_i(t)\mathbf{v}_i(t)\, \mathrm{e}^{i\mathbf{k}\cdot\mathbf{r}_i(t)} + \mathbf{F}_i\, \mathrm{e}^{i\mathbf{k}\cdot\mathbf{r}_i(t)}) \tag{3.139}$$

We can write the second term on the right-hand side of this equation in the form of the Fourier transform of a divergence by noting that

$$\begin{aligned}\sum_{i=1}^{N} \mathbf{F}_i\, \mathrm{e}^{i\mathbf{k}\cdot\mathbf{r}_i} &= \tfrac{1}{2}\sum_{i=1}^{N}\sum_{j=1}^{N} (\mathbf{F}_{ij}\, \mathrm{e}^{i\mathbf{k}\cdot\mathbf{r}_i} + \mathbf{F}_{ji}\, \mathrm{e}^{i\mathbf{k}\cdot\mathbf{r}_j}) \\ &= \tfrac{1}{2}\sum_{i=1}^{N}\sum_{j=1}^{N} \mathbf{F}_{ij}(\mathrm{e}^{i\mathbf{k}\cdot\mathbf{r}_i} - \mathrm{e}^{i\mathbf{k}\cdot\mathbf{r}_j}) \\ &= \tfrac{1}{2}\sum_{i=1}^{N}\sum_{j=1}^{N} \mathbf{F}_{ij}(\mathrm{e}^{-i\mathbf{k}\cdot\mathbf{r}_{ij}} - 1)\, \mathrm{e}^{i\mathbf{k}\cdot\mathbf{r}_j} \\ &= -i\mathbf{k}\cdot\tfrac{1}{2}\sum_{i=1}^{N}\sum_{j=1}^{N} \mathbf{r}_{ij}\mathbf{F}_{ij}\frac{\mathrm{e}^{-i\mathbf{k}\cdot\mathbf{r}_{ij}} - 1}{-i\mathbf{k}\cdot\mathbf{r}_{ij}}\, \mathrm{e}^{i\mathbf{k}\cdot\mathbf{r}_j} \end{aligned}\tag{3.140}$$

Combining (3.139) and (3.140) and performing an inverse Fourier transform we obtain the instantaneous analogue of equation (3.123). We could of course continue the analysis of Section 3.7 to remove the streaming contribution from the pressure tensor, but this is more laborious in **k** space than in real space and we will not give this here. We can use our instantaneous expression for the pressure tensor to describe fluctuations in an equilibrium system. In this case the streaming velocity is, of course, zero, and from (2.14),

$$\mathbf{P}(\mathbf{k},t)=\sum_{i=1}^{N} m\mathbf{v}_i(t)\mathbf{v}_i(t)\,\mathrm{e}^{i\mathbf{k}\cdot\mathbf{r}_i(t)}-\tfrac{1}{2}\sum_{i,j=1}^{N}\mathbf{r}_{ij}(t)\mathbf{F}_{ij}(t)\frac{\mathrm{e}^{-i\mathbf{k}\cdot\mathbf{r}_{ij}(t)}-1}{-i\mathbf{k}\cdot\mathbf{r}_{ij}(t)}\,\mathrm{e}^{i\mathbf{k}\cdot\mathbf{r}_j(t)} \tag{3.141}$$

The **k**-space analysis given provides a better understanding of the Irving–Kirkwood operator O_{ij}. In **k**-space it is not necessary to perform the apparently difficult operation of Taylor expanding delta functions.

Before we close this section we will try to make the equation for the momentum density, $\mathbf{J}(\mathbf{r},t)=\rho(\mathbf{r},t)\mathbf{u}(\mathbf{r},t)$, a little clearer. In **k**-space this equation is a convolution,

$$\mathbf{J}(\mathbf{k},t)=\int \mathrm{d}\mathbf{k}'\,\rho(\mathbf{k}-\mathbf{k}',t)\mathbf{u}(\mathbf{k}',t) \tag{3.142}$$

Does this definition of the streaming velocity, **u**, make good physical sense? One sensible definition for the streaming velocity, **u**, would be that velocity which minimizes the sum-of-squares of deviations from the particle velocities $\mathbf{v}_i$. For simplicity we set $t=0$, and let R be that sum of squares

$$R=\sum_{i=1}^{N}(\mathbf{v}_i-\mathbf{u}(\mathbf{r}_i))^2=\sum_{i=1}^{N}\left(\mathbf{v}_i-\sum_n \mathbf{u}(\mathbf{k}_n)\,\mathrm{e}^{-i\mathbf{k}_n\cdot\mathbf{r}_i}\right)^2 \tag{3.143}$$

If $\mathbf{u}(\mathbf{r})$ minimizes this sum-of-squares then the derivative of R with respect to each of the Fourier components $\mathbf{u}(\mathbf{k}_m)$, must be zero. Differentiating (3.143) we obtain

$$\frac{\partial R}{\partial \mathbf{u}(\mathbf{k}_m)}=2\sum_{i=1}^{N}\left(\mathbf{v}_i-\sum_n \mathbf{u}(\mathbf{k}_n)\,\mathrm{e}^{i\mathbf{k}_n\cdot\mathbf{r}_i}\right)\mathrm{e}^{i\mathbf{k}_m\cdot\mathbf{r}_i}=0 \tag{3.144}$$

This implies that

$$\sum_{i=1}^{N} m\mathbf{v}_i\,\mathrm{e}^{-i\mathbf{k}_m\cdot\mathbf{r}_i}=\sum_{i=1}^{N}\sum_n m\mathbf{u}(\mathbf{k}_n)\,\mathrm{e}^{-i\mathbf{k}_n\cdot\mathbf{r}_i}\,\mathrm{e}^{-i\mathbf{k}_m\cdot\mathbf{r}_i} \tag{3.145}$$

Both sides of this equation can be identified as **k**-space variables,

$$\mathbf{J}(-\mathbf{k}_m)=\sum_n \mathbf{u}(\mathbf{k}_n)\rho(-\mathbf{k}_m-\mathbf{k}_n) \tag{3.146}$$

So that

$$\mathbf{J}(\mathbf{k}_m) = \sum_n \mathbf{u}(\mathbf{k}_n)\rho(\mathbf{k}_m - \mathbf{k}_n) \tag{3.147}$$

This is the Fourier series version of equation (3.142).

We can use the same procedure to calculate an expression for the heat-flux vector. As we will see this procedure is very much simpler than the Irving–Kirkwood method described in Section 3.7. We begin by identifying the instantaneous expression for the instantaneous wavevector dependent energy density in a fluid at equilibrium,

$$\rho e(\mathbf{k}, t) = \sum_i \left(\tfrac{1}{2} m\mathbf{v}_i^2(t) + \tfrac{1}{2}\sum_j \phi_{ij}(\mathbf{r}_{ij}(t)) \right) \mathrm{e}^{i\mathbf{k}\cdot\mathbf{r}_i(t)} \tag{3.148}$$

This is an instantaneous, wavevector-dependent analogue of (3.112). To simplify notation in the following we will suppress the time argument for all phase variables. The time argument will always be t. If we calculate the rate of change of the energy density we find that

$$\begin{aligned}\frac{\partial \rho e(\mathbf{k}, t)}{\partial t} &= i\mathbf{k}\cdot\sum_i \mathbf{v}_i(\tfrac{1}{2}mv_i^2 + \phi_i)\,\mathrm{e}^{i\mathbf{k}\cdot\mathbf{r}_i} + \sum_i m\mathbf{v}_i\cdot\dot{\mathbf{v}}_i\,\mathrm{e}^{i\mathbf{k}\cdot\mathbf{r}_i} \\ &\quad + \tfrac{1}{2}\sum_{i,j}\left(\dot{\mathbf{r}}_i\cdot\frac{\partial \phi_{ij}}{\partial \mathbf{r}_i} + \dot{\mathbf{r}}_j\cdot\frac{\partial \phi_{ij}}{\partial \mathbf{r}_j} \right)\mathrm{e}^{i\mathbf{k}\cdot\mathbf{r}_i}\end{aligned} \tag{3.149}$$

where we use the notation $\phi_i = \tfrac{1}{2}\Sigma_j \phi_{ij}$. If we denote the energy of particle i as e_i and $\mathbf{F}_{ij}$ as the force exerted on particle i due to j then (3.149) can be rewritten as

$$\begin{aligned}\frac{\partial \rho e}{\partial t} &= i\mathbf{k}\cdot\sum_i \mathbf{v}_i e_i\,\mathrm{e}^{i\mathbf{k}\cdot\mathbf{r}_i} + \sum_i \mathbf{v}_i\cdot\mathbf{F}_i\,\mathrm{e}^{i\mathbf{k}\cdot\mathbf{r}_i} - \tfrac{1}{2}\sum_{i,j}(\mathbf{v}_i\cdot\mathbf{F}_{ij} - \mathbf{v}_j\cdot\mathbf{F}_{ij})\,\mathrm{e}^{i\mathbf{k}\cdot\mathbf{r}_i} \\ &= i\mathbf{k}\cdot\sum_i \mathbf{v}_i e_i\,\mathrm{e}^{i\mathbf{k}\cdot\mathbf{r}_i} + \tfrac{1}{2}\sum_{i,j}\mathbf{v}_i\cdot\mathbf{F}_{ij}(\mathrm{e}^{i\mathbf{k}\cdot\mathbf{r}_i} - \mathrm{e}^{i\mathbf{k}\cdot\mathbf{r}_j})\end{aligned} \tag{3.150}$$

This equation involves the same combination of exponents as we saw for the pressure tensor in (3.140). Expanding exponentials to first order in $\mathbf{k}$, and using (2.24), we find that the wavevector-dependent heat-flux vector can be written as

$$\mathbf{J}_Q(\mathbf{k} = 0) = \sum_i \mathbf{v}_i e_i - \tfrac{1}{2}\sum_{i,j}\mathbf{r}_{ij}\mathbf{F}_{ij}\cdot\mathbf{v}_i + O(k^2) \tag{3.151}$$

In $\mathbf{r}$ space rather than $\mathbf{k}$ space the expressions for the instantaneous pressure tensor and heat-flux vector become

$$\mathbf{P}(\mathbf{r},t) = \sum_{i=1}^{N}\left[m(\mathbf{v}_i(t)-\mathbf{u}(t))(\mathbf{v}_i(t)-\mathbf{u}(t)) - \tfrac{1}{2}\sum_{j=1}^{N}\mathbf{r}_{ij}(t)\mathbf{F}_{ij}(t)\right]\delta(\mathbf{r}_i(t)-\mathbf{r}) \tag{3.152}$$

$$\mathbf{J}_Q(\mathbf{r},t) = \sum_{i=1}^{N}\left[(\mathbf{v}_i(t)-\mathbf{u}(t))e_i(t) - \tfrac{1}{2}\sum_{j=1}^{N}\mathbf{r}_{ij}(t)(\mathbf{v}_i(t)-\mathbf{u}(t))\cdot\mathbf{F}_{ij}(t)\right]\delta(\mathbf{r}_i(t)-\mathbf{r}) \tag{3.153}$$

Our procedure for calculating microscopic expressions for the hydrodynamic densities and fluxes relies upon establishing a correspondence between the microscopic and macroscopic forms of the continuity equations. These equations refer only to the divergence of the pressure tensor and heat flux. Strictly speaking, therefore, we can only determine the divergences of the flux tensors. We can add any divergence-free quantity to our expressions for the flux tensors without affecting the identification process.

3.9 THE KINETIC TEMPERATURE

We obtain an instantaneous expression for the temperature by analysing the expression for the pressure tensor (3.150) for the case of an ideal gas at equilibrium. Thus if $n(\mathbf{r},t)$ is the local instantaneous number density,

$$\frac{3n(\mathbf{r},t)k_B T(\mathbf{r},t)}{2} = \sum_{i=1}^{N} m(\mathbf{v}_i(t)-\mathbf{u}(\mathbf{r},t))^2\delta(\mathbf{r}_i(t)-\mathbf{r}) \tag{3.154}$$

We will call this expression for the temperature the kinetic temperature. In using this expression for the temperature we are employing a number of approximations. Firstly, we are ignoring the number of degrees of freedom which are frozen by the instantaneous determination of $\mathbf{u}(\mathbf{r},t)$. Secondly, and more importantly, we are assuming that in a nonequilibrium system the kinetic temperature is identical to the thermodynamic temperature T_T,

$$T_T = \left.\frac{\partial E}{\partial S}\right|_V \tag{3.155}$$

This is undoubtedly an approximation. It would be true if the postulate of local thermodynamic equilibrium was exact. However, we know that the energy, pressure, enthalpy, etc., are all functions of the thermodynamic forces driving the system away from equilibrium. These are nonlinear effects which vanish in Newtonian fluids. Presumably the entropy is also a function of these driving forces. It is extremely unlikely that the field dependence of the entropy and the energy are precisely those required for the exact equivalence of the kinetic and thermodynamic temperatures for all nonequilibrium systems. Recent calculations of the entropy of systems very far from equilibrium support the hypothesis that the kinetic and thermodynamic temperatures are in fact different (Evans, 1989). Outside the linear (Newtonian) regime the kinetic temperature is a convenient operational (as opposed to thermodynamic) state variable. If a nonequilibrium system is in a steady state both the kinetic and the thermodynamic temperatures must be constant in time. Furthermore, we expect that, outside the linear regime in systems with a unique nonequilibrium steady state, the thermodynamic temperature should be a monotonic function of the kinetic temperature.

REFERENCES

Berne, B.J. (1977). In: *Statistical Mechanics Part B: Time Dependent Processes* (B.J. Berne ed.), Chap. 5. New York: Plenum Press.

Dyson, F.J. (1949). *Phys. Rev.*, **75**, 486. Feynman, R.P. (1951). *Phys. Rev.*, **84**, 108.

Eckmann, J.-P. and Ruelle, D. (1985). *Rev. Mod. Phys.*, **57**, 617.

Edberg, R., Evans, D.J. and Morriss, G.P. (1986). *J. Chem. Phys.*, **84**, 6933.

Evans, D.J. (1989). *J. Stat. Phys.*, **57**, 745.

Evans, D.J. and Morriss, G.P. (1984a). *Chem. Phys.*, **87**, 451.

Evans, D.J. and Morriss, G.P. (1984b). *J. Chem. Phys.*, **81**, 1984.

Evans, D.J., Hoover, W.G., Failor, B.H., Moran, B. and Ladd, A.J.C. (1983). *Phys. Rev. Ser. A*, **28**, 1016.

Fermi, E., Pasta, J.G. and Ulam, S.M. (1955). *LASL Report LA-1940*, Los Alamos Science Laboratory, North Mexico. See also *Collected Works of Enrico Fermi*, Vol. 2, p. 978. Chicago: University of Chicago Press.

Gauss, K.F. (1829). *J. Reine Angew. Math.*, **IV**, 232.

Goldstein, H. (1980). *Classical Mechanics*, 2nd edn. Reading, MA: Addison-Wesley.

Hoover, W.G. (1986). *Molecular Dynamics. Lecture Notes in Physics*, Vol. 258. Berlin: Springer.

Irving, J.H. and Kirkwood, J.G. (1950). *J. Chem. Phys.*, **18**, 817.

Magnus, W. (1954). *Commun. Pure Appl. Math.*, **7**, 649.

Oseledec, V.I. (1968), *Trudy Mosk. Mat. Obsc.*, [Moscow Math. Soc.], **19**, 179.

Pars, L.A. (1979). *A Treatise on Analytical Dynamics.* Woodbridge, Connecticut: Ox Bow.

Pechukas, P. and Light, J.C. (1966). *J. Chem. Phys.*, **44**, 3897.

Whittacker, E.T. (1904). *A Treatise on Analytical Dynamics of Particles and Rigid Bodies.* London: Macmillan.

4 The Green–Kubo relations

4.1 THE LANGEVIN EQUATION

In 1828 the botanist Robert Brown (1828a,b) observed the motion of pollen grains suspended in a fluid. Although the system was allowed to come to equilibrium, he observed that the grains seemed to undergo a kind of unending irregular motion. This motion is now known as Brownian motion. The motion of large pollen grains suspended in a fluid composed of much lighter particles can be modelled by dividing the accelerating force into two components: a slowly varying *drag* force, and a rapidly varying *random* force due to the thermal fluctuations in the velocities of the solvent molecules. The Langevin equation, as it is known, is conventionally written in the form

$$\frac{\mathrm{d}\mathbf{v}}{\mathrm{d}t} = -\zeta\mathbf{v} + \mathbf{F}_{\mathbf{R}} \tag{4.1}$$

Using the Navier–Stokes equations to model the flow around a sphere it is known that the friction coefficient $\zeta = 6\pi\eta d/m$, where η is the shear viscosity of the fluid, d is the diameter of the sphere and m is its mass. The random force per unit mass, $\mathbf{F}_{\mathbf{R}}$, is used to model the force on the sphere due to the bombardment of solvent molecules. This force is called random because it is assumed that $\langle \mathbf{v}(0) \cdot \mathbf{F}_{\mathbf{R}}(t) \rangle = 0$, $\forall t$. A more detailed investigation of the drag on a sphere which is forced to oscillate in a fluid shows that a non-Markovian generalization (see Section 2.4) of the Langevin equation (Langevin, 1908) is required to describe the time-dependent drag on a rapidly oscillating sphere,

$$\frac{\mathrm{d}\mathbf{v}(t)}{\mathrm{d}t} = -\int_0^t \mathrm{d}t'\, \zeta(t-t')\mathbf{v}(t') + \mathbf{F}_{\mathbf{R}}(t) \tag{4.2}$$

In this case the viscous drag on the sphere is not simply linearly proportional to the instantaneous velocity of the sphere as in (4.1). Instead it is linearly proportional to the velocity at all previous times. As we will see there are many transport processes which can be described by an equation of this form. We will refer to the equation

$$\frac{dA(t)}{dt} = -\int_0^t dt'\, K(t-t')A(t') + F(t) \tag{4.3}$$

as the generalized Langevin equation for the phase variable $A(\Gamma)$. $K(t)$ is the time-dependent transport coefficient that we seek to evaluate. We assume that the equilibrium canonical ensemble average of the random force and the phase variable A, vanishes for all times.

$$\langle A(0)F(t)\rangle = \langle A(t_0)F(t_1+t)\rangle = 0, \qquad \forall t \text{ and } t_0 \tag{4.4}$$

The time displacement by t_0 is allowed because the equilibrium time-correlation function is independent of the time origin. Multiplying both sides of (4.3) by the complex conjugate of $A(0)$ and taking a canonical average we see that,

$$\frac{dC(t)}{dt} = -\int_0^t dt'\, K(t-t')C(t') \tag{4.5}$$

where $C(t)$ is defined to be the equilibrium autocorrelation function,

$$C(t) \equiv \langle A(t)A^*(0)\rangle \tag{4.6}$$

Another function we will find useful is the flux autocorrelation function $\phi(t)$

$$\phi(t) = \langle \dot{A}(t)\dot{A}^*(0)\rangle \tag{4.7}$$

Taking a Laplace transform of (4,5) we see that there is an intimate relationship between the transport memory kernel $K(t)$ and the equilibrium fluctuations in A. The left-hand side of (4.5) becomes

$$\int_0^\infty dt\, e^{-st}\frac{dC(t)}{dt} = [e^{-st}C(t)]_0^\infty - \int_0^\infty dt(-s\,e^{-st})C(t) = s\tilde{C}(s) - C(0)$$

and as the right-hand side is the Laplace transform of a convolution,

$$s\tilde{C}(s) - C(0) = -\tilde{K}(s)\tilde{C}(s) \tag{4.8}$$

So that

$$\tilde{C}(s) = \frac{C(0)}{s+\tilde{K}(s)} \tag{4.9}$$

One can convert the A autocorrelation function into a flux autocorrelation function by realizing that,

$$\frac{d^2C(t)}{dt^2} = \frac{d}{dt}\left\langle \frac{dA(t)}{dt} A^*(0) \right\rangle = \frac{d}{dt}\langle [iLA(t)]A^*(0)\rangle$$

$$= \frac{d}{dt}\langle A(t)[-iLA^*(0)]\rangle = -\langle [iLA(t)][-iLA^*(0)]\rangle$$

$$= -\phi(t)$$

Then we take the Laplace transform of a second derivative to obtain

$$-\tilde{\phi}(s) = \int_0^\infty dt\, e^{-st}\frac{d^2C(t)}{dt^2} = \left[e^{-st}\frac{dC(t)}{dt}\right]_0^\infty + s\int_0^\infty dt\, e^{-st}\frac{dC(t)}{dt}$$

$$= s[e^{-st}C(t)]_0^\infty + s^2\int_0^\infty dt\, e^{-st}C(t) = s^2\tilde{C}(s) - sC(0) \tag{4.10}$$

Here we have used the result that $dC(0)/dt = 0$. Eliminating $\tilde{C}(s)$ between equations (4.9) and (4.10) gives

$$\tilde{K}(s) = \frac{\tilde{\phi}(s)}{C(0) - \dfrac{\tilde{\phi}(s)}{s}} \tag{4.11}$$

Rather than try to give a general interpretation of this equation, it may prove more useful to apply it to the Brownian motion problem. $C(0)$ is the time zero value of an equilibrium time-correlation function and can be easily evaluated as $k_B T/m$, and $d\mathbf{v}/dt = \mathbf{F}/m$ where $\mathbf{F}$ is the total force on the Brownian particle. Thus,

$$\tilde{\zeta}(s) = \frac{\tilde{C}^F(s)}{mk_B T - \dfrac{\tilde{C}^F(s)}{s}} \tag{4.12}$$

where

$$\tilde{C}^F(s) = \tfrac{1}{3}\langle \mathbf{F}(0)\cdot\tilde{\mathbf{F}}(s)\rangle \tag{4.13}$$

is the Laplace transform of the total force autocorrelation function. In writing (4.13) we have used the fact that the equilibrium ensemble average denoted $\langle \ldots \rangle$, must be isotropic. The average of any second-rank tensor, say $\langle \mathbf{F}(0)\mathbf{F}(t)\rangle$, must therefore be a scalar multiple of the second-rank identity tensor. That scalar must of course be $\frac{1}{3}\mathrm{tr}\{\langle \mathbf{F}(0)\mathbf{F}(t)\rangle\} = \frac{1}{3}\langle \mathbf{F}(0)\cdot\mathbf{F}(t)\rangle$.

In the so-called Brownian limit where the ratio of the Brownian particle mass to the mean square of the force becomes infinite,

$$\tilde{\zeta}(s) = \frac{\beta}{3m} \int_0^\infty \mathrm{d}t\, \mathrm{e}^{-st} \langle \mathbf{F}(t) \cdot \mathbf{F}(0) \rangle \tag{4.14}$$

For any finite value of the Brownian ratio, (4.12) shows that the integral of the force autocorrelation function is zero. This is seen most easily by solving (4.12) for C^{F} and taking the limit as $s \to 0$.

Equation (49), which gives the relationship between the memory kernel and the phase autocorrelation function, implies that the velocity autocorrelation function $Z(t) \equiv \frac{1}{3}\langle \mathbf{v}(0) \cdot \mathbf{v}(t) \rangle$ is related to the friction coefficient by the equation,

$$\tilde{Z}(s) = \frac{k_{\mathrm{B}} T/m}{s + \tilde{\zeta}(s)} \tag{4.15}$$

This equation is valid outside the Brownian limit. The integral of the velocity autocorrelation function is related to the growth of the mean-square displacement giving yet another expression for the friction coefficient

$$\tilde{Z}(0) = \lim_{t \to \infty} \int_0^t \mathrm{d}t'\, \tfrac{1}{3} \langle \mathbf{v}(0) \cdot \mathbf{v}(t') \rangle = \lim_{t \to \infty} \int_0^t \mathrm{d}t'\, \tfrac{1}{3} \langle \mathbf{v}(t) \cdot \mathbf{v}(t') \rangle$$

$$= \lim_{t \to \infty} \tfrac{1}{3} \langle \mathbf{v}(t) \cdot \Delta\mathbf{r}(t) \rangle = \lim_{t \to \infty} \frac{1}{6} \frac{\mathrm{d}}{\mathrm{d}t} \langle \Delta\mathbf{r}(t)^2 \rangle \tag{4.16}$$

Here the displacement vector $\Delta\mathbf{r}(t)$ is defined by

$$\Delta\mathbf{r}(t) = \mathbf{r}(t) - \mathbf{r}(0) = \int_0^t \mathrm{d}t'\, \mathbf{v}(t') \tag{4.17}$$

Assuming that the mean-square displacement is linear in time, in the long time limit it follows from (4.15) that the friction coefficient $\tilde{\zeta}(0)$, can be calculated from

$$\frac{k_{\mathrm{B}} T}{m \tilde{\zeta}(0)} \equiv D = \frac{1}{6} \lim_{t \to \infty} \frac{\mathrm{d}}{\mathrm{d}t} \langle \Delta\mathbf{r}(t)^2 \rangle = \frac{1}{6} \lim_{t \to \infty} \frac{\langle \Delta\mathbf{r}(t)^2 \rangle}{t} \tag{4.18}$$

This is the Einstein (1905) relation for the diffusion coefficient D. The relationship between the diffusion coefficient and the integral of the velocity autocorrelation function (4.16) is an example of a Green–Kubo relation (Green, 1954; Kubo, 1957).

It should be pointed out that the transport properties we have just evaluated are properties of systems at equilibrium. The Langevin equation describes the irregular Brownian motion of particles *in an equilibrium system.*

Similarly, the self-diffusion coefficient characterizes the random walk executed by a particle in an equilibrium system. The identification of the zero-frequency friction coefficient, $6\pi\eta d/m$, with the viscous drag on a sphere which is forced to move with constant velocity through a fluid, implies that equilibrium fluctuations can be modelled using nonequilibrium transport coefficients, in this case the shear viscosity of the fluid. This hypothesis is known as the Onsager regression hypothesis (Onsager, 1931). The hypothesis can be inverted: one can calculate transport coefficients from a knowledge of the equilibrium fluctuations. We will now discuss these relations in more detail.

4.2 MORI–ZWANZIG THEORY

We will show that for an arbitrary phase variable $A(\mathbf{\Gamma})$ evolving under equations of motion which preserve the equilibrium distribution function, one can always write down a Langevin equation. Such an equation is an exact consequence of the equations of motion. We will use the symbol iL, to denote the Liouvillean associated with these equations of motion. These *equilibrium* equations of motion could be field-free Newtonian equations of motion or they could be field-free thermostatted equations of motion such as Gaussian isokinetic or Nosé–Hoover equations (see Section 5.2). The equilibrium distribution could be microcanonical, canonical or even isothermal–isobaric provided that, if the latter is the case, suitable distribution preserving dynamics are employed. For simplicity we will compute equilibrium time-correlation functions over the canonical distribution function f_c

$$f_c(\mathbf{\Gamma}) = \frac{e^{-\beta H_0(\mathbf{\Gamma})}}{\int d\mathbf{\Gamma}\, e^{-\beta H_0(\mathbf{\Gamma})}} \tag{4.19}$$

We saw in the previous section that a key element of the derivation was that the correlation of the random force, $\mathbf{F}_R(t)$ with the Langevin variable A, vanishes for all time. We will now use the notation developed in Section 3.5, which treats phase variables, $A(\mathbf{\Gamma})$, $B(\mathbf{\Gamma})$, as vectors in $6N$-dimensional phase space with a scalar product defined by $\int d\mathbf{\Gamma}\, f_0(\mathbf{\Gamma})B(\mathbf{\Gamma})A^*(\mathbf{\Gamma})$, and denoted as (B, A^*). We will define a projection operator which will transform any phase variable B into a vector which has no correlation with the Langevin variable A. The component of B parallel to A is merely

$$PB(\mathbf{\Gamma}, t) \equiv \frac{(B(\mathbf{\Gamma}, t), A^*(\mathbf{\Gamma}))}{(A(\mathbf{\Gamma}), A^*(\mathbf{\Gamma}))} A(\mathbf{\Gamma}) \tag{4.20}$$

This equation defines the projection operator P.

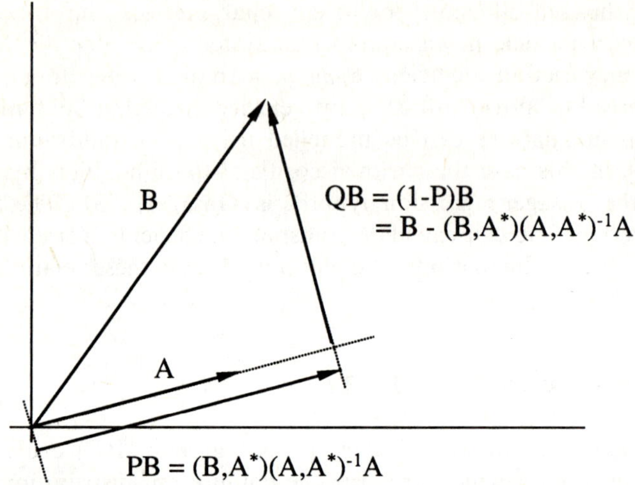

Fig. 4.1 The projection operator, P, operating on B produces a vector which is the component of B parallel to A.

The operator $Q = 1 - P$ is the complement of P and computes the component of B orthogonal to A (see Fig. 4.1).

$$(QB(t), A^*) = \left(B(t) - \frac{(B(t), A^*)}{(A, A^*)} A, A^* \right)$$
$$= (B(t), A^*) - \frac{(B(t), A^*)}{(A, A^*)} (A, A^*) = 0 \tag{4.21}$$

In more physical terms the projection operator Q computes that part of any phase variable which is random with respect to a Langevin variable A.

Other properties of the projection operators are that,

$$PP = P; \qquad QQ = Q; \qquad QP = PQ = 0 \tag{4.22}$$

Secondly, P and Q are Hermitian operators (like the Liouville operator itself). To prove this we note that,

$$(PB, C^*)^* = \frac{((B, A^*)A, C^*)^*}{(A, A^*)^*} = \frac{(B, A^*)^*(A, C^*)^*}{(A, A^*)}$$
$$= \frac{(B^*, A)(A^*, C)}{(A, A^*)} = \frac{(A, B^*)(C, A^*)}{(A, A^*)}$$
$$= \frac{((C, A^*)A, B^*)}{(A, A^*)} = (PC, B^*) \tag{4.23}$$

Furthermore, since $Q = 1 - P$ where 1 is the identity operator, and since both the identity operator and P are Hermitian, so is Q.

We will wish to compute the random and direct components of the propagator e^{iLt}. The random and direct parts of the Liouvillean iL are iQL and iPL, respectively. These Liouvilleans define the corresponding random and direct propagators, e^{iQLt} and e^{iPLt}. We can use the Dyson equation to relate these two propagators. If we take e^{iQLt} as the refernece propagator in (3.100) and e^{iLt} as the test propagator, then

$$e^{iLt} = e^{iQLt} + \int_0^t d\tau \, e^{iL(t-\tau)} iPL \, e^{iQL\tau} \tag{4.24}$$

The rate of change of $A(t)$, the Langevin variable at time t is

$$\frac{dA(t)}{dt} = e^{iLt} \, iLA = e^{iLt} \, i(Q + P)LA \tag{4.25}$$

But,

$$e^{iLt} \, iPLA = e^{iLt} \frac{(iLA, A^*)}{(A, A^*)} A = \frac{(iLA, A^*)}{(A, A^*)} e^{iLt} A \equiv i\Omega A(t) \tag{4.26}$$

This defines the frequency $i\Omega$ which is an equilibrium property of the system It only involves equal time averages. Substituting this equation into (4.25) gives

$$\frac{dA(t)}{dt} = i\Omega A(t) + e^{iLt} \, iQLA \tag{4.27}$$

Using the Dyson decomposition of the propagator given in equation (4.24), this leads to

$$\frac{dA(t)}{dt} = i\Omega A(t) + \int_0^t d\tau \, e^{iL(t-\tau)} \, iPL \, e^{iQL\tau} \, iQLA + e^{iQLt} \, iQLA \tag{4.28}$$

We identify $e^{iQLt} \, iQLA$ as the random force $F(t)$ because

$$(F(t), A^*) = (e^{iQLt} \, iQLA, A^*) = (QF(t), A^*) = 0 \tag{4.29}$$

where we have used (4.22). It is very important to remember that the propagator which generates $F(t)$ from $F(0)$ is not the propagator e^{iLt}, rather it is the random propagator e^{iQLt}. The integral in (4.28) involves the term

$$\begin{aligned} iPL \, e^{iQLt} \, iQLA &= iPLF(t) = iPLQF(t) \\ &= \frac{(iLQF(t), A^*)}{(A, A^*)} A \\ &= -\frac{(QF(t), (iLA)^*)}{(A, A^*)} A \end{aligned}$$

as L is Hermitian and i is anti-Hermitian, $(\mathrm{iL})^* = (\mathrm{d}/\mathrm{d}t)^* = (\mathrm{d}\Gamma/\mathrm{d}t\cdot\mathrm{d}/\mathrm{d}\Gamma)^* = \mathrm{d}/\mathrm{d}t = \mathrm{iL}$ (the equations of motion are real). Since Q is Hermitian,

$$\mathrm{i}P\mathrm{L}\,\mathrm{e}^{\mathrm{i}Q\mathrm{L}t}\,\mathrm{i}Q\mathrm{L}A = -\frac{(F(t),(Q\mathrm{iL}A)^*)}{(A,A^*)}A$$

$$= -\frac{(F(t),F(0)^*)}{(A,A^*)}A \equiv -K(t)A \tag{4.30}$$

where we have defined a memory kernel $K(t)$; it is basically the autocorrelation function of the random force. Substituting this definition into (4.28) gives

$$\frac{\mathrm{d}A(t)}{\mathrm{d}t} = i\Omega A(t) - \int_0^t \mathrm{d}\tau\, \mathrm{e}^{\mathrm{iL}(t-\tau)}K(\tau)A + F(t)$$

$$= i\Omega A(t) - \int_0^t \mathrm{d}\tau\, K(\tau)A(t-\tau) + F(t) \tag{4.31}$$

This shows that the *generalized Langevin equation* is an exact consequence of the equations of motion for the system (Zwanzig, 1961; Mori, 1965a,b). Since the random force is random with respect to A, multiplying both sides of (4.31) by $A^*(0)$ and taking a canonical average gives the memory function equation,

$$\frac{\mathrm{d}C(t)}{\mathrm{d}t} = i\Omega C(t) - \int_0^t \mathrm{d}\tau\, K(\tau)C(t-\tau) \tag{4.32}$$

This is essentially the same as equation (4.5).

As we mentioned in the introduction to this section, the generalized Langevin equation and the memory function equation are exact consequences of any dynamics which preserves the equilibrium distribution function. As such the equations therefore describe equilibrium fluctuations in the phase variable A, and the equilibrium autocorrelation function for A, namely $C(t)$.

However, the generalized Langevin equation bears a striking resemblance to a nonequilibrium constitutive relation. The memory kernel $K(t)$ plays the role of a transport coefficient. Onsager's regression hypothesis (1931) states that the equilibrium fluctuations in a phase variable are governed by the same transport coefficients as is the relaxation of that same phase variable to equilibrium. This hypothesis implies that the generalized Langevin equation can be interpreted as a linear, *nonequilibrium* constitutive relation with the memory function $K(t)$, given by the *equilibrium* autocorrelation function of the random force.

Onsager's hypothesis can be justified by the fact that in observing an *equilibrium* system for a time which is of the order of the relaxation time for the memory kernel, it is *impossible* to tell whether or not the system is at equilibrium. We could be observing the final stages of the relaxation of a nonequilibrium system towards equilibrium, or we could be simply observing the small time-dependent fluctuations in an equilibrium system. On a short time-scale there is simply no way of telling the difference between these two possibilities. When we interpret the generalized Langevin equation as a nonequilibrium constitutive relation, it is clear that it can only be expected to be valid close to equilibrium. This is because it is a linear constitutive equation.

4.3 SHEAR VISCOSITY

It is relatively straightforward to apply the Mori–Zwanzig formalism to the calculation of fluctuation expressions for linear transport coefficients. Our first application of the method will be the calculation of shear viscosity. Before we do this we will say a little more about constitutive relations for shear viscosity. The Mori–Zwanzig formalism leads naturally to a non-Markovian expression for the viscosity. Equation (4.31) refers to a memory function rather than a simple Markovian transport coefficient such as the Newtonian shear viscosity. We will thus be led to a discussion of viscoelasticity (see Section 2.4).

We choose our test variable A, to be the x component of the wavevector-dependent transverse momentum current $\mathbf{J}^{\perp}(\mathbf{k}, t)$. For simplicity, we define the coordinate system so that $\mathbf{k}$ is in the y direction and $\mathbf{J}^{\perp}$ is in the x direction.

$$J_x(k_y, t) = \sum m v_{xi}(t) \exp(ik_y y_i(t)) \tag{4.33}$$

In Section 3.8 we saw that

$$\dot{J} = ikP_{yx}(k, t) \tag{4.34}$$

where for simplicity we have dropped the Cartesian indices for J and k. We note that at zero wavevector the transverse momentum current is a constant of the motion, $\mathrm{d}J/\mathrm{d}t = 0$. The quantities we need in order to apply the Mori–Zwanzig formalism are easily computed.

The frequency matrix $i\Omega$, defined in (4.26), is identically zero. This is always so in the single-variable case as $-\langle A^* \,\mathrm{d}A/\mathrm{d}t \rangle = 0$, for any phase variable A. The norm of the transverse current is calculated

$$\begin{aligned}(J(k), J^*(k)) &= \left\langle \sum_{i=1}^{N} p_{xi}\, \mathrm{e}^{iky_i} \sum_{j=1}^{N} p_{xj}\, \mathrm{e}^{-iky_j} \right\rangle \\ &= N\langle p_{x1}^2 \rangle + N(N-1)\langle p_{x1} p_{x2}\, \mathrm{e}^{ik(y_1 - y_2)} \rangle \\ &= Nmk_{\mathrm{B}}T \end{aligned} \tag{4.35}$$

At equilibrium p_{xi} is independent of p_{x2} and $(y_1 - y_2)$. The values of $\langle p_{x1}\rangle$ and $\langle p_{x2}\rangle$ are identically zero. The random force, F, can also easily be calculated since, if we use (4.34),

$$PP_{yx}(k) = \frac{(P_{yx}(k), J(-k))}{\langle |J(k)|^2 \rangle} J = 0 \tag{4.36}$$

Thus we can write,

$$F(0) = \mathrm{i}QLJ = (1 - P)ikP_{yx}(k) = ikP_{yx}(k) \tag{4.37}$$

The time-dependent random force (see (4.29)) is

$$F(t) = \mathrm{e}^{\mathrm{i}QLt} ikP_{yx}(k) \tag{4.38}$$

A Dyson decomposition of e^{QiLt} in terms of e^{iLt} shows that,

$$\mathrm{e}^{\mathrm{i}Lt} = \mathrm{e}^{Q\mathrm{i}Lt} + \int_0^t \mathrm{d}s\, \mathrm{e}^{\mathrm{i}L(t-s)} P\mathrm{i}L\, \mathrm{e}^{Q\mathrm{i}Ls} \tag{4.39}$$

Now for any phase variable B,

$$P\mathrm{i}LB = \langle J^* \mathrm{i}LB \rangle \frac{J}{Nmk_\mathrm{B}T} = -\langle B(\mathrm{i}LJ)^* \rangle \frac{J}{Nmk_\mathrm{B}T}$$

$$= -ik\langle BP_{yx}(-k)\rangle \frac{J}{Nmk_\mathrm{B}T} \tag{4.40}$$

Substituting this observation into (4.39) shows that the difference between the propagators $\mathrm{e}^{Q\mathrm{i}Lt}$ and $\mathrm{e}^{\mathrm{i}Lt}$ is of order k, and can therefore be ignored in the zero wavevector limit.

From equation (4.30) the memory kernel $K(t)$ is $\langle F(t)F^*(0)\rangle / \langle AA^*\rangle$. Using equation (4.38) the small wavevector form for $K(t)$ becomes

$$K(t) = k^2 \frac{\langle P_{yx}(k,t) P_{yx}(-k,0)\rangle}{Nmk_\mathrm{B}T} \tag{4.41}$$

The generalized Langevin equation (the analogue of equation (4.31)) is

$$\lim_{k\to 0} \frac{\mathrm{d}J_x(k_y,t)}{\mathrm{d}t} = \frac{-k^2}{Nmk_\mathrm{B}T}\int_0^t \mathrm{d}s \langle P_{yx}(k_y,s)P_{yx}(-k_y,0)\rangle J_x(k_y,t-s) + ik_y P_{yx}(k_y,t) \tag{4.42}$$

where we have taken explicit note of the Cartesian components of the relevant functions. Now we know that the rate of change of the transverse current is $ikP_{yx}(k,t)$. This means that the left-hand side of (4.42) is related to equilibrium fluctuations in the shear stress. We also know from (3.142) that $J(k) = \int \mathrm{d}k'\, \rho(k'-k)u(k')$, so, close to equilibrium, the transverse momentum current (our Langevin variable A) is closely related to the wavevector-dependent strain rate $\gamma(k)$. In fact the wavevector-dependent strain rate $\gamma(k)$ is $-ikJ(k)/\rho(k=0)$. Putting these two observations together we see that the generalized Langevin equation for the transverse momentum current is essentially a relation between fluctuations in the shear stress and the strain rate—a constitutive relation. Ignoring the random force (constitutive relations are deterministic), (4.42) can be written in the form of the constitutive relation (2.76),

$$\lim_{k\to 0} P_{yx}(t) = -\int_0^t ds\, \eta(k=0, t-s)\gamma(k=0, s) \tag{4.43}$$

If we use the fact that, $P_{yx}V = \lim(k \to 0)P_{yx}(k)$, $\eta(t)$ is easily seen to be

$$\eta(t) = \beta V \langle P_{xy}(t) P_{xy}(0)\rangle \tag{4.44}$$

Equation (4.43) is identical to the viscoelastic generalization of Newton's law of viscosity equation (2.76).

The Mori–Zwanzig procedure has been used to derive a viscoelastic constitutive relation. No mention has been made of the shearing boundary conditions required for shear flow. Neither is there any mention of viscous heating or possible nonlinearities in the viscosity coefficient. Equation (4.42) is a description of equilibrium fluctuations. However, unlike the case for the Brownian friction coefficient or the self-diffusion coefficient, the viscosity coefficient refers to nonequilibrium rather than equilibrium systems.

The zero wavevector limit is subtle. We can imagine longer and longer wavelength fluctuations in the strain rate $\gamma(k)$. For an equilibrium system, however, $\gamma(k=0) \equiv 0$ and $\langle \gamma(k=0)\gamma^*(k=0)\rangle \equiv 0$. There are *no* equilibrium fluctuations in the strain rate at $k=0$. The zero wavevector strain rate is completely specified by the boundary conditions.

If we invoke Onsager's regression hypothesis we can obviously identify the memory kernel $\eta(t)$ as the memory function for planar (i.e. $k=0$) Couette flow. We observe that there is no fundamental way of knowing whether we are watching small equilibrium fluctuations at small but non-zero wavevector, or the *last* stages of relaxation toward equilibrium of a finite k, nonequilibrium disturbance. Provided the nonequilibrium system is sufficiently close to equilibrium, the Langevin memory function will be the nonequilibrium memory kernel. However, the Onsager regression hypothesis is additional to, and not part of, the Mori–Zwanzig theory. In Section 6.3 we prove that the nonequilibrium linear viscosity coefficient is given exactly by the infinite time integral of the stress fluctuations. In Section 6.3 we will not use the Onsager regression hypothesis.

At this stage one might legitimately ask the question: what happens to these equations if we do *not* take the zero wavevector limit? After all we have already defined a wavevector dependent shear viscosity in (2.77). It is not a simple matter to apply the Mori–Zwanzig formalism to the finite wavevector case. Instead we will use a method which makes a direct appeal to the Onsager regression hypothesis.

Provided the time and spatially dependent strain rate is of sufficiently small amplitude, the generalized viscosity can be defined as (2.77)

$$P_{yx}(k,t) = -\int_0^t \mathrm{d}s\, \eta(k,t-s)\gamma(k,s) \tag{4.45}$$

Using the fact that $\gamma(k,t) = -iku_x(k,t) = -ikJ(k,t)/\rho$, and that $\mathrm{d}J(k,t)/\mathrm{d}t = ikP_{yx}(k,t)$, we can rewrite (4.45) as

$$\dot{J}(k,t) = -\frac{k^2}{\rho}\int_0^t \mathrm{d}s\, \eta(k,t-s)J(k,s) \tag{4.46}$$

If we Fourier–Laplace transform both sides of this equation over time, and using Onsager's hypothesis, multiply both sides by $J(-k,0)$ and average with respect to the equilibrium canonical ensemble, we obtain

$$\tilde{C}(k,\omega) = \frac{C(k,0)}{i\omega + \dfrac{k^2\tilde{\eta}(k,\omega)}{\rho}} \tag{4.47}$$

where $C(k,t)$ is the equilibrium transverse current autocorrelation function $\langle J(k,t)J(-k,0)\rangle$ and the tilde notation denotes a Fourier–Laplace transform in time,

$$\tilde{C}(\omega) = \int_0^\infty \mathrm{d}t\, C(t)\,\mathrm{e}^{-i\omega t} \tag{4.48}$$

We call the autocorrelation function of the wavevector-dependent shear stress,

$$N(k,t) \equiv \frac{1}{Vk_{\mathrm{B}}T}\langle P_{yx}(k,t)P_{yx}(-k,0)\rangle \tag{4.49}$$

We can use the relation $\mathrm{d}J(k,t)/\mathrm{d}t = ikP_{yx}(k,t)$, to transform from the transverse current autocorrelation function $C(k,t)$ to the stress autocorrelation function $N(k,t)$, since

$$\begin{aligned}\frac{\mathrm{d}^2}{\mathrm{d}t^2}\langle J(k,t)J(-k,0)\rangle &= -\langle \dot{J}(k,t)\dot{J}(-k,0)\rangle \\ &= -k^2\langle P_{yx}(k,t)P_{yx}(-k,0)\rangle \end{aligned} \tag{4.50}$$

This derivation closely parallels that for equation (4.10) and (4.11) in Section 4.1. The reader should refer to that section for details. Using

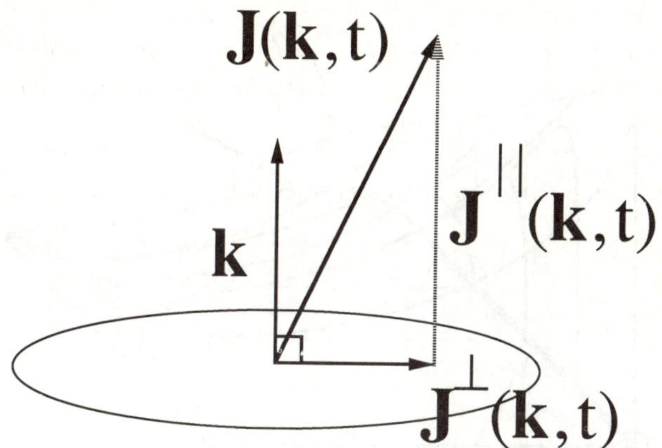

Fig. 4.2 We can resolve the wavevector-dependent momentum density into components which are parallel and orthogonal to the wavevector, $\mathbf{k}$: $\mathbf{J}(\mathbf{k},t) = \mathbf{J}^{\perp}(\mathbf{k},t) + \mathbf{J}^{\|}(\mathbf{k},t)$.

the fact that $\rho = Nm/V$, we see that

$$k^2 V k_B T \tilde{N}(k,\omega) = \omega^2 \tilde{C}(k,\omega) + i\omega C(k,0) \tag{4.51}$$

The equilibrium average $C(k,0)$ is given by equation (4.35). Substituting this equation into equation (4.47) gives us an equation for the frequency- and wavevector-dependent shear viscosity in terms of the stress autocorrelation function,

$$\tilde{\eta}(k,\omega) = \frac{\tilde{N}(k,\omega)}{1 - \dfrac{k^2 \tilde{N}(k,\omega)}{i\omega\rho}} \tag{4.52}$$

This equation is *not* of the Green–Kubo form. Green–Kubo relations are exceptional being only valid for infinitely slow processes. Momentum relaxation is only infinitely slow at *zero* wavevector. At finite wavevector momentum relaxation is a *fast* process. We can obtain the usual Green–Kubo form by taking the zero k limit of equation (4.52). In that case

$$\tilde{\eta}(0,\omega) = \lim_{k\to 0} \tilde{N}(k,\omega) \tag{4.53}$$

Because there are no fluctuations in the zero wavevector strain rate, the function $\tilde{N}(k,\omega)$ is discontinuous at the origin. For all nonzero values of k, $\tilde{N}(k,0) = 0$! Over the years many errors have been made as a result of this fact. Figure 4.3 illustrates these points schematically. The results for shear viscosity precisely parallel those for the friction constant of a Brownian

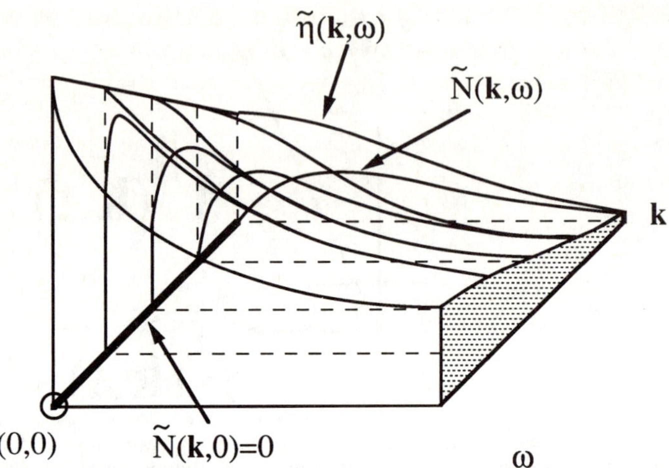

Fig. 4.3 Schematic diagram of the frequency- and wavevector-dependent viscosity and stress autocorrelation function. The relationship between the viscosity, $\tilde{\eta}(k, \omega)$, and the stress autocorrelation function, $\tilde{N}(k, \omega)$. At $k = 0$ both functions are identical. At $\omega = 0$ but $k \neq 0$, the stress autocorrelation function is identically zero. The stress autocorrelation function is discontinuous at the origin. The viscosity is continuous everywhere, but non-analytic at the origin (see Evans (1981)).

particle. Only in the Brownian limit is the friction constant given by the autocorrelation function of the Brownian force.

An immediate conclusion from the theory we have outlined is that *all* fluids are viscoelastic. Viscoelasticity is a direct result of the generalized Langevin equation which, in turn, is an exact consequence of the microscopic equations of motion.

4.4 GREEN–KUBO RELATIONS FOR NAVIER–STOKES TRANSPORT COEFFICIENTS

It is relatively straightforward to derive Green–Kubo relations for the other Navier–Stokes transport coefficients, namely bulk viscosity and thermal conductivity. In Section 6.3 when we describe the SLLOD equations of motion for viscous flow we will find a simpler way of deriving Green–Kubo relations for both viscosity coefficients. For now we simply state the Green–Kubo relation for bulk viscosity as (Zwanzig, 1965),

$$\eta_V = \frac{1}{Vk_B T}\int_0^\infty dt \langle [p(t)V(t) - \langle pV \rangle][p(0)V(0) - \langle pV \rangle] \rangle \quad (4.54)$$

The Green–Kubo relation for thermal conductivity can be derived by similar arguments to those used in the viscosity derivation. Firstly, we note from (2.26) that in the absence of a velocity gradient the internal energy per unit volume ρU obeys a continuity equation, $\rho\, \mathrm{d}U/\mathrm{d}t = -\boldsymbol{\nabla}\cdot\mathbf{J}_Q$. Secondly, we note that Fourier's definition of the thermal conductivity coefficient, λ, from equation (2.56a), is $\mathbf{J}_Q = -\lambda\boldsymbol{\nabla}T$. Combining these two results we obtain

$$\rho\frac{\mathrm{d}U}{\mathrm{d}t} = \lambda\nabla^2 T \tag{4.55}$$

Unlike the previous examples, both U and T have nonzero equilibrium values; namely, $\langle U\rangle$ and $\langle T\rangle$. A small change in the left-hand side of equation (4.55) can be written as $(\langle\rho\rangle + \Delta\rho)\,\mathrm{d}(\langle U\rangle + \Delta U)/\mathrm{d}t$. By definition $\mathrm{d}\langle U\rangle/\mathrm{d}t = 0$, so to first order in Δ, we have $\rho\,\mathrm{d}\Delta U/\mathrm{d}t$. Similarly, the spatial gradient of $\langle T\rangle$ does not contribute, so we can write

$$\rho\frac{\mathrm{d}\Delta U}{\mathrm{d}t} = \lambda\nabla^2\Delta T \tag{4.56}$$

The next step is to relate the variation in temperature ΔT to the variation in energy per unit volume $\Delta(\rho U)$. To do this we use the thermodynamic definition,

$$\left.\frac{1}{V}\frac{\partial E}{\partial T}\right|_V = \left.\frac{\partial(\rho U)}{\partial T}\right|_V = \rho c_V \tag{4.57}$$

where c_V is the specific heat per unit mass. We see from the second equality, that a small variation in the temperature ΔT is equal to $\Delta(\rho U)/\rho c_V$. Therefore,

$$\rho\Delta\dot{U} = \frac{\lambda}{\rho c_V}\nabla^2\rho\Delta U \tag{4.58}$$

If $D_T \equiv \lambda/\rho c_V$ is the thermal diffusivity, then in terms of the wavevector-dependent internal energy density, equation (4.58) becomes

$$\rho\Delta\dot{U}(\mathbf{k},t) \equiv -k^2 D_T\rho\Delta U(\mathbf{k},t) \tag{4.59}$$

If $C(k,t)$ is the wavevector-dependent internal energy density autocorrelation function,

$$C(k,t) \equiv \rho^2\langle\Delta U(\mathbf{k},t)\Delta U(-\mathbf{k},0)\rangle \tag{4.60}$$

then the frequency- and wavevector-dependent diffusivity is the memory function of energy density autocorrelation function,

$$\tilde{C}(k,\omega) = \frac{C(k,0)}{i\omega + k^2\tilde{D}_T(k,\omega)} \tag{4.61}$$

Using exactly the same procedures as in Section 4.1, we can convert (4.61) into an expression for the diffusivity in terms of a current correlation function. From (4.7) and (4.10), if $\phi = -\mathrm{d}^2 C/\mathrm{d}t^2$, then

$$\phi(k,t) = k^2 \langle J_{Qx}(k,t) J_{Qx}(-k,0)\rangle \tag{4.62}$$

Using equation (4.10), we obtain the analogue of (4.11)

$$k^2 \tilde{D}_T(k,\omega) = \frac{C(k,0) - i\omega\tilde{C}(k,\omega)}{\tilde{C}(k,\omega)} = \frac{\tilde{\phi}(k,\omega)}{C(k,0) - \dfrac{\tilde{\phi}(k,\omega)}{i\omega}} \tag{4.63}$$

If we define the analogue of (4.49), that is $\phi(k,t) \equiv k^2 N_Q(k,t)$, then (4.63) for the thermal diffusivity can be written in the same form as the wavevector-dependent shear viscosity equation (4.52). That is

$$\tilde{D}_T(k,\omega) = \frac{\tilde{N}_Q(k,\omega)}{C(k,0) - \dfrac{k^2}{i\omega}\tilde{N}_Q(k,\omega)} \tag{4.64}$$

Again we see that we must take the zero-wavevector limit *before* we take the zero-frequency limit, and using the canonical ensemble fluctuation formula for the specific heat,

$$\rho c_V = \frac{V}{k_B T^2} C(0,0) \tag{4.65}$$

we obtain the Green–Kubo expression for the thermal conductivity

$$\lambda = \frac{V}{k_B T^2} \int_0^\infty \mathrm{d}t \, \langle J_{Qx}(t) J_{Qx}(0)\rangle \tag{4.66}$$

This completes the derivation of the Green–Kubo formulae for thermal transport coefficients. These formulae relate thermal transport coefficients to equilibrium properties. In the next chapter we will develop nonequilibrium routes to the thermal transport coefficients.

REFERENCES

Brown, R. (1828a). *Ann. Phys. Chem.*, **14**, 294.
Brown, R. (1828b). *Phil. Mag.*, **4**, 161.
Einstein, A. (1905). *Ann. Phys.*, **17**, 549.
Evans, D.J. (1981). *Phys. Rev., Ser. A*, **23**, 2622.
Green, M.S. (1954). *J. Chem. Phys.*, **22**, 398.

Kubo, R. (1957). *J. Phys. Soc. Jpn.*, **12**, 570.
Langevin, P. (1908). *C.R. Acad. Sci. (Paris)*, **146**, 530.
Mori, H. (1965a). *Prog. Theor. Phys.*, **33**, 423.
Mori, H. (1965b). *Prog. Theor. Phys.*, **34**, 399.
Onsager, L. (1931). *Phys. Rev.*, **37**, 405; **38**, 2265.
Zwanzig, R. (1961). *Lectures in Theoretical Physics (Boulder)*, Vol. III, p. 135. New York: Wiley.
Zwanzig, R. (1965). *Ann. Rev. Phys. Chem.*, **16**, 67.

5 Linear Response Theory

5.1 ADIABATIC LINEAR RESPONSE THEORY

In this chapter we will discuss how an external field, F_e, perturbs an N-particle system. We assume that the field is sufficiently weak that only the linear response of the system need be considered. These considerations will lead us to equilibrium fluctuation expressions for *mechanical* transport coefficients such as the electrical conductivity. These expressions are formally identical to the Green–Kubo formulae that were derived in the last chapter. The difference is that the Green–Kubo formulae pertain to *thermal* transport processes where boundary conditions perturb the system away from equilibrium—all Navier–Stokes processes fall into this category. Mechanical transport coefficients, however, refer to systems where mechanical fields, which appear explicitly in the equations of motion for the system, drive the system away from equilibrium.

As we will see it is no coincidence that there is such close similarity between the fluctuation expressions for thermal- and mechanical-transport coefficients. In fact one can often mathematically transform the nonequilibrium boundary conditions for a thermal-transport process into a mechanical field. The two representations of the system are then said to be *congruent*.

A major difference between the derivations of the equilibrium fluctuation expressions for the two representations is that in the mechanical case one does not need to invoke Onsager's regression hypothesis. The linear mechanical response of a nonequilibrium system is analysed mathematically with resultant expressions for the response that involve equilibrium time-correlation functions. In the thermal case (Chapter 4) equilibrium fluctuations were studied and after invoking Onsager's hypothesis, the connection with nonequilibrium transport coefficients was made. Given a congruent mechanical representation of a thermal transport process, one can in fact prove the validity of Onsager's hypothesis.

The mechanical field, F_e, performs work on the system, preventing relaxation to equilibrium. This work is converted into heat. It is easy to show that the rate at which the field performs work on the system is, for small fields, proportional to F_e^2. This is, at least formally, a nonlinear effect.

This is why, in the complete absence of any thermostatting mechanism, Kubo (1957) was able to derive correct expressions for the linear response. However, in spite of heating being a nonlinear effect, a thermostatted treatment of linear response theory leads to a considerably more satisfying discussion. We will therefore include in this chapter a description of thermostats and isothermal linear response theory.

Consider a system of N atoms suddenly subject, at $t = 0$, to a time-dependent external field, F_e. The generalization of our discussion to vector or tensor fields is straightforward. For simplicity we will assume that the particles move in a three-dimensional Cartesian space. For times greater than zero the system is assumed to obey the dynamics given in the equations below.

$$\begin{aligned} \dot{\mathbf{q}}_i &= \frac{\mathbf{p}_i}{m} + \mathbf{C}_i F_e(t) \\ \dot{\mathbf{p}}_i &= \mathbf{F}_i + \mathbf{D}_i F_e(t) \end{aligned} \tag{5.1}$$

The phase variables $\mathbf{C}_i(\mathbf{\Gamma})$ and $\mathbf{D}_i(\mathbf{\Gamma})$ describe the coupling of the field to the system. We assume that the equations have been written in such a way that at equilibrium in the absence of the external field the *canonical* kinetic energy, K, satisfies the equipartition relation

$$\tfrac{3}{2} N k_B T = \left\langle \sum_{i=1}^{N} \frac{\mathbf{p}_i^2}{2m} \right\rangle \equiv \langle K \rangle \tag{5.2}$$

This implies that the canonical momenta give the peculiar velocities of each of the particles and that, therefore,

$$\sum_{i=1}^{N} \mathbf{p}_i = 0 \tag{5.3}$$

In this case H_0,

$$H_0(\mathbf{\Gamma}) \equiv \sum_{i=1}^{N} \frac{\mathbf{p}_i^2}{2m} + \Phi(\mathbf{q}) \tag{5.4}$$

is the instantaneous expression for the internal energy. We do not assume that a Hamiltonian exists which will generate the field-dependent equations of motion. In the absence of the external field and a thermostat, H_0 is the total energy, and is therefore a constant of the motion. The rate of change of internal energy due to the field is

$$\begin{aligned} \dot{H}_0(\mathbf{\Gamma}, t) &= \sum_{i=1}^{N} \left[\frac{\dot{\mathbf{p}}_i(t) \cdot \mathbf{p}_i}{m} - \dot{\mathbf{q}}_i(t) \cdot \mathbf{F}_i \right] \\ &= - \sum_{i=1}^{N} \left[-\mathbf{D}_i \cdot \frac{\mathbf{p}_i}{m} + \mathbf{C}_i \cdot \mathbf{F}_i \right] F_e(t) \\ &\equiv -J(\mathbf{\Gamma}) F_e(t) \end{aligned} \tag{5.5}$$

where $J(\mathbf{\Gamma})$, is called the *dissipative* flux.

The response of the system to the external field can be assessed by monitoring the average response of an arbitrary phase variable $B(\mathbf{\Gamma})$ at some later time t. The *average* response of the system is the response that is obtained by perturbing an ensemble of initial phases. It is usual to select the starting states from the equilibrium canonical ensemble, thus

$$f(\mathbf{\Gamma},0) = f_c(\mathbf{\Gamma}) = \frac{\exp[-\beta H_0(\mathbf{\Gamma})]}{\int \mathrm{d}\mathbf{\Gamma}' \exp[-\beta H_0(\mathbf{\Gamma}')]} = \frac{\mathrm{e}^{-\beta H_0(\mathbf{\Gamma})}}{Z(\beta)} \tag{5.6}$$

The average response $\langle B(t) \rangle$ can be calculated from the expression

$$\langle B(t) \rangle = \int \mathrm{d}\mathbf{\Gamma}\, B(\mathbf{\Gamma}) f(\mathbf{\Gamma},t) \tag{5.7}$$

This is the Schrödinger representation for the response of B. The problem of determining the response then reduces to determining the perturbed distribution function $f(t)$. The rate of change in the perturbed distribution function is given by the Liouville equation

$$\frac{\partial}{\partial t} f(\mathbf{\Gamma},t) = -iLf(\mathbf{\Gamma},t) = -\left[\left(\frac{\partial}{\partial \mathbf{\Gamma}} \cdot \dot{\mathbf{\Gamma}}(t)\right) + \dot{\mathbf{\Gamma}}(t) \cdot \frac{\partial}{\partial \mathbf{\Gamma}}\right] f(\mathbf{\Gamma},t) \tag{5.8}$$

The $\dot{\mathbf{\Gamma}}(t)$ in these equations is given by the first-order form of the equations of motion (5.1) with the external field evaluated at the current time, t.

If the equations of motion are derivable from a Hamiltonian it is easy to show that $\partial(\mathrm{d}\mathbf{\Gamma}/\mathrm{d}t)/\partial\mathbf{\Gamma} = 0$ (Section 3.3). We will assume that even in the case where no Hamiltonian exists which can generate the equations of motion (5.1), that $\partial(\mathrm{d}\mathbf{\Gamma}/\mathrm{d}t)/\partial\mathbf{\Gamma} = 0$. We refer to this condition as the *adiabatic incompressibility of phase space* (AIΓ). A sufficient, but not necessary, condition for this to hold is that the unthermostatted or *adiabatic* equations of motion are derivable from a Hamiltonian. It is of course possible to pursue the theory without this condition, but in practise it is rarely necessary to do so (the only known exception is discussed in MacGowan and Evans (1986)).

Thus in the adiabatic case, if AIΓ holds, we know that the Liouville operator is Hermitian (see Sections 3.3 and 3.5) and, therefore,

$$iLA = \frac{\partial}{\partial \mathbf{\Gamma}} \cdot \dot{\mathbf{\Gamma}} A = \dot{\mathbf{\Gamma}} \cdot \frac{\partial}{\partial \mathbf{\Gamma}} A = \mathrm{i}\mathrm{L}A \tag{5.9}$$

If we denote the Liouvillean for the field-free equations of motion as iL_0, and we break up the total Liouvillean into its field-free and field-dependent parts, equation (5.8) becomes,

$$\frac{\partial}{\partial t}(f_c + \Delta f(\mathbf{\Gamma},t)) = -(\mathrm{iL}_0 + \mathrm{i}\Delta\mathrm{L}(t))(f_c + \Delta f(\mathbf{\Gamma},t)) \tag{5.10}$$

where the distribution function $f(\boldsymbol{\Gamma}, t)$, is written as $f_c + \Delta f(\boldsymbol{\Gamma}, t)$. Since H_0 is a constant of the motion for the field-free adiabatic equations of motion; iL_0 therefore *preserves* the canonical ensemble,

$$\mathrm{iL}_0 f_c = \dot{\boldsymbol{\Gamma}} \cdot \frac{\partial}{\partial \boldsymbol{\Gamma}} f_c = \sum_{i=1}^{N} \left(\dot{\mathbf{q}}_i \cdot \frac{\partial}{\partial \mathbf{q}_i} + \dot{\mathbf{p}}_i \cdot \frac{\partial}{\partial \mathbf{p}_i} \right) \frac{\mathrm{e}^{-\beta H_0(\boldsymbol{\Gamma})}}{Z(\beta)}$$

$$= \frac{-\beta\, \mathrm{e}^{-\beta H_0}}{Z(\beta)} \sum_{i=1}^{N} \left(\dot{\mathbf{q}}_i \cdot \frac{\partial H_0}{\partial \mathbf{q}_i} + \dot{\mathbf{p}}_i \cdot \frac{\partial H_0}{\partial \mathbf{p}_i} \right) = 0 \tag{5.11}$$

Substituting (5.11) into equation (5.10) we see that

$$\frac{\partial}{\partial t} \Delta f(\boldsymbol{\Gamma}, t) + \mathrm{iL}_0 \Delta f(\boldsymbol{\Gamma}, t) = -\mathrm{i}\Delta \mathrm{L}(t) f_c(\boldsymbol{\Gamma}) + O(\Delta)^2 \tag{5.12}$$

In (5.12) we ignore perturbations to the distribution function which are second order in the field. (The Schrödinger–Heisenberg equivalence (Section 3.3) proves that these second-order terms for the distribution are identical to the second-order trajectory perturbations.) In Section 7.8 we discuss the nature of this linearization procedure in some detail. To linear order, the solution of equation (5.12) is

$$\Delta f(\boldsymbol{\Gamma}, t) = -\int_0^t \mathrm{d}s \exp(-\mathrm{iL}_0(t-s))\mathrm{i}\Delta \mathrm{L}(s) f_c(\boldsymbol{\Gamma}) + O(\Delta^2) \tag{5.13}$$

The correctness of this solution can easily be checked by noting that at $t = 0$, (5.13) has the correct initial condition, ($\Delta f(\boldsymbol{\Gamma}, t = 0) = 0$) and that the solution for $\Delta f(\boldsymbol{\Gamma}, t)$ given in (5.13) satisfies (5.12) for all subsequent times.

We will now operate on the canonical distribution function with the operator, $\mathrm{iL}(t)$. We again use the fact that iL_0 preserves the canonical distribution function.

$$\mathrm{i}\Delta \mathrm{L}(t) f_c(\boldsymbol{\Gamma}) = \mathrm{iL}(t) f_c(\boldsymbol{\Gamma}) = \dot{\boldsymbol{\Gamma}} \cdot \frac{\partial}{\partial \boldsymbol{\Gamma}} \frac{\exp[-\beta H_0(\boldsymbol{\Gamma})]}{Z(\beta)}$$

$$= -\beta f_c(\boldsymbol{\Gamma}) \dot{\boldsymbol{\Gamma}} \cdot \frac{\partial H_0}{\partial \boldsymbol{\Gamma}}$$

$$= -\beta \dot{H}_0^{\mathrm{ad}} f_c(\boldsymbol{\Gamma}) \tag{5.14}$$

The adiabatic time derivative of H_0 is given by the dissipative flux (5.5), so

$$\mathrm{i}\Delta \mathrm{L}(s) f_c(\boldsymbol{\Gamma}) = -\beta \dot{H}_0^{\mathrm{ad}}(s) f_c(\boldsymbol{\Gamma}) = \beta J(\boldsymbol{\Gamma}) F_e(s) f_c(\boldsymbol{\Gamma}) \tag{5.15}$$

The time argument associated with $\mathrm{i}\Delta\mathrm{L}(s)$ is the time argument of the external field.

Substituting (5.15) into (5.13) and in turn into (5.7), the linear response of the phase variable B is given by

$$\langle B(t)\rangle = \langle B(0)\rangle + \int d\mathbf{\Gamma}\, B(\mathbf{\Gamma})\Delta f(\mathbf{\Gamma}, t)$$

$$= \langle B(0)\rangle - \int_0^t ds \int d\mathbf{\Gamma}\, B(\mathbf{\Gamma}) \exp(-\mathrm{i}L_0(t-s))\beta J(\mathbf{\Gamma})F_e(s)f_c(\mathbf{\Gamma})$$

$$= \langle B(0)\rangle - \beta \int_0^t ds \int d\mathbf{\Gamma}\, B(\mathbf{\Gamma}; t-s)J(\mathbf{\Gamma}; 0)f_c(\mathbf{\Gamma})F_e(s) \qquad (5.16)$$

In deriving the third line of this equation from the second we have *unrolled* the propagator from the dissipative flux onto the response variable B. Note that the propagator has no effect on either the canonical distribution function (which is preserved by it), or on the external field $F_e(t)$ which is not a phase variable.

It is usual to express the result in terms of a linear susceptibility, χ_{BJ}, which is defined in terms of the equilibrium time-correlation function of B and J

$$\chi_{BJ}(t) \equiv \beta\langle B(t)J(0)\rangle \qquad (5.17)$$

To linear order, the canonical ensemble averaged linear response for $B(t)$ is

$$\lim_{F_e \to 0} \langle B(t)\rangle = \langle B(0)\rangle - \int_0^t ds\, \chi_{BJ}(t-s)F_e(s) \qquad (5.18)$$

This equation is very similar to the response functions we met in Chapter 4 when we discussed viscoelasticity and constitutive relations for thermal transport coefficients. The equation shows that the linear response is non-Markovian. *All* systems have memory. All N-body systems remember the field history over the decay time of the relevant time-correlation function, $\langle B(t)J(0)\rangle$. Markovian behaviour is only an idealization brought about by a lack of sensitivity in our measurements of the time-resolved many-body response.

There are a number of deficiencies in the derivation we have just given. Suppose that by monitoring $\langle B(t)\rangle$ for a family of external fields F_e, we wish to deduce the susceptibility $\chi(t)$. One cannot blindly use equation (5.18). This is because as the system heats up through the conversion of work into heat, the system temperature will change with time. This effect is quadratic in the magnitude of the external field. If χ increases with temperature, the long-time limiting value of $\langle B(t)\rangle$ will be infinite. If χ decreases with increasing temperature the limiting value of $\langle B(t)\rangle$ could well be zero. This is simply a reflection of the fact that in the absence of a thermostat there is no steady state. The linear steady state value for the response can only be obtained if we take the field strength to zero *before* we let time go to infinity. This procedure will inevitably lead to difficulties in both the experimental and numerical determination of the linear susceptibilities.

Another difficulty with the derivation is that if adiabatic linear response theory is applied to computer simulation, one would prefer not to use canonical averaging. This is because a single Newtonian equilibrium trajectory cannot generate or span the canonical ensemble. A single Newtonian trajectory can at most span a microcanonical subset of the canonical ensemble of states. A canonical evaluation of the susceptibility therefore requires an ensemble of trajectories if one is using Newtonian dynamics. This is inconvenient and very expensive in terms of computer time.

One cannot simply extend this adiabatic theory to the microcanonical ensemble. Kubo (1982) has shown that if one subjects a cumulative microcanonical ensemble (all states less than a specified energy have the same probability) to a mechanical perturbation, then the linear susceptibility is given by the equilibrium correlation of the test variable B and the dissipative flux J averaged over the delta microcanonical ensemble (all states with a precisely specified energy have the same probability). When the equilibrium ensemble of starting states is not identical to the equilibrium ensemble used to compute the susceptibilities, we say that the theory is *ergodically inconsistent*. We will now show how both of these difficulties can be resolved.

5.2 THERMOSTATS AND EQUILIBRIUM DISTRIBUTION FUNCTIONS

The Gaussian isokinetic thermostat

Thermostats were first introduced as an aid to performing nonequilibrium computer simulations. Only later was it realized that these devices have a fundamental role in the statistical mechanics of many-body systems. The first deterministic method for thermostatting molecular dynamics simulations was proposed simultaneously and independently by Hoover and Evans (Hoover *et al.*, 1982; Evans, 1983). Their method employs a damping or friction term in the equations of motion. Initially, the use of such damping terms had no theoretical justification. Later it was realized (Evans *et al.*, 1983) that these equations of motion could be derived using Gauss' principle of least constraint (Section 3.1). This systematized the extension of the method to other constraint functions.

Using Gauss' principle (Chapter 3), the isokinetic equations of motion for a system subject to an external field can be written as,

$$\begin{aligned} \dot{\mathbf{q}}_i &= \frac{\mathbf{p}_i}{m} + \mathbf{C}_i F_e(t) \\ \dot{\mathbf{p}}_i &= \mathbf{F}_i + \mathbf{D}_i F_e(t) - \alpha \mathbf{p}_i \end{aligned} \tag{5.19}$$

This is the thermostatted generalization of (5.1) where the thermostatting term $\alpha \mathbf{p}_i$ has been added. In writing these equations we are assuming:

(1) that the equations have been written in a form in which the canonical momenta are peculiar with respect to the streaming velocities of the particles;
(2) that $\Sigma\, \mathbf{p}_i = 0$; and
(3) that H_0 is the phase variable which corresponds to the internal energy.

In order to know that these three conditions are valid, we must know quite a lot about the possible flows induced in the system by the external field. This means that if we are considering shear flow, for example, the Reynolds number must be small enough for laminar flow to be stable. Otherwise we cannot specify the streaming component of a particles motion ($\mathbf{C}_i$ must contain the local hydrodynamic flow field $\mathbf{u}(\mathbf{r}, t)$) and we cannot expect condition (1) to be valid.

The isokinetic expression for the multiplier is easily seen to be

$$\alpha = \alpha_0 + \alpha_1 F_e(t) = \frac{\sum_i \frac{\mathbf{F}_i}{m_i} \cdot \mathbf{p}_i}{\sum_i \frac{1}{m_i} \mathbf{p}_i^2} + \frac{\sum_i \frac{\mathbf{D}_i}{m_i} \cdot \mathbf{p}_i}{\sum_i \frac{1}{m_i} \mathbf{p}_i^2} F_e(t) \tag{5.20}$$

It is instructive to compare this result with the corresponding field-free multiplier given in (3.32). It is important to keep in mind that the expression for the multiplier depends explicitly on the external field and, therefore, on time. This is why we define the time- and field-independent phase variables α_0 and α_1.

It is easy to show that if Gauss' principle is used to fix the internal energy H_0, then the equations of motion take on exactly the same form (Evans, 1983), except that the multiplier is

$$\alpha = \frac{\sum \mathbf{D}_i \cdot \mathbf{p}_i - \mathbf{C}_i \cdot \mathbf{F}_i}{\sum \frac{\mathbf{p}_i^2}{m}} F_e(t) \tag{5.21}$$

It may seem odd that the form of the field-dependent equations of motion is independent of whether we are constraining the kinetic or the total energy. This occurs because the vector character of the constraint force is the same for both forms of constraint (see Section 3.1). In the isoenergetic case it is clear that the multiplier vanishes when the external field is zero. This is as expected, since in the absence of an external field Newton's equations conserve the total energy.

Gaussian thermostats remove heat from the system at the rate $\mathrm{d}Q/\mathrm{d}t \equiv -(\mathrm{d}H_0/\mathrm{d}t)^{\mathrm{therm}}$

$$\dot{Q}(t) = \alpha(t) \sum_{i=1}^{N} \frac{\mathbf{p}_i^2}{m_i} \tag{5.22}$$

by applying a force of constraint which is parallel to the peculiar velocity of each particle in the system.

We will now discuss the equilibrium properties of Gaussian isokinetic systems in more detail. At equilibrium, the Gaussian isokinetic equations become

$$\begin{aligned} \dot{\mathbf{q}}_i &= \frac{\mathbf{p}_i}{m} \\ \dot{\mathbf{p}}_i &= \mathbf{F}_i - \alpha \mathbf{p}_i \end{aligned} \tag{5.23}$$

with the multiplier given by equation (5.20) with $F_e = 0$. Clearly the average value of the multiplier is zero at equilibrium with fluctuations in its value being precisely those required to keep the kinetic energy constant. Following our assumption that the initial value of the total linear momentum is zero, it is trivial to see that, like the kinetic energy, the total linear momentum is a constant of the motion.

The ergodically generated equilibrium distribution function $f_T(\mathbf{\Gamma})$ can be obtained by solving the Liouville equation for these equations of motion. It is convenient to consider the total time derivative of f. From the Liouville equation (3.42), we see that

$$\frac{\mathrm{d}f}{\mathrm{d}t} = -f \frac{\partial}{\partial \mathbf{\Gamma}} \cdot \dot{\mathbf{\Gamma}} = -f \sum_{i=1}^{N} \frac{\partial}{\partial \mathbf{p}_i} \cdot \dot{\mathbf{p}}_i = f \sum_{i=1}^{N} \frac{\partial}{\partial \mathbf{p}_i} \cdot \alpha \mathbf{p}_i \tag{5.24}$$

In computing the final derivative in this equation we get $3N$ identical intensive terms from the $3N$ derivatives, $\alpha \partial \mathbf{p}_i \cdot / \partial \mathbf{p}_i$. We also get $3N$ terms from $\mathbf{p}_i \cdot \partial \alpha / \partial \mathbf{p}_i$ which sum to give $-\alpha$. Since we are interested in statistical mechanical systems, we will ignore terms of relative order $1/N$ in the remaining discussion. It is certainly possible to retain these terms, but this would add considerably to the algebraic complexity without revealing any new physics. This being the case, equation (5.24) becomes

$$\frac{\mathrm{d}f}{\mathrm{d}t} = 3N\alpha f + O(1) f \tag{5.25}$$

From (5.20) it can be shown that

$$\frac{\mathrm{d}f}{\mathrm{d}t} = -\frac{3N}{2K} f\dot{\Phi} \tag{5.26}$$

or

$$\frac{\mathrm{d}\ln f}{\mathrm{d}t} = -\frac{3N}{2K}\frac{\mathrm{d}\Phi}{\mathrm{d}t} \tag{5.27}$$

Integrating both sides with respect to time enables us to evaluate the time-independent equilibrium distribution function,

$$f_{\mathrm{T}}(\boldsymbol{\Gamma}) = \frac{\exp[-\beta\Phi(\boldsymbol{\Gamma})]\delta(K(\boldsymbol{\Gamma}) - K_0)}{\int \mathrm{d}\boldsymbol{\Gamma}\, \exp[-\beta\Phi(\boldsymbol{\Gamma})]\delta(K(\boldsymbol{\Gamma}) - K_0)} \tag{5.28}$$

where the constant, $\beta = 3N/2K_0$. We call this distribution function the isokinetic distribution f_{T} (Evans and Morriss, 1983a,b). It has a very simple form: the kinetic degrees of freedom are distributed microcanonically, and the configurational degrees of freedom are distributed canonically. The thermodynamic temperatures $(\partial E/\partial S)_{N,\mathrm{V}} = T$) of these two subsystems are, of course, identical.

If one retains terms of order $1/N$ in the above derivation, the result is the same except that $\beta = (3N - 4)/2K_0$. Such a result could have been anticipated in advance because in our Gaussian isokinetic system, four degrees of freedom are frozen, one by the kinetic-energy constraint, and three because the linear momentum is fixed.

One can check that the isokinetic distribution is an equilibrium solution of the equilibrium Liouville equation. Clearly, $\mathrm{d}f_{\mathrm{T}}/\mathrm{d}t \neq 0$. As one follows the streaming motion of an evolving point in phase space $\boldsymbol{\Gamma}(t)$, the streaming derivative of the comoving local density is

$$\frac{\mathrm{d}f_{\mathrm{T}}}{\mathrm{d}t} = \frac{3N}{2K(\boldsymbol{\Gamma})}\dot{\Phi}(\boldsymbol{\Gamma}) f_{\mathrm{T}}(\boldsymbol{\Gamma}) \neq 0 \tag{5.29}$$

This is a direct consequence of the fact that, for a Gaussian isokinetic system, phase space is compressible. It is clear, however, that in the absence of external fields $\langle \mathrm{d}f_{\mathrm{T}}/\mathrm{d}t \rangle = 0$, because the mean value of $\mathrm{d}\Phi/\mathrm{d}t$ is zero. If we sit at a fixed point in phase space and ask whether, under Gaussian isokinetic dynamics, the isokinetic distribution function changes, then the answer is no. The isokinetic distribution is the equilibrium distribution function. It is preserved by the dynamics. Substitution into the Liouville equation gives

$$\frac{\partial f_{\mathrm{T}}}{\partial t} = -\dot{\boldsymbol{\Gamma}} \cdot \frac{\partial f_{\mathrm{T}}}{\partial \boldsymbol{\Gamma}} - f_{\mathrm{T}} \frac{\partial}{\partial \boldsymbol{\Gamma}} \cdot \dot{\boldsymbol{\Gamma}} = (\beta\dot{\Phi} + 3N\alpha) f_{\mathrm{T}} = 0 \tag{5.30}$$

The proof that the last two terms sum to zero is easily given using the fact that $\beta = 3N/2K$ and that $K = \Sigma \mathbf{p}^2/2m$ is a constant of the motion.

$$\beta \dot{\Phi} + 3N\alpha = -\beta \sum \mathbf{F}_i \cdot \frac{\mathbf{p}_i}{m} + 3N \frac{\sum \mathbf{F}_i \cdot \frac{\mathbf{p}_i}{m}}{\sum \frac{\mathbf{p}_i^2}{m}} = 0 \tag{5.31}$$

If the equilibrium isokinetic system is ergodic, a single trajectory in phase space will eventually generate the isokinetic distribution. However, a single isokinetic trajectory cannot ergodically generate a canonical distribution. We can, however, ask whether isokinetic dynamics will preserve the canonical distribution. If we integrate the equations of motion for an ensemble of systems which are initially distributed canonically, will that distribution be preserved by isokinetic dynamics? Clearly,

$$\begin{aligned}\frac{\partial f_c}{\partial t} &= f_c(3N\alpha + \beta \dot{K} + \beta \dot{\Phi}) \\ &= f_c\left(\beta - \frac{3N}{2K(\boldsymbol{\Gamma})}\right)\dot{\Phi}(\boldsymbol{\Gamma}) = f_c \Delta(\beta)\dot{\Phi}(\boldsymbol{\Gamma})\end{aligned} \tag{5.32}$$

is *not* identically zero. In this expression K is a phase variable and not a constant, and $\mathrm{d}\Phi/\mathrm{d}t$ is equal to zero on average. K would only be a constant if all members of the ensemble had identical kinetic energies. The mean value of $3N/2K$ is of course β.

Consider the time derivative of the canonical average of an arbitrary *extensive* phase variable, B, where the dynamics is Gaussian isokinetic.

$$\frac{\mathrm{d}}{\mathrm{d}t}\langle B(t)\rangle = \int \mathrm{d}\boldsymbol{\Gamma}\, B \frac{\partial f_c}{\partial t} = \int \mathrm{d}\boldsymbol{\Gamma}\, B \Delta(\beta)\dot{\Phi} f_c \tag{5.33}$$

The time derivative of the ensemble average is

$$\begin{aligned}\frac{\mathrm{d}}{\mathrm{d}t}\langle B(t)\rangle &= \left\langle B\left(\frac{3N}{2K} - \beta\right)\dot{\Phi}\right\rangle \\ &= \frac{\beta}{K_0}\langle B \Delta K \dot{\Phi}\rangle + O(\Delta^2)\end{aligned} \tag{5.34}$$

where $\Delta(K) \equiv K - \langle K\rangle = K - K_0$. Equation (5.34) can be written as a product of three extensive, zero-mean variables.

$$\begin{aligned}\frac{\mathrm{d}}{\mathrm{d}t}\langle B(t)\rangle &= \frac{\beta}{K_0}[\langle B\rangle\langle \Delta k \dot{\Phi}\rangle + \langle \Delta B \Delta K \dot{\Phi}\rangle] \\ &= \frac{\beta}{K_0}\langle \Delta B \Delta K \dot{\Phi}\rangle = O(1)\end{aligned} \tag{5.35}$$

In deriving these equations we have used the fact that $\langle \Delta(K)\, \mathrm{d}\Phi/\mathrm{d}t \rangle = 0$, and that the ensemble average of the product of three extensive, zero-mean phase variables is of order N.

The above equation shows that, although B is extensive, the change in $\langle B(t) \rangle$ with time (as the ensemble changes from canonical at $t = 0$, to whatever for the Gaussian isokinetic equations generate as $t \to \infty$) is of order 1 and, therefore, can be ignored relative to the average of B itself. In the thermodynamic limit the canonical distribution is *preserved* by Gaussian isokinetic dynamics.

Nosé–Hoover thermostat—canonical ensemble

The Gaussian thermostat generates the isokinetic ensemble by a *differential* feedback mechanism. The kinetic temperature is constrained precisely by setting its time derivative to zero. Control theory provides a range of alternative feedback processes. After the Gaussian thermostat was developed, Nosé (1984a,b) utilized an *integral* feedback mechanism. As we will see, the Nosé thermostat, especially after a simplifying reformulation by Hoover (1985), provides a simple and direct way of ergodically generating the canonical ensemble.

The original Nosé method considers an extended system with an additional degree of freedom, s, which acts like an external reservoir, interacting with the system of interest by scaling all the velocities of the particles, $\mathbf{v}_i = s\, \mathrm{d}\mathbf{q}_i/\mathrm{d}t$. The new potential energy that Nosé chose to associate with this new degree of freedom was $(g + 1)k_\mathrm{B} T \ln s$, where g is related to the number of degrees of freedom of the system and T is the desired value of the temperature. It is essentially the choice of the potential for s which leads to a dynamics which generates the canonical ensemble.

The equivalent Hoover formulation of the Nosé thermostat uses equations of motion with the same form as the Gaussian equations. The difference being that the thermostatting multiplier, α, is determined by a time integral of the difference of the actual kinetic temperature from its desired value. All present applications of the Nosé thermostat use the Hoover reformulation rather than the original, somewhat cumbersome approach.

The Nosé Hamiltonian for the extended system is

$$H_\mathrm{N}(\mathbf{q}, \mathbf{p}, s, p_s) = \sum_{i=1}^{N} \frac{\mathbf{p}_i^2}{2ms^2} + \Phi(\mathbf{q}) + \frac{p_s^2}{2Q} + (g + 1)k_\mathrm{B} T \ln s \qquad (5.36)$$

where Q is effectively the *mass* associated with the heat bath (s is dimensionless so the parameter Q does not have the dimensions of mass). The equations of motion generated by this Hamiltonian are

$$\dot{\mathbf{q}}_i = \frac{\mathbf{p}_i}{ms^2} \qquad \dot{\mathbf{p}}_i = \mathbf{F}_i$$
$$\dot{s} = \frac{p_s}{Q} \qquad \dot{p}_s = \sum_{i=1}^{N} \frac{\mathbf{p}_i^2}{ms^3} - \frac{(g+1)k_B T}{s} \tag{5.37}$$

If we eliminate the variable p_s from the equations of motion, we replace the last two equations with a single second-order differential equation for s.

$$\frac{d^2 s}{dt^2} = \frac{1}{Q}\left(\sum_{i=1}^{N} \frac{\mathbf{p}_i^2}{ms^3} - \frac{(g+1)k_B T}{s}\right) \tag{5.38}$$

If the system is at equilibrium, the average force on the s coordinate must be zero, so that

$$\left\langle \sum_{i=1}^{N} \frac{\mathbf{p}_i^2}{ms^3} \right\rangle = \left\langle \sum_{i=1}^{N} \frac{m^2 s^4 \dot{\mathbf{q}}_i^2}{ms^3} \right\rangle = (g+1)k_B T \left\langle \frac{1}{s} \right\rangle \tag{5.39}$$

Suppose we interpret the time appearing in (5.37) to be a non-Galilaean fictitious time, and the *real velocities* to be $\mathbf{v}_i = s(d\mathbf{q}_i/dt)$. The instantaneous temperature is related to $\Sigma_i m\mathbf{v}_i^2$, and its time-averaged value is equal to $(g+1)k_B T$, where $g+1$ is the number of degrees of freedom. This is consistent with a non-Galilaean time average which is given by

$$\langle A \rangle_t = \frac{\displaystyle\int_0^T dt \, \frac{A(t)}{s}}{\displaystyle\int_0^T dt \, \frac{1}{s}} \tag{5.40}$$

This is an unusual definition of a time average as it implies that equal intervals of non-Galilaean time, dt, correspond to unequal intervals in *real time* of dt/s. Large values of s can be understood as corresponding to a rapid progress of fictitious time t. Division by s in the time averages appearing in (5.40) cancels out the uneven passage of fictitious time restoring the averages to their correct Galilaean values.

To calculate the equilibrium distribution function corresponding to the Nosé Hamiltonian we use the fact that for an ergodic system the equilibrium distribution function for Nosé's extended system is microcanonical. The microcanonical partition function for the extended system is

$$Z = \frac{1}{N!} \int d\mathbf{p} \, d\mathbf{q} \, dp_s \, ds \, \delta(H_N(\mathbf{q}, \mathbf{p}, s, p_s) - E) \tag{5.41}$$

$$Z = \frac{1}{N!} \int d\mathbf{p} \, d\mathbf{q} \, dp_s \, ds \, \delta\left(\sum_{i=1}^{N} \frac{\mathbf{p}_i^2}{2ms^2} + \Phi(\mathbf{q}) + \frac{p_s^2}{2Q} + (g+1)k_B T \ln s - E\right) \tag{5.42}$$

where $\mathbf{q}$ and $\mathbf{p}$ are $3N$-dimensional vectors, $\mathbf{q} \equiv (\mathbf{q}_1, \ldots, \mathbf{q}_N)$ and $\mathbf{p} \equiv (\mathbf{p}_1, \ldots, \mathbf{p}_N)$. If we change variables from $\mathbf{p}$ to $\mathbf{p}'$, where $\mathbf{p}'_i = \mathbf{p}_i/s$ for all i, then

$$Z = \frac{1}{N!} \int \mathrm{d}\mathbf{p}' \, \mathrm{d}\mathbf{q} \, \mathrm{d}p_s \, \mathrm{d}s \, s^{3N} \delta\left(H'_0(\mathbf{q}, \mathbf{p}') + \frac{p_s^2}{2Q} + (g+1)k_\mathrm{B} T \ln s - E \right) \tag{5.43}$$

where H'_0 is the usual N-particle Hamiltonian $\Sigma_i \mathbf{p}_i'^2/2m_i + \Phi(\mathbf{q})$ (the prime indicates that H'_0 is a function of $\mathbf{p}'$). The integral over s can be performed as the only contributions come from the zeros of the argument of the delta function. If $G(s) = H'_0 + p_s^2/2Q + (g+1)k_\mathrm{B} T \ln s - E$, then G has only one zero, which is

$$s_0 = \exp\left[-\frac{H'_0 + \dfrac{p_s^2}{2Q} - E}{(g+1)k_\mathrm{B} T} \right] \tag{5.44}$$

Using the identity $\delta(G(s)) = \delta(s - s_0)/\mathrm{d}G(s)/\mathrm{d}s$ it is easy to show that integrating over s gives

$$Z = \frac{1}{N!} \int \mathrm{d}\mathbf{p}' \, \mathrm{d}\mathbf{q} \, \mathrm{d}p_s \frac{1}{(g+1)k_\mathrm{B} T} \exp\left[-\frac{3N+1}{(g+1)k_\mathrm{B} T} \left(H'_0 + \frac{p_s^2}{2Q} - E \right) \right] \tag{5.45}$$

The integral over p_s is the infinite integral of a Gaussian and the result is

$$Z = \frac{1}{N!} \sqrt{\frac{2\pi Q}{(g+1)(3N+1)k_\mathrm{B} T}} \int \mathrm{d}\mathbf{p}' \, \mathrm{d}\mathbf{q} \exp\left[-\frac{3N+1}{(g+1)k_\mathrm{B} T} (H'_0 - E) \right] \tag{5.46}$$

If we *choose* $g = 3N$ then this partition function is simply related to the canonical partition function

$$Z = \frac{1}{3N+1} \sqrt{\frac{2\pi Q}{k_\mathrm{B} T}} \frac{1}{N!} \int \mathrm{d}\mathbf{p}' \, \mathrm{d}\mathbf{q} \exp\left[-\frac{H'_0 - E}{k_\mathrm{B} T} \right] \tag{5.47}$$

If the variables $\mathbf{q}$, $\mathbf{p}$, s and p_s are distributed microcanonically, then the variables $\mathbf{p}'$ and $\mathbf{q}$ are canonically distributed. The notion of non-Galilaean time makes this formulation of the Nosé thermostat rather cumbersome to use and difficult to interpret.

The next step in the development of this method was made by Hoover (1985) who realized that if one's interest lies solely in computing averages over $\mathbf{q}, \mathbf{p}'$ in real time then you may as well rewrite the equations of motion in terms of $\mathbf{q}, \mathbf{p}'$ and real time, t', and eliminate the variables $\mathbf{p}$, s, p_s and t

entirely. He used the time transformation

$$t' = \int_0^t \frac{d\tau}{s} \tag{5.48}$$

so that $dt' = dt/s$, and rewrote the Nosé equations of motion as

$$\begin{aligned} &\frac{d\mathbf{q}_i}{dt'} = \frac{\mathbf{p}_i'}{m} \qquad \frac{d\mathbf{p}_i'}{dt'} = \mathbf{F}_i - \zeta \mathbf{p}_i' \\ &\frac{d\zeta}{dt'} = \frac{1}{Q}\left(\sum_{i=1}^{N} \frac{\mathbf{p}_i'^2}{m} - (g+1)k_B T \right) = \frac{1}{\tau^2}\left(\frac{K(\mathbf{p}')}{K_0} - 1 \right) \\ &\frac{ds}{dt} = \zeta s \end{aligned} \tag{5.49}$$

where K_0 is the value of the kinetic energy corresponding to the required value of the temperature $K_0 = (g+1)k_B T/2$, $K(\mathbf{p}')$ is the instantaneous value of the kinetic energy, τ is a relaxation time which is related to the *mass* of the s degree of freedom ($\tau^2 = Q/2K_0$) and $\zeta = p_s/Q$. The motion of the system of interest can now be determined without reference to s. It is an irrelevant variable which can be ignored. The variable $d\zeta/dt$ is a function of $\mathbf{p}'$ only, so the complete description of the system can be given in terms of the variables $\mathbf{q}$ and $\mathbf{p}'$.

An important result, obtained from this time transformation by Evans and Holian (1985), is that time averages in terms of the variables $\mathbf{q}$, $\mathbf{p}'$ and t' take their usual form, i.e.

$$\langle A(\mathbf{q}, \mathbf{p}') \rangle_{t'} = \frac{1}{T'} \int_0^{T'} dt'\, A(\mathbf{q}, \mathbf{p}', t') = \langle A(\mathbf{q}, \mathbf{p}') \rangle_{\text{canonical}} \tag{5.20}$$

To obtain this result we start by considering the Nosé–Hoover phase variable Liouvillean $iL_{NH}(\mathbf{q}, \mathbf{p}', s, p_s; t')$ and relating it to the Nosé Liouvillean $iL_N(\mathbf{q}, \mathbf{p}, s, p_s; t)$.

$$\begin{aligned} iL_N(\mathbf{q}, \mathbf{p}, s, p_s, t) &= \dot{\mathbf{q}} \cdot \frac{\partial}{\partial \mathbf{q}} + \dot{\mathbf{p}} \cdot \frac{\partial}{\partial \mathbf{p}} + \dot{s} \frac{\partial}{\partial s} + \dot{p}_s \frac{\partial}{\partial p_s} \\ &= \frac{\mathbf{p}}{ms^2} \cdot \frac{\partial}{\partial \mathbf{q}} + \mathbf{F} \cdot \left.\frac{\partial}{\partial \mathbf{p}}\right|_s + \frac{p_s}{Q} \left.\frac{\partial}{\partial s}\right|_{\mathbf{p}} + \left(\sum_{i=1}^{N} \frac{\mathbf{p}_i^2}{ms^2} - (g+1)k_B T \right) \frac{1}{s} \frac{\partial}{\partial p_s} \end{aligned} \tag{5.51}$$

Using the results:

$$\left.\frac{\partial}{\partial s}\right|_{\mathbf{p}} = \left.\frac{\partial}{\partial s}\right|_{\mathbf{p}'} - \mathbf{p}' \cdot \left.\frac{\partial}{\partial \mathbf{p}}\right|_s \tag{5.52}$$

and

$$\left.\frac{\partial}{\partial \mathbf{p}}\right|_s = \frac{1}{s}\left.\frac{\partial}{\partial \mathbf{p}'}\right|_s \tag{5.53}$$

equation (5.51) becomes

$$\begin{aligned}
\mathrm{iL_N} &= \frac{\mathbf{p}}{ms^2}\cdot\frac{\partial}{\partial \mathbf{q}} + \left(\mathbf{F} - \mathbf{p}'\frac{p_s}{Q}\right)\frac{1}{s}\cdot\left.\frac{\partial}{\partial \mathbf{p}'}\right|_s + \left.\frac{p_s}{Q}\frac{\partial}{\partial s}\right|_{\mathbf{p}'} + \left(\sum_{i=1}^{N}\frac{\mathbf{p}_i^2}{ms^2} - (g+1)k_\mathrm{B}T\right)\frac{1}{s}\frac{\partial}{\partial p_s} \\
&= \frac{\mathbf{p}'}{ms}\cdot\frac{\partial}{\partial \mathbf{q}} + (\mathbf{F} - \zeta\mathbf{p}')\frac{1}{s}\cdot\left.\frac{\partial}{\partial \mathbf{p}'}\right|_s + \left.\zeta\frac{\partial}{\partial s}\right|_{\mathbf{p}'} + \left(\sum_{i=1}^{N}\frac{\mathbf{p}_i'^2}{m} - (g+1)k_\mathrm{B}T\right)\frac{1}{s}\frac{\partial}{\partial p_s} \\
&= \frac{1}{s}\left[\frac{\mathbf{p}'}{m}\cdot\frac{\partial}{\partial \mathbf{q}} + (\mathbf{F} - \zeta\mathbf{p}')\cdot\left.\frac{\partial}{\partial \mathbf{p}'}\right|_s + \left.s\zeta\frac{\partial}{\partial s}\right|_{\mathbf{p}'} + \left(\sum_{i=1}^{N}\frac{\mathbf{p}_i'^2}{m} - (g+1)k_\mathrm{B}T\right)\frac{\partial}{\partial p_s}\right] \\
&= \frac{1}{s}\mathrm{iL_{NH}}(\mathbf{q}, \mathbf{p}', s, p_s, t')
\end{aligned} \tag{5.54}$$

If A is an arbitrary phase variable, then the Liouvillean describes the rate of change of A. If we consider A to be a function of $\mathbf{q}$ and $\mathbf{p}$, then the rate of change of A with respect to time t is

$$\frac{\mathrm{d}A(\mathbf{q},\mathbf{p})}{\mathrm{d}t} = \mathrm{iL_N}(\mathbf{q}, \mathbf{p}, s, p_s, t)A(\mathbf{q}, \mathbf{p}) \tag{5.55}$$

Since $\mathrm{iL_N}$ contains no explicit time dependence, integrating with respect to time gives

$$\begin{aligned}
A(\mathbf{q}, \mathbf{p}, t) &= \exp\left[\int_0^t \mathrm{d}\tau\, \mathrm{iL_N}(\mathbf{q}, \mathbf{p}, s, p_s)\right]A(\mathbf{q}, \mathbf{p}, 0) \\
&= \exp[\mathrm{iL_N}(\mathbf{q}, \mathbf{p}, s, p_s)t]A(\mathbf{q}, \mathbf{p}, 0)
\end{aligned} \tag{5.56}$$

In a similar fashion, we can consider A to be a function of $\mathbf{q}$ and $\mathbf{p}'$. In this case it is natural to ask for the value of A at t'.

$$\begin{aligned}
A(\mathbf{q}, \mathbf{p}', t') &= \exp\left[\int_0^{t'} \mathrm{d}\tau'\, \mathrm{iL_{NH}}(\mathbf{q}, \mathbf{p}')\right]A(\mathbf{q}, \mathbf{p}', 0) \\
&= \exp[\mathrm{iL_{NH}}(\mathbf{q}, \mathbf{p}')t']A(\mathbf{q}, \mathbf{p}', 0)
\end{aligned} \tag{5.57}$$

Now A is a function of the reduced phase space only, so the dependence on s and p_s can be ignored. These two different representations of the phase variable can be equated. To do this consider the time transformation (5.48). It implies,

$$t' = \int_0^{t'} \mathrm{d}\tau' = \int_0^t \frac{\mathrm{d}\tau}{s} \tag{5.58}$$

So that $d\tau' = d\tau/s$, and

$$A(\mathbf{q},\mathbf{p},t) = \exp\left[\left[\int_0^t d\tau\, iL_N(\mathbf{q},\mathbf{p})\right]A(0) = \exp\left[\left[\int_0^t \frac{d\tau}{s} iL_{NH}(\mathbf{q},\mathbf{p}')\right]A(0)\right.\right.$$

$$= \exp\left[\left[\int_0^{t'} d\tau'\, iL_{NH}(\mathbf{q},\mathbf{p}')\right]A(0) = A(\mathbf{q},\mathbf{p}',t')\right. \tag{5.59}$$

Using (5.59) and the time transformation (5.48) we find that $T' = \int_0^{T'} dt' = \int_0^T dt\, 1/s$ so that we can derive (5.40),

$$\langle A\rangle_t = \frac{\displaystyle\int_0^{T'} dt'\, A(\mathbf{q},\mathbf{p},t')}{\displaystyle\int_0^{T'} dt'} = \frac{\displaystyle\int_0^T dt\, \frac{A(\mathbf{q},\mathbf{p},t)}{s}}{\displaystyle\int_0^T dt\, \frac{1}{s}} = \langle A\rangle \tag{5.60}$$

So the time average over t is equal to the time average over t'. Using the variables $\mathbf{q}$, $\mathbf{p}'$ and t' the time average over equal intervals of t' takes the usual form. The time average over $\mathbf{q}$, $\mathbf{p}$ and t, however, involves the scaling variable s, or equivalently a time average over unequal intervals of t.

One can of course dispense with the original form of Nosé's equations entirely. There is now no need to consider the notion of non-Galilaean time. We simply repeat the derivation we gave for the isokinetic distribution based on the Gaussian isokinetic equations of motion, for the Nosé–Hoover equations. Since there is no need to refer to non-Galilaean time we refer to $\mathbf{q}, \mathbf{p}', t'$ simply as, $\mathbf{q}, \mathbf{p}, t$ (dropping the prime). The N-particle distribution function $f(\boldsymbol{\Gamma}, \zeta)$ generated by the Nosé–Hoover equations of motion can be obtained by solving the Liouville equation for the equations of motion written in terms of $\mathbf{q}$, $\mathbf{p}$ and t. It is convenient to consider the total derivative of f which from the Liouville equation is

$$\frac{df}{dt} = -f\left(\frac{\partial}{\partial \boldsymbol{\Gamma}}\cdot\dot{\boldsymbol{\Gamma}} + \frac{\partial}{\partial \zeta}\dot{\zeta}\right) \tag{5.61}$$

From the equations of motion (5.49), dropping the primes, it is easy to see that $d\zeta/dt$ is a function of $\mathbf{q}$ and $\mathbf{p}$, and hence independent of ζ. The only nonzero contribution to the right-hand side comes from the $\mathbf{p}$ dependence of $d\mathbf{p}/dt$, so that

$$\frac{df}{dt} = 3N\zeta f \tag{5.62}$$

Consider the time derivative of the quantity $H_0 + Q\zeta^2/2$

$$\frac{\mathrm{d}}{\mathrm{d}t}\left(H_0 + \frac{Q\zeta^2}{2}\right) = \dot{H}_0 + Q\zeta\dot{\zeta}$$

$$= -\zeta \sum_{i=1}^{N} \frac{\mathbf{p}_i^2}{m} + \zeta\left(\sum_{i=1}^{N} \frac{\mathbf{p}_i^2}{m} - (g+1)k_B T\right)$$

$$= -\zeta(g+1)k_B T \tag{5.63}$$

If we take $g = 3N - 1$ we find that

$$\frac{\mathrm{d}}{\mathrm{d}t} \ln f = -\beta \frac{\mathrm{d}}{\mathrm{d}t}\left(H_0 + \frac{Q\zeta^2}{2}\right) \tag{5.64}$$

So that the equilibrium distribution function is the extended canonical distribution f_c

$$f_c(\mathbf{\Gamma}, \zeta) = \frac{\exp[-\beta(H_0 + \frac{1}{2}Q\zeta^2)]}{\int \mathrm{d}\mathbf{\Gamma}\, \mathrm{d}\zeta \exp[-\beta(H_0 + \frac{1}{2}Q\zeta^2)]} \tag{5.65}$$

In the Hoover representation of the equations of motion, the scaling variable s has essentially been eliminated so that the number of degrees of freedom of the system changes from $3N + 1$ to $3N$ and g changes from $3N$ to $3N - 1$.

5.3 ISOTHERMAL LINEAR RESPONSE THEORY

In Section 5.2 we considered two forms of thermostatted dynamics—the Gaussian isokinetic dynamics and the Nosé–Hoover canonical ensemble dynamics. Both these thermostatted equations of motion can add or remove energy from the system to control its temperature. It is particularly important to incorporate thermostatted dynamics when the system is perturbed by an external field. This allows the heat produced irreversibly to be removed continuously, and the system maintained in a steady, nonequilibrium state. We now generalize the adiabatic linear response theory of Section 5.1 in order to treat the perturbed thermostatted systems which we described in Section 5.2. We consider (Morriss and Evans, 1985) an N-particle system evolving under the Gaussian isokinetic dynamics for $t < 0$, but subject to an external field, F_e, for all times $t > 0$. The equations of motion are given by

$$\begin{aligned}\dot{\mathbf{q}}_i &= \frac{\mathbf{p}_i}{m} + \mathbf{C}_i F_e(t) \\ \dot{\mathbf{p}}_i &= \mathbf{F}_i + \mathbf{D}_i F_e(t) - \alpha \mathbf{p}_i\end{aligned} \tag{5.66}$$

The term $\alpha \mathbf{p}_i$ couples the system to a thermostat and we shall take

$$\alpha = \alpha_0 + \alpha_1 F_e(t) = \frac{\sum_i \frac{\mathbf{F}_i}{m_i} \cdot \mathbf{p}_i}{\sum_i \frac{1}{m_i} \mathbf{p}_i^2} + \frac{\sum_i \frac{\mathbf{D}_i}{m_i} \cdot \mathbf{p}_i}{\sum_i \frac{1}{m_i} \mathbf{p}_i^2} F_e(t) \tag{5.67}$$

so that the peculiar kinetic energy, $K(\mathbf{\Gamma}) = \Sigma_i \mathbf{p}_i^2/2m = K_0$, is a constant of the motion. In the absence of the field these equations of motion ergodically generate the isokinetic distribution function, f_T, equation (5.28), with $\beta = 3N/2K_0$. As we have seen, the isokinetic distribution function, f_T, is preserved by the field-free isokinetic equations of motion and that

$$\frac{\partial f_T}{\partial t} = -iL_T f_T = 0 \tag{5.68}$$

We use iL_T for the zero-field, isokinetic Liouvillean.

To calculate the linear thermostatted response we need to solve the linearized Liouville equation for thermostatted systems. Following the same arguments used in the adiabatic case (equations (5.8)–(5.12)), the linearized Liouville equation is

$$\frac{\partial}{\partial t} \Delta f(\mathbf{\Gamma}, t) + iL_T \Delta f(\mathbf{\Gamma}, t) = -\Delta iL(t) f_T(\mathbf{\Gamma}) + O(\Delta^2) \tag{5.69}$$

where $iL(t)$ is the external-field dependent, isokinetic Liouvillean and $\Delta iL(t) = iL(t) - iL_T$. Its solution is the analogue of (5.13) namely

$$\Delta f(\mathbf{\Gamma}, t) = -\int_0^t \mathrm{d}s \exp(-iL_T(t-s)) \Delta iL(s) f_T(\mathbf{\Gamma}) + O(\Delta^2) \tag{5.70}$$

Using equations (5.8), (5.28) and (5.66), and the fact that $\beta = 3N/2K_0$, it is easy to show that

$$\begin{aligned} \Delta iL(t) f_T(\mathbf{\Gamma}) &= iL(t) f_T(\mathbf{\Gamma}) - iL_T f_T(\mathbf{\Gamma}) = iL(t) f_T(\mathbf{\Gamma}) \\ &= \left[\dot{\mathbf{\Gamma}}(t) \cdot \frac{\partial}{\partial \mathbf{\Gamma}} + \left(\frac{\partial}{\partial \mathbf{\Gamma}} \cdot \dot{\mathbf{\Gamma}}(t) \right) \right] f_T(\mathbf{\Gamma}) \\ &= -\beta \dot{\Phi} f_T(\mathbf{\Gamma}) - f_T(\mathbf{\Gamma}) \sum_{i=1}^N \frac{\partial}{\partial \mathbf{p}_i} \cdot (\alpha \mathbf{p}_i) \end{aligned} \tag{5.71}$$

There is one subtle point in deriving the last line of (5.71),

$$\begin{aligned}\sum_{i=1}^{N} \dot{\mathbf{p}}_i \cdot \frac{\partial f_{\mathrm{T}}}{\partial \mathbf{p}_i} &= \sum_{i=1}^{N} \dot{\mathbf{p}}_i \cdot \frac{\partial}{\partial \mathbf{p}_i} \frac{\delta(K(\mathbf{p}) - K_0)\,\mathrm{e}^{-\beta\Phi(\mathbf{q})}}{Z_{\mathrm{T}}(\beta)} \\ &= \frac{\mathrm{e}^{-\beta\Phi(\mathbf{q})}}{Z_{\mathrm{T}}(\beta)} \sum_{i=1}^{N} \dot{\mathbf{p}}_i \cdot \frac{\partial K(\mathbf{p})}{\partial \mathbf{p}_i} \frac{\partial \delta(K(\mathbf{p}) - K_0)}{\partial K(\mathbf{p})} \\ &= \frac{\mathrm{e}^{-\beta\Phi}}{Z_{\mathrm{T}}(\beta)} \dot{K}(\mathbf{p}) \frac{\partial \delta(K(\mathbf{p}) - K_0)}{\partial K(\mathbf{p})} = 0 \end{aligned} \tag{5.72}$$

The last line follows because $K(\mathbf{p})$ is a constant of the motion for the Gaussian isokinetic equations of motion. We have also assumed that the only contribution to the phase-space compression factor comes from the thermostatting term $\alpha \mathbf{p}_i$. This means that in the absence of a thermostat, as in the adiabatic case, the phase space is incompressible and

$$\frac{\partial}{\partial \mathbf{\Gamma}} \cdot \dot{\mathbf{\Gamma}}(t)^{\mathrm{ad}} = \sum_{i=1}^{N} \left(\frac{\partial}{\partial \mathbf{q}_i} \cdot \mathbf{C}_i + \frac{\partial}{\partial \mathbf{p}_i} \cdot \mathbf{D}_i \right) = 0 \tag{5.73}$$

This assumption or condition, is known as the *adiabatic incompressibility of phase space* ($AI\Gamma$). A sufficient, but not necessary, condition for it to hold is that the adiabatic equations of motion should be derivable from a Hamiltonian. It is important to note that $AI\Gamma$ does not imply that the phase space for the thermostatted system should be incompressible. Rather, it states that if the thermostat is *removed* from the field-dependent equations of motion, the phase space is incompressible. It is essentially a condition on the external field coupling terms $\mathbf{C}_i(\mathbf{q}, \mathbf{p})$ and $\mathbf{D}_i(\mathbf{q}, \mathbf{p})$. It is not necessary that $\mathbf{C}_i$ be independent of $\mathbf{q}$, and $\mathbf{D}_i$ be independent of $\mathbf{p}$. Indeed in Section 6.3 we show that this is not the case for planar Couette flow, but the combination of partial derivatives in (5.73) is zero. It is possible to generalize the theory to treat systems where $AI\Gamma$ does not hold but this generalization has proved to be unnecessary.

Using equation (5.67) for the multiplier α, we have to first order in N

$$\begin{aligned}\Delta iL(t) f_{\mathrm{T}}(\mathbf{\Gamma}) &= -(\beta \dot{\Phi}(t) + 3N\alpha) f_{\mathrm{T}}(\mathbf{\Gamma}) \\ &= \beta \sum_{i=1}^{N} \left[\mathbf{C}_i \cdot \mathbf{F}_i - \mathbf{D}_i \cdot \frac{\mathbf{p}_i}{m} \right] F_{\mathrm{e}}(t) f_{\mathrm{T}}(\mathbf{\Gamma}) \\ &= \beta J(\mathbf{\Gamma}) F_{\mathrm{e}}(t) f_{\mathrm{T}}(\mathbf{\Gamma}) \end{aligned} \tag{5.74}$$

This equation shows that $\Delta iL(t) f(\mathbf{\Gamma})$ is independent of thermostatting. Equations (5.74) and (5.15) are essentially identical. This is why the dissipative flux J is defined in terms of the *adiabatic* derivative of the internal energy.

Interestingly, the kinetic part of the dissipative flux, $J(\boldsymbol{\Gamma})$, comes from the multiplier α, while the potential part comes from the time derivative of Φ.

Substituting (5.74) into (5.70), the change in the isokinetic distribution function is given by

$$\Delta f(\boldsymbol{\Gamma}, t) = -\beta \int_0^t \mathrm{d}s \exp(-iL(t-s)) J(\boldsymbol{\Gamma}) F_e(s) f_T(\boldsymbol{\Gamma}) \tag{5.75}$$

Using this result to calculate the mean value of $B(t)$, the isothermal linear response formula corresponding to equation (5.16) is

$$\begin{aligned}
\langle B(t)\rangle_T - \langle B(0)\rangle_T &= \int \mathrm{d}\boldsymbol{\Gamma}\, B(\boldsymbol{\Gamma}) \Delta f(\boldsymbol{\Gamma}, t) \\
&= -\beta \int_0^t \mathrm{d}s \int \mathrm{d}\boldsymbol{\Gamma}\, B(\boldsymbol{\Gamma}) \exp(-iL(t-s)) J(\boldsymbol{\Gamma}) F_e(s) f_T(\boldsymbol{\Gamma}) \\
&= -\beta \int_0^t \mathrm{d}s \int \mathrm{d}\boldsymbol{\Gamma}\, J(\boldsymbol{\Gamma}) f_T(\boldsymbol{\Gamma}) \exp(iL(t-s)) B(\boldsymbol{\Gamma}) F_e(s) \\
&= -\beta \int_0^t \mathrm{d}s \int \mathrm{d}\boldsymbol{\Gamma}\, B(t-s) J(\boldsymbol{\Gamma}) f_T(\boldsymbol{\Gamma}) F_e(s) \\
&= -\beta \int_0^t \mathrm{d}s \langle B(t-s) J(0)\rangle_{T,0} F_e(s)
\end{aligned} \tag{5.76}$$

Equation (5.76) is very similar in form to the adiabatic linear response formula derived in Section 5.1. The notation $\langle \ldots \rangle_{T,0}$ signifies that a field-free (0), isokinetic (T) ensemble average should be taken. Differences from the adiabatic formula are that:

(1) the field-free Gaussian isokinetic propagator governs the time evolution in the equilibrium time correlation function $\langle B(t-s)J(0)\rangle_{T,0}$;
(2) the ensemble averaging is Gaussian isokinetic rather than canonical;
(3) because both the equilibrium and nonequilibrium motions are thermostatted, the long-time limit of $\langle B(t)\rangle_T$ on the left-hand side of (5.76), is finite; and
(4) the formula is ergodically consistent. There is only one ensemble referred to in the expression, the Gaussian isokinetic distribution. The dynamics used to calculate the time evolution of the phase variable B in the equilibrium time-correlation function, ergodically generates the ensemble of time zero starting states $f_T(\boldsymbol{\Gamma})$. We refer to this as *ergodically consistent linear response theory*.

The last point means that time averaging rather than ensemble averaging can be used to generate the time zero starting states for the equilibrium time-correlation function on the right-hand side of equation (5.76).

It can be useful, especially for theoretical treatments, to use ergodically inconsistent formulations of linear response theory. It may be convenient to employ canonical rather than isokinetic averaging, for example. For the canonical ensemble, assuming $AI\Gamma$, we have in place of equation (5.71),

$$\begin{aligned} \Delta iL(t) f_c(\Gamma) &= -(\beta \Delta \dot{\Phi} + 3N\Delta\alpha) f_c(\Gamma) \\ &= \left(\beta \sum \mathbf{F}_i \cdot \mathbf{C}_i - \frac{3N}{2K} \sum \frac{\mathbf{p}_i}{m} \cdot \mathbf{D}_i \right) F_e f_c(\Gamma) \\ &= \beta J(\Gamma) F_e f_c(\Gamma) + \beta \frac{\Delta K}{\langle K \rangle_{c,0}} \sum \frac{\mathbf{p}_i}{m} \cdot \mathbf{D}_i F_e f_c(\Gamma) + 0(\Delta^2) \end{aligned} \tag{5.77}$$

where $\Delta\, \mathrm{d}\Phi/\mathrm{d}t$ is the difference between the rate of change of Φ with the external field turned on and with the field turned off ($\mathrm{d}\Phi(F_e)/\mathrm{d}t - \mathrm{d}\Phi(F_e = 0)/\mathrm{d}t$). Similarly, $\Delta\alpha = \alpha(F_e) - \alpha(F_e = 0) = \alpha_1 F_e$ (see (5.67)). To first order in $\Delta K \equiv K - \langle K \rangle_{c,0}$, we have,

$$\begin{aligned} \langle B(t) \rangle_c = \langle B(0) \rangle_c &- \beta \int_0^t \mathrm{d}s \langle B(t-s) J(0) \rangle_{c,0} F_e(s) \\ &- \beta \int_0^t \mathrm{d}s \left\langle B(t-s) \frac{\Delta K(0)}{\langle K \rangle_{c,0}} \sum \frac{\mathbf{p}_i(0)}{m} \cdot \mathbf{D}_i(0) \right\rangle_{c,0} F_e(s) \end{aligned} \tag{5.78}$$

Using the same methods as those used in deriving equation (5.35), we can show that if B is extensive, the second integral in equation (5.78) is of order 1 and can therefore be ignored.

Thus for a canonical ensemble of starting states and thermostatted Gaussian isokinetic dynamics, the response of an extensive variable, B, is given by

$$\langle B(t_T) \rangle_c = \langle B(0) \rangle_c - \beta \int_0^t \mathrm{d}s \langle B((t-s)_T) J(0) \rangle_{c,0} F_e(s) \tag{5.79}$$

Like the isokinetic ensemble formula, the response, $\langle B(t_T) \rangle_c$, possesses a well-defined steady-state limit.

It is straightforward to apply the linear response formalism to a wide variety of combinations of statistical mechanical ensembles and equilibrium dynamics. The resultant susceptibilities are shown in Table 5.1. It is important to appreciate that the dissipative flux, $J(\Gamma)$, is *determined* by both the choice of equilibrium ensemble of starting states *and* the choice of the equilibrium dynamics.

Table 5.1 Linear susceptibilities expressed as equilibrium time-correlation functions.‡

Adiabatic response of canonical ensemble

$$\chi = \beta \langle B(t_N) J(0) \rangle_c \quad \text{(T.5.1)}$$

Isothermal response of canonical or isothermal ensemble

$$\chi = \beta \langle B(t_T) J(0) \rangle_{c,T} \quad \text{(T.5.2)}$$

Isoenergetic response of canonical or microcanonical ensembles (Evans and Morriss, 1984b)

$$\chi(t) = \beta \langle B(t_N) J(0) \rangle_{c,E} \quad \text{(T.5.3)}$$

Isoenthalpic response of isoenthalpic ensemble

$$\chi = \beta \langle B(t_I) J(0) \rangle_I \quad \text{(T.5.4)}$$

$-JF_e \equiv dI/dt$, isoenthalpic dynamics defined in Evans and Morriss (1984b)

Nosé dynamics of the canonical ensemble§

$$\chi = \beta \langle B(t_c) J(0) \rangle_c \quad \text{(T.5.5)}$$

---

‡ Equilibrium dynamics: t_N, Newtonian; t_T, Gaussian isokinetic; t_I, Gaussian isoenthalpic; t_c, Nosé–Hoover. Ensemble averaging: $\rangle_c$, canonical; $\rangle_T$, isokinetic; $\rangle_E$, microcanonical; $\rangle_I$, isoenthalpic.

§ Proof of (T.5.5) can be found in Holian and Evans (1983).

5.4 THE EQUIVALENCE OF THERMOSTATTED LINEAR RESPONSES

We shall now address the important question of how the various linear susceptibilities described in Table 5.1 relate to one another. For simplicity, let us assume that the initial unperturbed ensemble is canonical. In this case the only difference between the adiabatic, the isothermal, the isoenergetic and the Nosé susceptibilities is in the respective field-free propagators used to generate the equilibrium time-correlation functions. We will now discuss the differences between the adiabatic and isothermal responses; however, the analysis of the other cases involves similar arguments. Without loss of generality, we shall assume that the dissipative flux J and the response phase variable B are both extensive and have mean values which vanish at equilibrium. The susceptibility is of order N.

The only difference between (T.5.1) and (T.5.2) is in the time propagation of the phase variable B,

$$B(t_T) = U_T(t) B(\mathbf{\Gamma}) = \exp(iL_T t) B(\mathbf{\Gamma}) \quad (5.80)$$

and

$$B(t_N) = U_N(t)B(\boldsymbol{\Gamma}) = \exp(\mathrm{i}L_N t)B(\boldsymbol{\Gamma}) \tag{5.81}$$

In equations (5.80) and (5.81) the Liouvillean $\mathrm{i}L_N$ is the Newtonian Liouvillean, and $\mathrm{i}L_T$ is the Gaussian isokinetic Liouvillean obtained from the equations of motion (5.23), with α given by the $F_e \to 0$ limit of equation (5.20). In both cases there is no explicit time dependence in the Liouvillean. We note that the multiplier, α, is intensive.

We can now use the Dyson equation (3.102) to calculate the difference between the isothermal and adiabatic susceptibilities for the canonical ensemble. If $\Rightarrow$ denotes the isothermal propagator and $\rightarrow$ the Newtonian, the difference between the two relevant equilibrium time-correlation functions is

$$\langle J \Rightarrow B \rangle - \langle J \rightarrow B \rangle = \langle J \Rightarrow \Delta \rightarrow B \rangle \equiv \delta \langle J \Rightarrow B \rangle \tag{5.82}$$

where we have used the Dyson equation (3.102). Now the difference between the isothermal and Newtonian Liouvillean is

$$\Delta = \mathrm{i}L_T - \mathrm{i}L_N = \Delta\dot{\boldsymbol{\Gamma}} \cdot \frac{\partial}{\partial \boldsymbol{\Gamma}} = -\alpha \sum_{i=1}^{N} \mathbf{p}_i \cdot \frac{\partial}{\partial \mathbf{p}_i} \tag{5.83}$$

Thus

$$\delta \langle J \Rightarrow B \rangle = -\int_0^t \mathrm{d}s \left\langle J \exp(\mathrm{i}L_T s)\alpha \sum_{i=1}^{N} \mathbf{p}_i \cdot \frac{\partial}{\partial \mathbf{p}_i} \exp(\mathrm{i}L_N(t-s))B \right\rangle \tag{5.84}$$

where α is the field-free Gaussian multiplier appearing in the isothermal equation of motion. We assume that it is possible to define a new phase variable B' by

$$\exp(\mathrm{i}L_N t)B' = \sum_{i=1}^{N} \mathbf{p}_i \cdot \frac{\partial}{\partial \mathbf{p}_i} \exp(\mathrm{i}L_N t)B \tag{5.85}$$

This is a rather unusual definition of a phase variable, but if B is an analytic function of the momenta, then an extensive phase variable B' always exists. First, we calculate the average value of $B'(t)$

$$\begin{aligned}
\langle B'(t_N) \rangle &= \left\langle \sum_{i=1}^{N} \mathbf{p}_i \cdot \frac{\partial}{\partial \mathbf{p}_i} B(t_N) \right\rangle = \int \mathrm{d}\boldsymbol{\Gamma}\, f_c(\boldsymbol{\Gamma}) \sum_{i=1}^{N} \mathbf{p}_i \cdot \frac{\partial}{\partial \mathbf{p}_i} B(t_N) \\
&= -\int \mathrm{d}\boldsymbol{\Gamma}\, B(t_N) \sum_{i=1}^{N} \frac{\partial}{\partial \mathbf{p}_i} \cdot (\mathbf{p}_i f_c(\boldsymbol{\Gamma})) \\
&= -3N \langle B(t_N) \rangle + 2\beta \langle B(t_N) K(0) \rangle \\
&= 2\beta \langle B(t_N)[K(0) - \langle K(0) \rangle] \rangle
\end{aligned} \tag{5.86}$$

Unless B is trivially related to the kinetic energy K, $\langle B'(t_N)\rangle = 0$. Typically, B will be a thermodynamic flux such as the heat-flux vector of the symmetric traceless part of the pressure tensor. In these cases $\langle B'(t_N)\rangle$ vanishes because of Curie's principle (Section 2.3).

Assuming, without loss of generality, that $\langle B(t_N)\rangle = 0$, then we can show that

$$\begin{aligned}\delta\langle J \Rightarrow B\rangle &= -\int_0^t \mathrm{d}s\, \langle J \exp(\mathrm{iL_T}s)\alpha \exp(\mathrm{iL_N}(t-s))B'\rangle \\ &= -\int_0^t \mathrm{d}s\, \langle J(-s_T)\alpha(0)B'(t_N - s_N)\rangle \end{aligned} \tag{5.87}$$

Because J, B and B' are extensive and α is intensive, equation (5.87) can be expressed as the product of three zero-mean extensive quantities divided by N. The average of three local, zero-mean quantities is extensive, and thus the quotient is intensive. Therefore, except in the case where B is a scalar function of the kinetic energy, the difference between the susceptibilities computed under Newton's equations and under Gaussian isokinetic equations is of order $1/N$ compared with the magnitude of the susceptibilities themselves. This means that in the large-system limit the adiabatic and isokinetic susceptibilities are identical. Similar arguments can be used to show the thermodynamic equivalence of the adiabatic and Nosé susceptibilities. It is pleasing to be able to prove that the mechanical response is independent of the thermostatting mechanism and so only depends upon the thermodynamic state of the system.

Two further comments can be made at this stage: firstly, there is a simple reason why the differences in the respective susceptibilities is significant in the case where B is a scalar function of the kinetic energy. This is simply a reflection of the fact that in this case B is intimately related to a constant of the motion for Gaussian isokinetic dynamics. One would expect to see a difference in the susceptibilities in this case. Secondly, in particular cases one can use Dyson decomposition techniques (in particular equation (3.107)), to examine the differences between the adiabatic and isokinetic susceptibilities systematically. Evans and Morriss (1984a) used this approach to calculate the differences, evaluated using Newtonian and isokinetic dynamics, between the correlation functions for the Navier–Stokes transport coefficients. The results showed that the equilibrium time correlation functions for the shear viscosity, for the self-diffusion coefficient and for the thermal conductivity, are independent of thermostatting in the large-system limit.

REFERENCES

Evans, D.J. (1983). *J. Chem. Phys.*, **78**, 3297.

Evans, D.J. and Holian, B.L. (1985). *J. Chem. Phys.*, **83**, 4069.

Evans, D.J., Hoover, W.G., Failor, B.H., Moran, B. and Ladd, A.J.C. (1983). *Phys. Rev., Ser. A*, **8**, 1016.

Evans, D.J. and Morriss, G.P. (1983a). *Phys. Lett.* **98A**, 433.

Evans, D.J. and Morriss, G.P. (1983b). *Chem. Phys.*, **77**, 63.

Evans, D.J. and Morriss, G.P. (1984a). *Chem. Phys.*, **87**, 451.

Evans, D.J. and Morriss, G.P. (1984b). *Comput. Phys. Rep.*, **1**, 297.

Holian, B.L. and Evans, D.J. (1983). *J. Chem. Phys.*, **78**, 5147.

Hoover, W.G. (1985). *Phys. Rev., Ser. A*, **31**, 1695.

Hoover, W.G., Ladd, A.J.C. and Moran, B. (1982). *Phys. Rev. Lett.*, **48**, 1818.

Kubo, R. (1957). *J. Phys. Soc. Jpn.*, **12**, 570.

Kubo, R. (1982). *Int. J. Quantum Chem.*, **16**, 25.

MacGowan, D. and Evans, D.J. (1986). *Phys. Lett.*, **117A**, 414.

Morriss, G.P. and Evans, D.J. (1985). *Mol. Phys.*, **54**, 629.

Nosé, S. (1984a). *J. Chem. Phys.*, **81**, 511.

Nosé, S. (1984b). *Mol. Phys.*, **52**, 255.

6 Computer Simulation Algorithms

6.1 INTRODUCTION

We will now show how linear response theory can be used to design computer simulation algorithms for the calculation of transport coefficients. There are two types of transport coefficients: mechanical and thermal. In this chapter we will show how thermal transport coefficients can be calculated using mechanical methods.

In nature, nonequilibrium systems may respond essentially adiabatically, or, depending upon circumstances, they may respond in an approximately isothermal manner—the quasi-isothermal response. No natural systems can be precisely adiabatic or isothermal. There will always be some transfer of the dissipative heat produced in nonequilibrium systems towards thermal boundaries. This heat may be radiated, convected or conducted to the boundary reservoir. Provided this heat transfer is slow on a microscopic time-scale and provided that the temperature gradients implicit in the transfer process lead to negligible temperature differences on a microscopic length scale, we call the system *quasi-isothermal.* We assume that quasi-isothermal systems can be modelled on a microscopic scale in computer simulations as isothermal systems.

In view of the robustness of the susceptibilities and equilibrium time-correlation functions to various thermostatting procedures (see Sections 5.2 and 5.4), we expect that quasi-isothermal systems may be modelled using Gaussian or Nosé–Hoover thermostats or enostats. Furthermore, since heating effects are *quadratic* functions of the thermodynamic forces, the *linear* response of nonequilibrium systems can always be calculated by analysing the adiabatic, the isothermal or the isoenergetic response.

Because of the fundamental relations between the linear nonequilibrium response and time-dependent equilibrium fluctuations (Table 6.1) we have two ways of calculating the susceptibilities. We could perform an equilibrium simulation and calculate the appropriate equilibrium time-correlation functions. The principal advantage of this method is that all possible transport

Table 6.1 Green–Kubo relations for Navier–Stokes transport coefficients.

Self-diffusion

$$D = \tfrac{1}{3} \int_0^\infty \mathrm{d}t \, \langle \mathbf{v}_i(t) \cdot \mathbf{v}_i(0) \rangle \tag{T.6.1}$$

Thermal conductivity

$$\lambda = \frac{V}{3k_B T^2} \int_0^\infty \mathrm{d}t \, \langle \mathbf{J}_Q(t) \cdot \mathbf{J}_Q(0) \rangle \tag{T.6.2}$$

Shear viscosity

$$\eta = \frac{V}{k_B T} \int_0^\infty \mathrm{d}t \, \langle P_{xy}(t) P_{xy}(0) \rangle \tag{T.6.3}$$

Bulk viscosity

$$\eta_v = \frac{1}{V k_B T} \int_0^\infty \mathrm{d}t \, \langle (p(t)V(t) - \langle pV \rangle)(p(0)V(0) - \langle pV \rangle) \rangle \tag{T.6.4}$$

coefficients can, in principle, be calculated from a single molecular-dynamics run. This approach is, however, very expensive in computer time with poor signal-to-noise ratios, and gives results that often depend strongly and non-monotonically upon the size of the system being simulated. A frequently more useful approach is to perform a nonequilibrium simulation of the transport process. For mechanical transport process we apply an external field, F_e, and calculate the transport coefficient, L, from a linear constitutive relation:

$$L = \int_0^\infty \mathrm{d}t \, \chi(t) = \lim_{F_e \to \infty} \lim_{t \to \infty} \frac{\langle B(t) \rangle}{F_e} \tag{6.1}$$

The use of (6.1) necessitates a thermostat, otherwise the work done on the system would be transformed continuously into heat and no steady state could be achieved (the limit, $t \to \infty$, would not exist). This method, known as non-equilibrium molecular dynamics (NEMD), has the added advantage that it can, in principle, be used to calculate nonlinear as well as linear transport coefficients. These coefficients can be calculated as a function of external-field strength, frequency or wavevector. The most efficient, number independent way to calculate *mechanical* transport coefficients is to ignore the elegant results of response theory and to duplicate the transport process, essentially as it occurs in nature.

Thermal transport processes are, in principle, much more difficult to simulate on a computer. A thermal-transport process is one which is driven by boundary conditions rather than mechanical fields. For thermal processes we cannot utilize time-dependent perturbation theory because no external field appears in the Hamiltonian which could be used as a perturbation variable. In spite of this difference, susceptibilities for thermal processes show many similarities to their mechanical counterparts (compare (5.73) with the results of Chapter 4). If J is the flux of some conserved quantity (mass, momentum of energy) and if X is a gradient in the density of that conserved quantity, then a linear Navier–Stokes transport coefficient is defined by a constitutive relation of the form

$$J = LX \tag{6.2}$$

In Chapter 4 we showed that each of the Navier–Stokes transport coefficients, L, is related to equilibrium fluctuations by Green–Kubo relations. These relations are set out in Table 6.1. Remarkably, Navier–Stokes thermal-transport coefficients are related to equilibrium time-correlation functions in essentially the same way as are mechanical-transport coefficients. We must stress, however, that this close formal similarity between thermal- and mechanical-transport coefficients only applies to Navier–Stokes thermal-transport processes. If fluxes of non-conserved variables are involved, then Green–Kubo relations must be generalized (see equation (4.12) and Section 4.3).

The ensemble averages employed in Table 6.1, are usually taken to be canonical, while the time dependence of the correlation functions is generated by field-free Newtonian equations of motion. In Section 5.4 we proved that, except for bulk viscosity, thermostatted equations of motion can also be used to generate the equilibrium time-correlation functions. For bulk viscosity the correlation function involves functions of the kinetic energy of the system. Therefore we *cannot* use Gaussian isokinetic equations of motion (see (5.86) and (5.87)), because for these equations the kinetic energy is a constant of the motion.

To calculate thermal-transport coefficients using computer simulation, we have the same two options that are available to us in the mechanical case. We could use equilibrium molecular dynamics to calculate the appropriate equilibrium time-correlation functions, or we could mimic experiment as closely as possible and calculate the transport coefficients from their defining constitutive relations. Perhaps surprisingly, the first technique to be used was equilibrium molecular dynamics (Alder and Wainwright, 1956). Much later the more efficient nonequilibrium approach was pioneered by Hoover and Ashurst (1975). Although the realistic nonequilibrium approach proved more efficient than equilibrium simulations it was still far from ideal. This was because for thermal transport processes appropriate boundary conditions are needed to drive the system away from equilibrium—moving walls or walls maintained at different temperatures. These boundary conditions necessarily make the system inhomogeneous. In dense fluids, particles pack against these walls giving rise to significant number dependence and interpretative difficulties.

The most effective way to calculate thermal-transport coefficients exploits the formal similarities between susceptibilities for thermal and mechanical transport coefficients. We *invent* a *fictitious* external field which interacts with the system in such a way as to precisely mimic the linear thermal-transport process. The general procedure is outlined in Table 6.2. These methods are called 'synthetic' because the invented mechanical perturbation does not exist in nature. It is our invention and its purpose is to produce a precise mechanical analogue of a thermal-transport process.

With regard to step (3) in Table 6.2, it is not absolutely necessary to invent equations of motion which satisfy $AI\Gamma$ (see Section 5.3). One can generalize response theory so that $AI\Gamma$ is not required. However, it is simpler and more convenient to require $AI\Gamma$ and thus far it has always proved possible to generate algorithms which satisfy $AI\Gamma$. Although $AI\Gamma$ is satisfied, most sets of equations of motion used in synthetic NEMD are not derivable from a Hamiltonian. The preferred algorithms for thermal conductivity and shear viscosity are not derivable from Hamiltonians. In the case of thermal

Table 6.2 Synthetic NEMD.

(1) For the transport coefficient of interest L_{ij}, $J_i \equiv L_{ij} X_j$. Identify the Green–Kubo relation for the transport coefficient,

$$L_{ij} = \int_0^\infty \mathrm{d}t \, \langle J_i(t) J_j(0) \rangle$$

(2) Invent a fictitious field, F_e, and its coupling to the system such that the dissipative flux is J_j.

$$\dot{H}_0^{\mathrm{ad}} = -J_j F_e$$

(3) Ensure that $AI\Gamma$ is satisfied, that the equations of motion are homogeneous and that they are consistent with periodic boundary conditions

(4) Apply a thermostat

(5) Couple F_e to the system isothermally or isoenergetically and compute the steady state average, $\langle J_i(t) \rangle$, as a function of the external field, F_e. Linear response theory then proves,

$$L_{ij} = \lim_{F_e \to 0} \lim_{t \to \infty} \frac{\langle J_i(t) \rangle}{F_e}$$

conductivity the Hamiltonian approach must be abandoned because of conflicts with the periodic boundary condition convention used in simulations. For shear viscosity the breakdown of the Hamiltonian approach occurs for more complicated reasons.

Equations of motion generated by this procedure are not unique, and it is usually not possible to predict *a priori* which particular algorithm will be most efficient. It is important to realize that the algorithms generated by this procedure are only guaranteed to lead to the correct *linear* ($\lim F_e \to 0$) transport coefficients. We have said nothing so far about generating the correct nonlinear response.

Many discussions of the relative advantages of NEMD and equilibrium molecular dynamics revolve around questions of efficiency. For large fields, NEMD is *orders of magnitude* more efficient than equilibrium molecular dynamics. However, one can always make NEMD arbitrarily inefficient by choosing a sufficiently small field. At fields which are small enough for the response to be linear, there is no simple answer to the question of whether NEMD is more efficient than equilibrium molecular dynamics. The number dependence of errors for the two methods are very different—compared with equilibrium molecular dynamics the relative accuracy of NEMD can be made arbitrarily great by increasing the system size.

These discussions of efficiency ignore two major advantages of NEMD over equilibrium molecular dynamics. Firstly, by simulating a *nonequilibrium* system one can visualize and study the microscopic physical mechanisms that are important to the transport processes (this is true both for synthetic and *realistic* NEMD). One can readily study the distortions of the local molecular structure of nonequilibrium systems. For molecular systems under shear flow, one can watch the shear-induced processes of molecular alignment, rotation and conformational change (Edberg *et al.*, 1987). Obtaining this sort of information from equilibrium time-correlation functions is possible but it is so difficult that no one has yet attempted the task. It is likely that no one ever will. Secondly, the NEMD opens the door to studying the nonlinear response of systems far from equilibrium.

We will now give an extremely brief description of how one performs molecular dynamics simulations. We refer the reader to more detailed treatments which can be found in the excellent monograph by Allen and Tildesley (1987) and in the review of NEMD by the present authors (Evans and Morriss, 1984a).

Consider the potential energy, Φ, of a system of N interacting particles. The potential energy can always be expanded into a sum of pair, triplet, etc., interactions:

$$\Phi(r) = \frac{1}{2!}\sum \phi^{(2)}(\mathbf{r}_i, \mathbf{r}_j) + \frac{1}{3!}\sum \phi^{(3)}(\mathbf{r}_i, \mathbf{r}_j, \mathbf{r}_k) + \dots \tag{6.3}$$

For the inert-gas fluids it is known that the total potential energy can be reasonably accurately written as a sum of effective pair interactions with an effective pair-interaction potential denoted by $\phi(\mathbf{r}_i, \mathbf{r}_j)$. The Lennard–Jones potential, ϕ^{LJ}, is frequently used as an effective pair potential,

$$\phi^{\mathrm{LJ}}(\mathbf{r}_i, \mathbf{r}_j) = \phi^{\mathrm{LJ}}(r_{ij}) \equiv 4\varepsilon\left[\left(\frac{\sigma}{r_{ij}}\right)^{12} - \left(\frac{\sigma}{r_{ij}}\right)^{6}\right] \tag{6.4}$$

The potential energy of the two particles i,j is solely a function of their separation distance r_{ij} and is independent of the relative orientation of their separation vector $\mathbf{r}_{ij}$. The Lennard–Jones potential is characterized by a well depth ε, which controls the energy of the interaction, and a distance σ, which is the distance at which the potential energy of the pair changes sign due to the cancellation of the van der Waals attractive forces by the short-range quantum repulsive forces. If $\varepsilon/k_B = 119.8$ K and $\sigma = 3.405$ Å, the Lennard–Jones potential forms a surprisingly accurate representation of liquid argon (Hansen and Verlet, 1969). For proper scaling during simulations, all calculations are performed in reduced units where $\varepsilon/k_B = \sigma = m = 1$. This amounts to measuring all distances in units of σ, all temperatures in units of ε/k_B and all masses in units of m. The Lennard–Jones potential is often truncated at a distance, $r_c = 2.5\sigma$. Other potentials that are commonly used include the Weeks–Chandler–Andersen potential, usually written as WCA, which is the Lennard–Jones potential truncated at the position of minimum potential energy ($2^{1/6}\sigma$) and then shifted up so that the potential is zero at the cutoff.

$$\begin{aligned} \phi^{\mathrm{WCA}}(r_{ij}) &= 4\varepsilon\left[\left(\frac{\sigma}{r_{ij}}\right)^{12} - \left(\frac{\sigma}{r_{ij}}\right)^{6}\right] + 1; \qquad r < 2^{1/6}\sigma \\ &= 0 \qquad ; \qquad r > 2^{1/6}\sigma \end{aligned} \tag{6.5}$$

The main advantage of this potential is its extremely short range of interaction. This permits simulations to be carried out much more quickly than is possible with the longer range Lennard–Jones potential. Another short-range potential that is often used is the soft-sphere potential which omits the r^{-6} term from the Lennard–Jones poential. The soft-sphere potential is often truncated at 1.5σ.

In molecular dynamics one simply solves the equations of motion for a system of ($N \simeq 100$–$100\,000$) interacting particles. The force on particle i, due to particle j, $\mathbf{F}_{ij}$, is evaluated from the equation

$$\mathbf{F}_{ij} = -\frac{\partial \phi_{ij}}{\partial \mathbf{r}_i} \tag{6.6}$$

The N interacting particles are placed in a cubic cell which is surrounded by an infinite array of identical cells—so-called periodic boundary conditions. To compute the force on a given particle in the primitive cell, one locates the closest (or minimum) image positions of the other $N - 1$ particles. The minimum image of particle i may be within the primitive cell, or in one of the surrounding image cells (see Fig. 6.1). One then finds all the minimum images particles for i that lie within the potential cutoff distance r_c and uses (6.6) to compute the contributions to the force on i, $\mathbf{F}_i = \Sigma\,\mathbf{F}_{ij}$.

Finally, one solves Newton's or Hamilton's equations of motion for the system

$$\begin{aligned}\dot{\mathbf{r}}_i &= \frac{\mathbf{p}_i}{m} \\ \dot{\mathbf{p}}_i &= \mathbf{F}_i\end{aligned} \tag{6.7}$$

If, during the course of the motion particle i leaves the primitive cell, it will be replaced under the periodic boundary condition convention by an image of itself, travelling with exactly the same momentum, one lattice vector distant. We prefer to use Hamilton's form for the equations of motion because this form is much more convenient than the Newtonian form both for NEMD

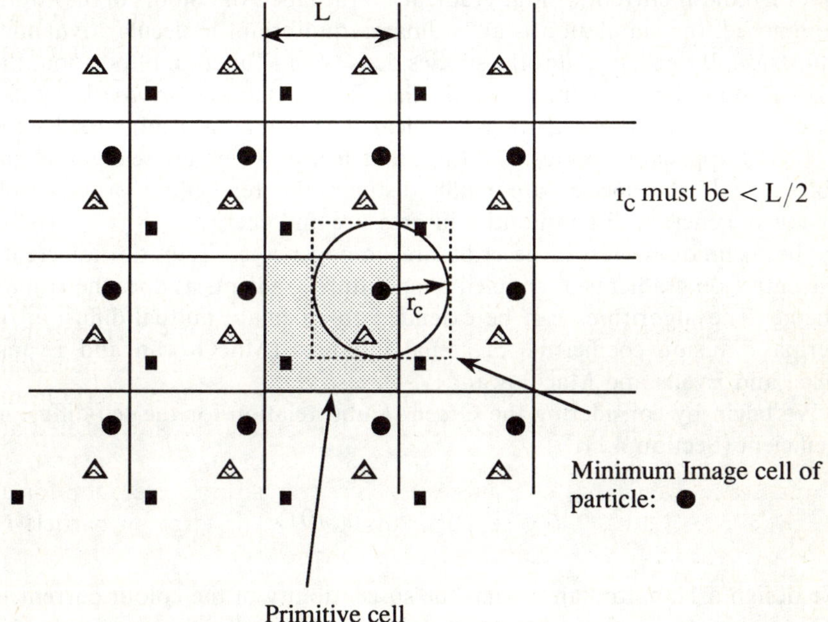

Fig. 6.1 Orthogonal periodic boundary conditions.

and for equilibrium molecular dynamics with velocity-dependent forces (such as thermostats). We often solve these equations of motion using a fifth-order Gear predictor–corrector method. In studies of the transient response of systems to external fields we use the less-efficient Runge–Kutta methods. Unlike the Gear algorithms, Runge–Kutta methods are self-starting, achieving full accuracy in the first timestep.

We will now give a summary of some of the synthetic NEMD algorithms that have been used to calculate Navier–Stokes transport coefficients.

6.2 SELF-DIFFUSION

The first nonequilibrium molecular dynamics (NEMD) algorithm for self-diffusion was devised by Holian (Erpenbeck and Wood, 1977). In this elegant scheme the self-diffusion coefficient was evaluated as the limiting value of the mutual-diffusion coefficient as the two species become identical. In this limit the two species differ only by a colour label which plays no role in their subsequent dynamics but which is reset in a probabilistic fashion as particles cross a labelling plane. A concentration gradient in coloured species is set up and the mutual-diffusion coefficient is calculated from the constitutive relation (colour current/colour gradient). If the labels or colours of the atoms are ignored, the simulation is an ordinary equilibrium molecular dynamics simulation. If one calculates the species density as a function of position, the periodic boundary conditions imply that it is a periodic *sawtooth* profile. Exactly how sharp the teeth are is not clear. The technique is inhomogeneous and is not applicable to mutual diffusion of species which are really different molecules. If the species are really distinct, the relabelling process will obviously generate discontinuities in pressure and energy.

The techniques we will describe are homogeneous. They do not create concentration gradients or coupled temperature gradients as does the Holian scheme. The algorithms can be extended to calculate mutual-diffusion or thermal-diffusion coefficients of actual mixtures (MacGowan and Evans, 1986a and Evans and MacGowan, 1987).

We begin by considering the Green–Kubo relation for the self-diffusion coefficient (Section 4.1):

$$D = \int_0^\infty \mathrm{d}t \, \langle v_{xi}(t) v_{xi}(0) \rangle \tag{6.8}$$

We design a Hamiltonian so that the susceptibility of the colour current to the magnitude of the perturbing colour field is closely related to the single-particle velocity autocorrelation function (6.8). Consider the *colour*

Hamiltonian (Evans *et al.*, 1983)

$$H = H_0 - \sum_{i=1}^{N} c_i x_i F(t), \qquad t > 0 \tag{6.9}$$

where H_0 is the unperturbed Hamiltonian. The c_i are called colour changes. We call this property colour rather than charge to emphasize the fact that H_0 is independent of the set of colour charges $\{c_i\}$. At equilibrium, in the absence of the colour field, the dynamics is colour blind. For simplicity we consider an even number of particles N, with

$$c_i = (-1)^i \tag{6.10}$$

The response we consider is the colour current density J_x

$$J_x = \frac{1}{V} \sum_{i=1}^{N} c_i \dot{x}_i \tag{6.11}$$

Since we are dealing with a Hamiltonian system, $AI\Gamma$ (Section 5.3) is automatically satisfied. The dissipation function is

$$\dot{H}_0^{\text{ad}} = F(t) \sum_{i=1}^{N} c_i v_{xi} = F(t) J_x V \tag{6.12}$$

Linear response theory therefore predicts that (Sections 5.1 and 5.3)

$$\langle J_x(t) \rangle = \beta V \int_0^t \mathrm{d}s \, \langle J_x(t-s) J_x(0) \rangle_0 F(s) \tag{6.13}$$

where the propagator implicit in $J_x(t-s)$ is the field-*free* equilibrium propagator. (If we were considering electrical rather than colour conductivity, equation (6.13) would give the Kubo expression for the electrical conductivity.) To obtain the diffusion coefficient we need to relate the colour current autocorrelation function to the single-particle velocity autocorrelation function. This relation, as we shall see, depends slightly on the choice of the equilibrium ensemble. If we choose the canonical ensemble, then

$$\langle J_x(t) J_x(0) \rangle_c = \frac{1}{V^2} \sum_{i,j}^{N} c_i c_j \langle v_{xi}(t) v_{xj}(0) \rangle_c \tag{6.14}$$

In the thermodynamic limit, for the canonical ensemble, if $j \neq i$, then $\langle v_{xi}(t) v_{xj}(0) \rangle = 0, \forall t$. This is clear since if c is the speed of sound, $v_{xj}(0)$ can only be correlated with other particles within its *sound cone* (i.e. a volume with radius, ct). In the thermodynamic limit there will always be infinitely more particles outside the sound cone than within it. Since the particles outside this cone cannot possibly be correlated with particle i, we find that

$$\langle J_x(t) J_x(0) \rangle_c = \frac{1}{V^2} \sum_{i=1}^{N} c_i^2 \langle v_{xi}(t) v_{xi}(0) \rangle_c = \frac{N}{V^2} \langle v_x(t) v_x(0) \rangle_c \tag{6.15}$$

Combining this equation with the Green–Kubo relation for self-diffusion gives

$$D = \frac{1}{\beta\rho} \lim_{t\to\infty} \lim_{F\to 0} \frac{\langle J_x(t)\rangle}{F} \tag{6.16}$$

If we are working within the *molecular dynamics* ensemble in which the total linear momentum of the system is zero, then v_{xi} is not independent of v_{xj}. In this case there is a correction of order N^{-1} to this equation and the self-diffusion coefficient becomes (Evans *et al.*, 1983),

$$D = \frac{N-1}{N} \frac{1}{\beta\rho} \lim_{t\to\infty} \lim_{F\to 0} \frac{\langle J_x(t)\rangle}{F} \tag{6.17}$$

In the absence of a thermostat the order of the limits in (6.17) and (6.10) is important. They cannot be reversed. If a thermostat is applied to the system, a trivial application of the results in Section 5.3 allows the limits to be taken in either order.

As an example of the use of thermostats we will now derive the Gaussian isokinetic version of the colour diffusion algorithm. Intuitively it is easy to see that as the heating effect is nonlinear (that is $O(F^2)$), it does not effect the linear response. The equations of motion we employ are:

$$\dot{\mathbf{q}}_i = \frac{\mathbf{p}_i}{m} \tag{6.18}$$

and

$$\dot{\mathbf{p}}_i = \mathbf{F}_i + \mathbf{i}c_i F - \alpha\left(\mathbf{p}_i - \mathbf{i}c_i \frac{J_x}{\rho}\right) \tag{6.19}$$

where the Gaussian multiplier required to thermostat the system is obtained from the constraint equation

$$\frac{1}{m}\sum_{i=1}^{N}\left(\mathbf{p}_i - \mathbf{i}c_i\frac{J_x}{\rho}\right)^2 = 3Nk_B T \tag{6.20}$$

In this definition of the temperature we calculate the peculiar particle velocities relative to the streaming velocity of *each species*. If one imagined that the two species are physically separated, then this definition of the temperature is independent of the bulk velocity of the two species. In the absence of this definition of the peculiar kinetic energy, the thermostat and the colour field would work against each other and the temperature would have an explicit quadratic dependence on the colour current. Combining

(6.12) and (6.13) we identify the thermostatting multiplier as

$$\alpha = \frac{\sum m\mathbf{F}_i \cdot \left(\mathbf{p}_i - \mathbf{i}c_i \frac{mJ_x}{\rho} \right)}{\sum \mathbf{p}_i \cdot \left(\mathbf{p}_i - \mathbf{i}c_i \frac{mJ_x}{\rho} \right)} \tag{6.21}$$

In the original paper (Evans *et al.*, 1983), the thermostat was only applied to the components of the velocity which were orthogonal to the colour field. It can be shown that the *linear* response of these two systems is identical, provided that the systems are at the same state point.

The algorithm is homogeneous since if we translate particle i and its interacting neighbours, the total force on i remains unchanged. The algorithm is also consistent with ordinary periodic boundary conditions (Fig. 6.1). There is no change in the colour charge of particles if they enter or leave the primitive cell. It may seem paradoxical that we can measure diffusion coefficients without the presence of concentration gradients; however, we have replaced the chemical-potential gradient which drives real diffusion with a fictitious colour field. A gradient in chemical potential implies a composition gradient and a coupled temperature gradient. Our colour field acts homogeneously and leads to no temperature or density gradients. Linear response theory, when applied to our fictitious colour field, tells us how the transport properties of our fictitious mechanical system relate to the thermal-transport process of diffusion.

By applying a sinusoidal colour field $F(t) = F_0 e^{i\omega t}$, we can calculate the entire equilibrium velocity autocorrelation function. Noting the amplitude and the relative phase of the colour current, we can calculate the complex frequency-dependent susceptibility

$$\chi(\omega) = \int_0^\infty dt\, e^{-i\omega t} \chi(t) = \lim_{F \to 0} \frac{J(\omega)}{F(\omega)} \tag{6.22}$$

An inverse Fourier–Laplace transform of $\chi(t)$ gives the velocity autocorrelation function.

Figure 6.2 shows the results of computer simulations of the diffusion coefficient for the 108-particle Lennard–Jones fluid at a reduced temperature of 1.08 and a reduced density of 0.85. The open circles were obtained using the algorithm outlined in this section (Evans *et al.*, 1983) which is based on equation (6.17). In Fig. 6.2 the colour conductivity (left y axis) and the diffusion coefficient (right y axis) are plotted as functions of the colour current. The self-diffusion coefficient is obtained by extrapolating the current to zero.

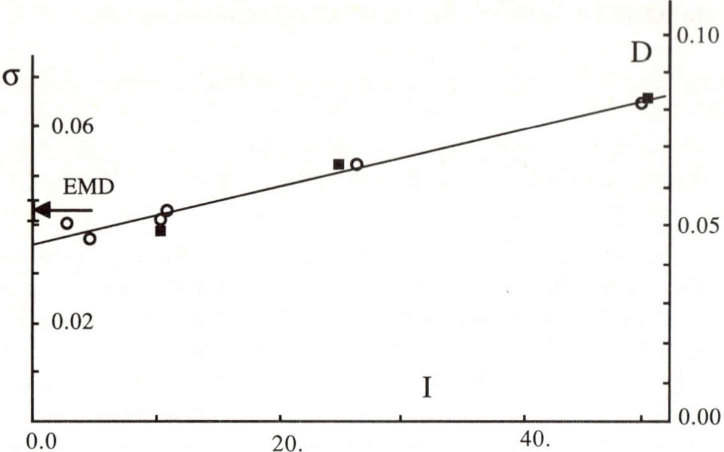

Fig. 6.2 The results (■) of nonequilibrium simulations which were performed at constant colour current rather than constant applied colour field (○). The constant-current methods are described in more detail in Section 6.8. Briefly, one treats the colour field $F(t)$ as a Lagrange multiplier whose value is chosen in such a way that the colour current is a constant of the motion. It is clear from the diagram that the constant-current and constant-colour-field simulations are also in statistical agreement with each other.

The arrow labelled 'EMD' shows the results of equilibrium molecular dynamics where the diffusion coefficient was obtained (Levesque and Verlet, 1970) by integrating the velocity autocorrelation function (Section 4.1). The nonequilibrium and nonequilibrium simulations are in statistical agreement with each other.

Also shown in Fig. 6.2 are the results of simulations performed at constant colour current, rather than constant applied colour field. We will return to this matter when we describe Norton ensemble methods in Section 6.6.

In terms of computational efficiency, the self-diffusion coefficient, being a single-particle property, is far more efficiently computed from equilibrium simulations than from the algorithm given above. The algorithm outlined above is useful for pedagogical reasons; it is the simplest NEMD algorithm and is also the basis for developing algorithms for the mutual-diffusion coefficients of mixtures (Evans and MacGowan, 1987). The mutual-diffusion coefficient, being a collective transport property, is difficult to calculate using equilibrium molecular dynamics (Erpenbeck, 1989). If the two coloured species are distinct electrically charged species, the colour conductivity is actually the electrical conductivity, and the algorithm given above provides a simple means for its calculation.

6.3 COUETTE FLOW AND SHEAR VISCOSITY

We now describe a homogeneous algorithm for calculating the shear viscosity. Among the Navier–Stokes transport processes, shear viscosity is unique in that a steady, homogeneous, algorithm is possible using only the periodic boundary conditions to drive the system to a nonequilibrium state. Apart from the possible presence of a thermostat, the equations of motion can be simple Newtonian equations of motion. We will begin by describing how to adapt periodic boundary conditions for planar Couette flow. We will assume that the reader is familiar with the use of fixed orthogonal periodic boundary conditions in equilibrium molecular dynamics simulations (Allen and Tildesley, 1987). Because shearing periodic boundaries alone can be used to drive shear flow, an understanding of the so-called *Lees and Edwards boundary conditions* (Lees and Edwards, 1972) is sufficient to define an algorithm for planar Couette flow. This algorithm is called the *boundary-driven* algorithm. As this algorithm is based simply on the adaption of periodic boundary conditions to simulations of shear flow, the algorithm is exactly arbitrarily far from equilibrium.

From a theoretical point of view, the boundary-driven algorithm is difficult to work with. Because there is no explicit external field appearing in the equations of motion one cannot employ response theory to link the results obtained from these simulations with, say, the Green–Kubo relations for shear viscosity. This algorithm also has some disadvantages from a numerical point of view. This will lead us to a discussion of the so-called SLLOD algorithm. This algorithm still employs Lees–Edwards boundary conditions, but it eliminates all the disadvantages of the simple boundary-drive method. The SLLOD algorithm is also exactly arbitrarily far from equilibrium.

Lees–Edwards shearing periodic boundaries

Figure 6.3 shows one way of representing planar Couette flow in a periodic system. In the figure we only employ two particles per unit cell, although in an actual computer simulation this number typically ranges from about 100 to, possibly, several tens of thousands. As the particles move under Newton's equations of motion they feel the interatomic forces exerted by the particles within the unit cell and by the image particles whose positions are determined by the instantaneous lattice vectors of the periodic array of cells. The motion of the image cells defines the strain rate, $\gamma \equiv \partial u_x/\partial y$, for the flow. The motion of the cell images is such that their individual origins move with an x velocity which is proportional to the y coordinate of the particular cell origin.

$$\mathbf{u}(\mathbf{r}, t) = \mathbf{i}\gamma y \tag{6.23}$$

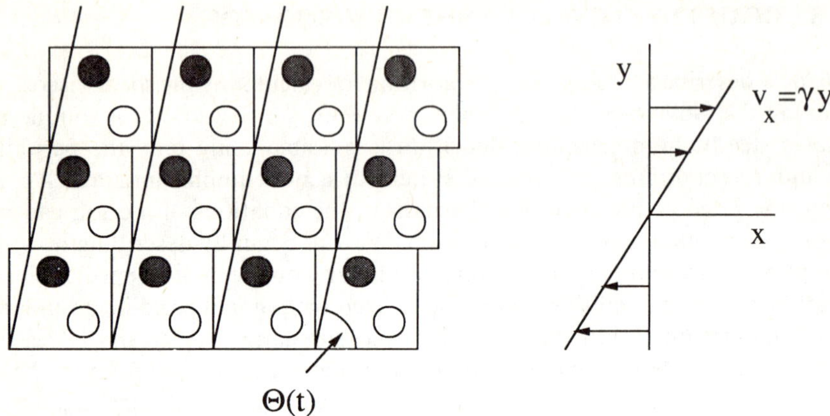

Fig. 6.3 Lees–Edwards periodic boundary conditions for planar Couette flow.

If the Reynolds number is sufficiently small and turbulence does not occur, we expect that the motion of image particles above and below any given cell will, *in time*, induce a linear streaming velocity, $\mathbf{u}(\mathbf{r})$, on each particle within the cell.

If, during the course of time, a given particle moves out of a cell, it will be replaced by its periodic image. If the particle moves through a y face of a cell (that is, through the planes $y = 0$ or $y = L$), the replacing image particle will not have the same laboratory velocity, nor necessarily the same x coordinate. This movement of particles into and out of the primitive cell promotes the generation of a stable linear streaming velocity profile.

Although there are jump discontinuities in both the laboratory coordinates and the laboratory velocities of particles between cells, there is no way in which the particles can actually sense the boundaries of any given cell—they are merely book-keeping devices. The system is spatially homogeneous. As we shall see, those components of particle velocity and position which are discontinuous have *no* thermodynamic meaning.

We have depicted the Lees–Edwards boundary conditions in the so-called *sliding brick* representation. There is a completely equivalent *deforming cube* representation that can be used if one so prefers (see Fig. 6.4). We will mainly use the language of the sliding-brick representation—however, our choice is completely arbitrary.

We will now consider the motion of particles under Lees–Edwards boundary conditions in more detail. Consider a simulation cube of side L, located so that the streaming velocity at the cube origin is zero (that is the cube $0 < \{x, y, z\} < L$). The laboratory velocity of particle i is then the

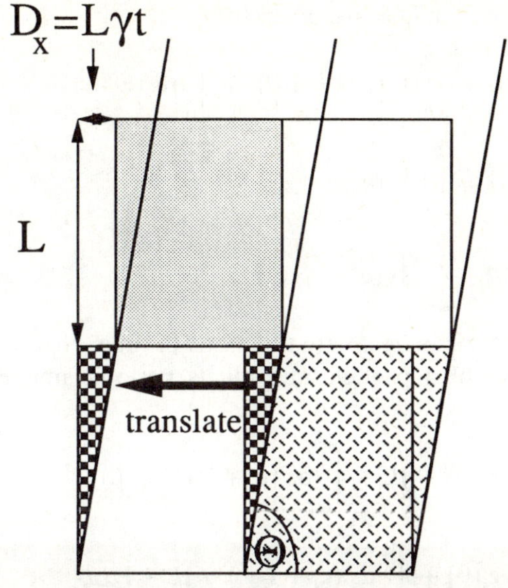

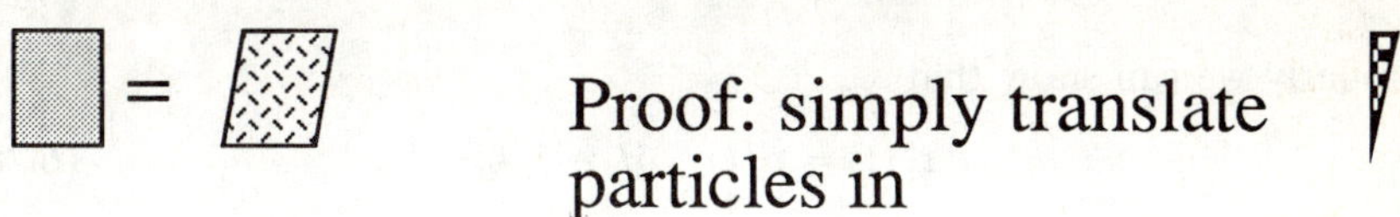

Fig. 6.4 The sliding-brick and deforming-cube representations of Lees–Edwards boundary conditions are equivalent.

sum of two parts; a peculiar or thermal velocity $\mathbf{c}_i$, and a streaming velocity $\mathbf{u}(\mathbf{r}_i)$, so

$$\dot{\mathbf{r}}_i = \mathbf{c}_i + \mathbf{u}(\mathbf{r}_i) \tag{6.24}$$

Imagine that at $t = 0$ we have the usual periodic replication of the simulation cube where the boundary condition is

$$\mathbf{r}_i = (\mathbf{r}_i)_{\text{mod L}} \tag{6.25}$$

(with the modulus of a vector defined to be the vector of the moduli of the elements). As the streaming velocity is a function of y only, we need to consider explicitly boundary crossings in the y direction. At $t = 0$, $\mathbf{r}_i$ has images at $\mathbf{r}_i' + \mathbf{j}L$, and $\mathbf{r}_i''$ has images at $\mathbf{r}_i - \mathbf{j}L$. After time t the positions of particle i and these two images are given by

$$\mathbf{r}_i(t) = \mathbf{r}_i(0) + \int_0^t ds\, \dot{\mathbf{r}}_i(s) = \mathbf{r}_i(0) + \int_0^t ds\, (\mathbf{c}_i + \mathbf{i}\gamma y_i)$$

$$\mathbf{r}'_i(t) = \mathbf{r}'_i(0) + \int_0^t ds\, (\mathbf{c}'_i + \mathbf{i}\gamma y'_i) \tag{6.26}$$

$$\mathbf{r}''_i(t) = \mathbf{r}''_i(0) + \int_0^t ds\, (\mathbf{c}''_i + \mathbf{i}\gamma y''_i)$$

where $\mathbf{c}_i$ and y_i (and their images) are functions of time. Now, by definition, the peculiar velocities of a particle and all its periodic images are equal, $\mathbf{c}_i = \mathbf{c}'_i = \mathbf{c}''_i$, so that

$$\begin{aligned}\mathbf{r}'_i(t) &= \mathbf{r}_i(0) + \mathbf{j}L + \int_0^t ds\, (\mathbf{c}_i + \mathbf{i}\gamma(y_i + L)) \\ &= \mathbf{r}_i(0) + \int_0^t ds\, (\mathbf{c}_i + \mathbf{i}\gamma y_i) + \mathbf{j}L + \mathbf{i}\gamma L t \\ &= \mathbf{r}_i(t) + \mathbf{j}L + \mathbf{i}\gamma L t\end{aligned} \tag{6.27}$$

Similarly we can show that

$$\mathbf{r}''_i(t) = \mathbf{r}_i(t) - \mathbf{j}L - \mathbf{i}\gamma L t \tag{6.28}$$

If $\mathbf{r}_i(t)$ moves out the bottom of the simulation cube, it is replaced by the image particle at $\mathbf{r}'_i(t)$ (Fig. 6.5)

$$\mathbf{r}_i^{\mathrm{new}} = (\mathbf{r}'_i)_{\mathrm{mod\,L}} = (\mathbf{r}_i + \mathbf{i}\gamma L t)_{\mathrm{mod\,L}} \tag{6.29}$$

or if $\mathbf{r}_i(t)$ moves out of the top of the simulation cube, it is replaced by the image particle at $\mathbf{r}_i(t)''$

$$\mathbf{r}_i^{\mathrm{new}} = (\mathbf{r}''_i)_{\mathrm{mod\,L}} = (\mathbf{r}_i - \mathbf{i}\gamma L t)_{\mathrm{mod\,L}} \tag{6.30}$$

The change in the laboratory velocity of a particle is given by the time derivative of equations (6.29) and (6.30). These rules for imaging particles and their velocities are shown schematically in Fig. 6.5.

There is a major difficulty with the boundary-driven algorithm. The way in which the boundaries induce a shearing motion to the particles takes time to occur, approximately given by the sound-traversal time for the primitive cell. This is the minimum time taken for the particles to realize that the shear is taking place. The boundary-driven method as described above, therefore *cannot* be used to study time-dependent flows. The most elegant solution to this problem introduces the SLLOD algorithm. We will defer a discussion

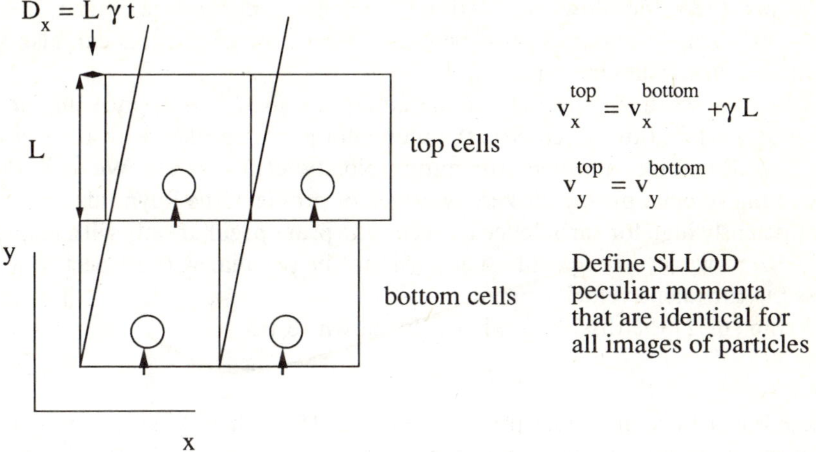

Fig. 6.5 A particle moving out of the top of a cell is replaced by its image from the cell below.

of thermostats and the evaluation of thermodynamic properties until after we have discussed the SLLOD algorithm.

The SLLOD algorithm

The boundary-driven shear-flow algorithm has a number of disadvantages, the principal one being its lack of contact with response theory. We will now describe two synthetic field algorithms for simulating any form of flow deformation. Historically, the first fictitious force method proposed for viscous-flow calculations was the DOLLS tensor method (Hoover *et al.*, 1980). This method can be derived from the DOLLS tensor Hamiltonian,

$$H = H_0 + \sum_{i=1}^{N} \mathbf{q}_i \mathbf{p}_i : (\nabla \mathbf{u}(t))^{\mathrm{T}} \tag{6.31}$$

It generates the following equations of motion

$$\begin{aligned} \dot{\mathbf{q}}_i &= \frac{\mathbf{p}_i}{m} + \mathbf{q}_i \cdot \nabla \mathbf{u} \\ \dot{\mathbf{p}}_i &= \mathbf{F}_i - \nabla \mathbf{u} \cdot \mathbf{p}_i \end{aligned} \tag{6.32}$$

These equations of motion *must* be implemented with compatible periodic boundary conditions. If the strain-rate tensor has only one nonzero element and it is off-diagonal, the deformation is planar Couette flow and Lees–Edwards boundary conditions must be used. If the strain-rate tensor is isotropic then the flow is dilational and the appropriate variation of Lees–Edwards boundaries must be used. Other flow geometries can also be simulated using these equations.

One can see from the first of the equations (6.32), that since $d\mathbf{q}_i/dt$ is obviously a laboratory velocity, the momenta $\mathbf{p}_i$ are *peculiar* with respect to the low Reynolds number streaming velocity $\mathbf{u}(\mathbf{r}) = \mathbf{r}\cdot\nabla\mathbf{u}$. We call this streaming velocity profile the zero wavevector profile. If the Reynolds number is sufficiently high for turbulence to occur, the $\mathbf{p}_i$ are peculiar only with respect to the zero wavevector profile. They will not be peculiar with respect to any possible turbulent velocity profiles.

From (6.32) the dissipation is easily shown to be

$$\dot{H}_0^{\mathrm{ad}} = -\nabla\mathbf{u}:\mathbf{P}V \tag{6.33}$$

where $\mathbf{P}$ is the instantaneous pressure tensor (3.152), whose kinetic component is given in terms of the peculiar momenta $\mathbf{p}_i$. Since the DOLLS tensor equations of motion are derivable from a Hamiltonian, the $AI\Gamma$ condition is clearly satisfied and we see immediately from equations (6.33) and (5.73) that, in the linear regime, close to equilibrium, the shear and bulk viscosities are related to equilibrium fluctuations via the Green–Kubo formula (T.6.3). This *proves* that the DOLLS tensor algorithm is correct for the limiting linear regime. The linear response of the pressure tensor is, therefore,

$$\langle \mathbf{P}(t)\rangle = -\beta V \int_0^t \mathrm{d}s\, \langle \mathbf{P}(t-s)\mathbf{P}\rangle : \nabla\mathbf{u}(s) \tag{6.34}$$

The DOLLS tensor method has now been replaced by the SLLOD algorithm (Evans and Morriss, 1984b). The only difference between the SLLOD algorithm and the DOLLS tensor equations of motion involves the equation of motion for the momenta. The Cartesian components that couple to the strain-rate tensor are transposed. Unlike the DOLLS tensor equations, the SLLOD equations of motion *cannot* be derived from a Hamiltonian.

$$\begin{aligned} \dot{\mathbf{q}}_i &= \frac{\mathbf{p}_i}{m} + \mathbf{q}_i\cdot\nabla\mathbf{u} \\ \dot{\mathbf{p}}_i &= \mathbf{F}_i - \mathbf{p}_i\cdot\nabla\mathbf{u} \end{aligned} \tag{6.35}$$

It is easy to see that the dissipation function for the SLLOD algorithm is precisely the same as for the DOLLS tensor equations of motion. In spite of the absence of a generating Hamiltonian, the SLLOD equations also satisfy $AI\Gamma$. This means that the linear response for both systems is identical and is given by (6.34). By taking the limit $\gamma \to 0$, followed by the limit $t \to \infty$, we see that the linear shear viscosity can be calculated from a nonequilibrium simulation, evolving under either the SLLOD or the DOLLS tensor equations of motion. With, $\nabla\mathbf{u} = \mathbf{ji}(\partial u_x/\partial y)$, and calculating the ratio of stress to strain rate we calculate

$$\eta = \lim_{t\to\infty} \lim_{\gamma\to 0} \frac{-\langle P_{xy}(t)\rangle}{\gamma} \tag{6.36}$$

From (6.34) we see that the susceptibility is precisely the Green–Kubo expression for the shear viscosity (Table 6.1). Because the linear response of the SLLOD and DOLLS tensor algorithms are related to equilibrium fluctuations by the Green–Kubo relations, these algorithms can be used to calculate the reaction of systems to time-varying strain rates. If the shear rate is a sinusoidal function of time, then the Fourier transform of the susceptibility gives the complex, frequency-dependent shear viscosity measured in viscoelasticity (Sections 2.4 and 4.3).

If the strain-rate tensor is isotropic then the equations of motion describe adiabatic dilation of the system. If this dilation rate is sinusoidal then the limiting small field bulk viscosity can be calculated by monitoring the amplitude and phase of the pressure response and extrapolating both the amplitude and frequency to zero (Hoover *et al.*, 1980). It is again easy to see from (6.34) that the susceptibility for the dilation-induced pressure change is precisely the Green–Kubo transform of the time-dependent equilibrium fluctuations in the hydrostatic pressure (Table 6.1).

Although the DOLLS tensor and SLLOD algorithms have the same dissipation and give the correct *linear* behaviour, the DOLLS tensor algorithm begins to yield incorrect results at quadratic order in the strain rate. These errors show up first as errors in the normal stress differences. For irrotational flows ($\nabla\mathbf{u} = (\nabla\mathbf{u})^{\mathrm{T}}$) so the SLLOD and DOLLS tensor methods are identical, as can easily be seen from their equations of motion.

We will now show that the SLLOD algorithm gives an exact description of shear flow *arbitrarily far from equilibrium.* This method is also correct in the high Reynolds number regime in which laminar flow is unstable. Consider superimposing a linear velocity profile on a canonical ensemble of N-particle systems. This will generate the *local equilibrium* distribution function for Couette flow, f_1

$$f_1 = \frac{\exp\left[-\beta\left(\frac{m(\mathbf{v}_i + \mathbf{i}\gamma y_i)^2}{2} + \Phi\right)\right]}{\int \mathrm{d}\Gamma \exp\left[-\beta\left(\frac{m(\mathbf{v}_i + \mathbf{i}\gamma y_i)^2}{2} + \Phi\right)\right]} \tag{6.37}$$

Macroscopically, such an ensemble is described by a linear streaming velocity profile,

$$\mathbf{u}(\mathbf{r}, t) = \mathbf{i}\gamma y \tag{6.38}$$

so that the second-rank strain-rate tensor, $\nabla\mathbf{u}$, has only one nonzero element, $(\nabla\mathbf{u})_{yx} = \gamma$. The local equilibrium distribution function is *not* the same as the steady-state distribution. This is easily seen when we realize that the shear stress evaluated for f_l is zero. The local distribution function is no more than a canonical distribution with a superimposed linear velocity profile. No molecular relaxation has yet taken place.

If we allow this relaxation to take place by advancing time using Newton's equations (possibly supplemented with a thermostat) the system will go on shearing forever. This is because the linear velocity profile of the local distribution generates a *zero* wavevector transverse momentum current. As we saw in Section 3.8, the zero wavevector momentum density is conserved. The transverse momentum current will persist forever, at least for an infinite system.

Now let us see what happens under the SLLOD equations of motion (6.34), when the strain-rate tensor is given by (6.38). Differentiating the first equation, then substituting for $d\mathbf{p}_i/dt$ using the second equation gives

$$m\ddot{\mathbf{q}}_i = \mathbf{F}_i - \mathbf{i}\gamma p_{yi} + \mathbf{i}(\gamma p_{yi} + \mathbf{m}\dot{\gamma} y_i) = \mathbf{F}_i + \mathbf{i}m\dot{\gamma} y_i \tag{6.39}$$

If the strain rate γ is switched on at time zero, and remains steady thereafter,

$$\gamma(t) = \gamma\Theta(t) \Rightarrow \dot{\gamma} = \gamma\delta(t) \tag{6.40}$$

Thus $d\gamma/dt$ is a delta function at $t = 0$. Now consider subjecting a canonical ensemble to these transformed SLLOD equations of motion (6.39). If we integrate the velocity of particle i over an infinitesimal time interval about zero, we see that

$$\mathbf{v}_i(0^+) - \mathbf{v}_i(0) = \int_0^{0^+} ds\, \dot{\mathbf{v}}(s) = \mathbf{i}\gamma y_i \tag{6.41}$$

So at time 0^+ the x velocity of every particle is incremented by an amount proportional to the product of the strain rate times its y coordinate. At time 0^+, the other components of the velocity and positions of the particles are unaltered because there are no delta function singularities in their equations of motion. Applying (6.41) to a canonical ensemble of systems will clearly generate the local equilibrium distribution for planar Couette flow.

The application of SLLOD dynamics to the canonical ensemble is thus seen to be equivalent to applying Newton's equations to the local distribution function. The SLLOD equations of motion have therefore succeeded in transforming the boundary condition, expressed in the form of the local distribution function, into the form of a smooth mechanical force which appears as a *mechanical* perturbation in the equations of motion. This property is unique to SLLOD dynamics. It is not satisfied by the DOLLS tensor equations of motion. Since one cannot really call into question the validity of the application of Newtonian dynamics to the local distribution as a correct description of Couette flow, we are led to the conclusion that the adiabatic application of SLLOD dynamics to the canonical ensemble gives an exact description of Couette flow (Fig. 6.6).

Knowing that the SLLOD equations are exact, and that they generate Green–Kubo expressions for the shear and bulk viscosities provides a *proof* of the validity of the Green–Kubo expressions themselves. The SLLOD transformation of a thermal-transport process into a mechanical one provides us with a direct route to the Green–Kubo relations for the viscosity coefficients. From equation (6.35) we see that we already have these relations for both the shear and bulk viscosity coefficients. We also see that these expressions are identical to those derived in Chapter 4, using the generalized Langevin equation. It is clear that the present derivation is simpler and gives greater physical insight into the processes involved.

Compared with the boundary-driven methods, the advantages of using the SLLOD algorithm in computer simulations are many. Under periodic boundaries the SLLOD momenta, which are peculiar with respect to the zero wavevector velocity field, are continuous functions of time and space. This is not so for the laboratory velocities $\mathbf{v}_i$. The internal energy and the pressure tensor of the system are more simply expressed in terms of SLLOD momenta rather than laboratory momenta. The internal energy, E, is given as

$$E(T, \rho, N, \gamma) = \langle H_0 \rangle = \left\langle \sum_{i=1}^{N} \frac{\mathbf{p}_i^2}{2m} + \tfrac{1}{2} \sum_{i,j}^{N} \phi_{ij} \right\rangle \tag{6.42}$$

$$\frac{d\mathbf{q}_i}{dt} = \frac{\mathbf{p}_i}{m} + \mathbf{i}\gamma y_i$$

$$\frac{d\mathbf{p}_i}{dt} = \mathbf{F}_i - \mathbf{i}\gamma p_{yi}$$

are equivalent to:

$$m\frac{d^2\mathbf{q}_i}{dt^2} = \mathbf{F}_i + \mathbf{i}\, m\frac{d\gamma}{dt} y_i$$

If the strain rate is a step function then

$\frac{d\gamma}{dt} = \gamma\,\delta(t)$ and the t=0 velocities are incremented

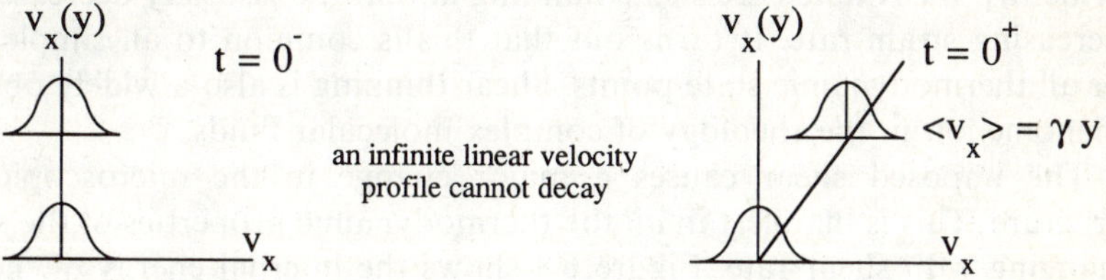

Fig. 6.6 SLLOD equations of motion give an exact representation of planar Couette flow.

while the ensemble averaged pressure tensor is

$$\mathbf{P}(T,\rho,N,\gamma)V = \left\langle \sum_{i=1}^{N} \frac{\mathbf{p}_i\mathbf{p}_i}{m} - \tfrac{1}{2}\sum_{i,j}^{N} \mathbf{r}_{ij}\mathbf{F}_{ij} \right\rangle \tag{6.43}$$

For simulations of viscoelasticity special measures must be taken in the boundary-driven algorithm to ensure that the time-varying strain rate is actually what it is expected to be. In the SLLOD method no special techniques are required for simulations of time-dependent flows; one simply has to solve the equations of motion with a time-dependent strain rate and ensure that the periodic boundary conditions are precisely consistent with the strain derived by integrating the imposed strain rate $\gamma(t)$.

Since the SLLOD momenta are peculiar with respect to the zero wavevector velocity profile, the obvious way of thermostatting the algorithm is to use the equations

$$\begin{aligned}\dot{\mathbf{q}}_i &= \frac{\mathbf{p}_i}{m} + \mathbf{i}\gamma y_i \\ \dot{\mathbf{p}}_i &= \mathbf{F}_i - \mathbf{i}\gamma p_{yi} - \alpha \mathbf{p}_i\end{aligned} \tag{6.44}$$

The thermostatting multiplier, α, is calculated in the usual way by ensuring that $\mathrm{d}(\Sigma\, p_i^2)/\mathrm{d}t = 0$.

$$\alpha = \frac{\sum(\mathbf{F}_i \cdot \mathbf{p}_i - \gamma p_{xi} p_{yi})}{\sum p_i^2} \tag{6.45}$$

The temperature is assumed to be related to the peculiar kinetic energy. In these equations it is *assumed* that a linear velocity profile is stable. However, as we have mentioned a number of times, the linear velocity profile is only stable at low Reynolds number, ($\mathrm{Re} = \rho m \gamma L^2/\eta$).

Figure 6.7 shows the shear viscosity of 2048 WCA particles as a function of strain rate. The fluid is close to the Lennard–Jones triple point. The reduced temperature and density are 0.722 and 0.8442, respectively. The simulations were carried out using the Gaussian isokinetic SLLOD algorithm. We see that there is a substantial change in the viscosity with shear rate. Evidently WCA fluids are shear thinning in that the viscosity decreases with increasing strain rate. It turns out that this is common to all simple fluids for all thermodynamic state points. Shear thinning is also a widely observed phenomenon in the rheology of complex molecular fluids.

The imposed shear causes a major change in the microscopic fluid structure. This is manifest in *all* the thermodynamic properties of the system changing with shear rate. Figure 6.8 shows the internal energy of the fluid plotted as a function of strain rate. For reduced strain rates in the range 0–1.5 we see that both the shear viscosity and the internal energy change by

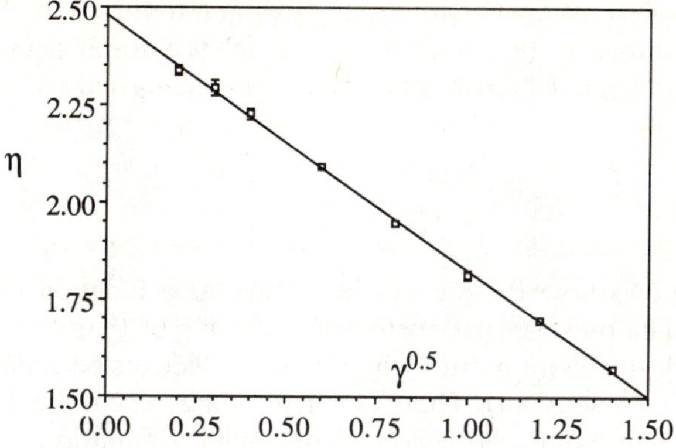

Fig. 6.7 Viscosity of the $N = 2048$ WCA fluid.

approximately 50% compared with their equilibrium values. Furthermore, the viscosity coefficient appears to vary as the square root of the strain rate, while the energy appears to change with a 1.5 power of the strain rate. Over the range of strain rates studied, the maximum deviation from the functional forms is 2.5% for the viscosity and 0.1% for the internal energy. Recently there has been much discussion of the relation of these apparently nonanalytic dependences to mode-coupling theory (see: Yamada and Kawasaki, 1973; Kawasaki and Gunton, 1973; Ernst *et al.*, 1978; Evans, 1983; Kirkpatrick, 1984; van Beijeren, 1984; de Schepper *et al.*, 1986). It is clear that the final resolution of this matter is still a long way off.

One of the most interesting and subtle rheological effects concerns the diagonal elements of the pressure tensor. For Newtonian fluids (i.e. fluids characterized by a strain-rate-independent and frequency-independent viscosity)

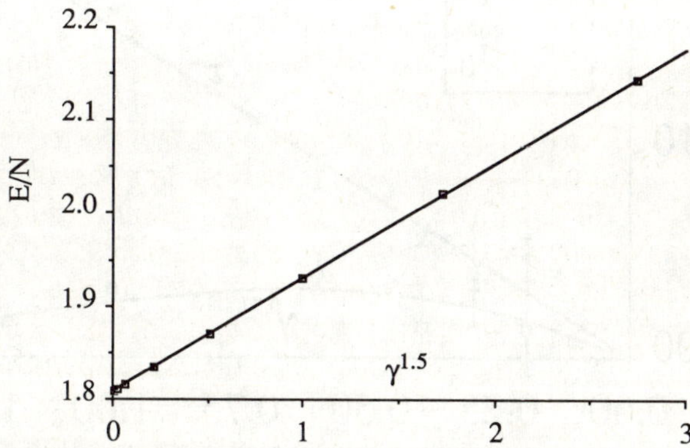

Fig. 6.8 The internal energy of a fluid plotted as a function of the strain rate.

the diagonal elements are equal to each other and to their equilibrium values. Far from equilibrium, this is not true. We define normal stress coefficients, η_0 and η_- (the so-called out-of-plane and in-plane normal stress coefficients) as

$$\eta_0 \equiv -P_{zz} - \tfrac{1}{2}(P_{xx} + P_{yy})/2\gamma \tag{6.46}$$

$$\eta_- \equiv -(P_{xx} - P_{yy})/2\gamma \tag{6.47}$$

Figure 6.9 shows how these coefficients vary as a function of $\gamma^{0.5}$ for the WCA fluid. The out-of-plane coefficient is far larger than the in-plane one, except at very small strain rates where both coefficients become zero (i.e. the fluid becomes Newtonian). These coefficients are very difficult to compute accurately. They require both larger and longer simulations to achieve an accuracy comparable with that for the shear viscosity. In terms of the macroscopic hydrodynamics of non-Newtonian fluids, these normal stress differences are responsible for a wide variety of interesting phenomena (e.g. the Weissenberg effect; see Rainwater and Hanley (1985) and Rainwater *et al.* (1985)).

If one allows the strain rate to be a sinusoidal function of time and extrapolates the system response to zero amplitude, one can calculate the linear viscoelastic response of a fluid. Figure 6.10 shows complex frequency-dependent shear viscosity for the Lennard–Jones fluid (Evans, 1980) at its triple point.

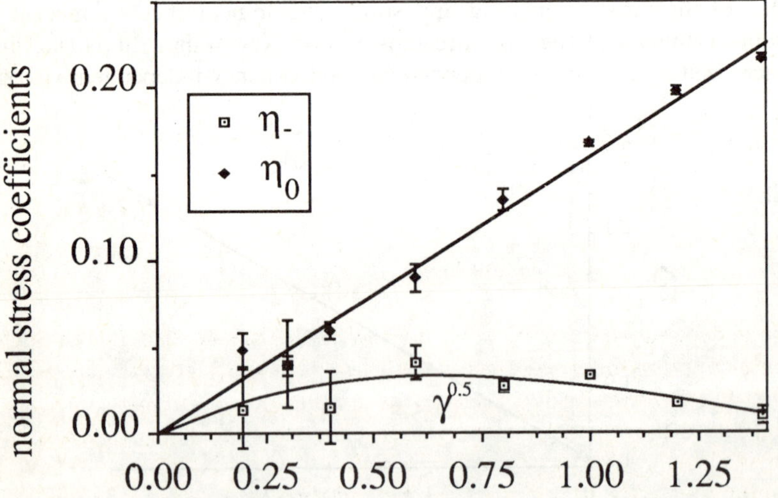

Fig. 6.9 Normal stress coefficients for the $N = 2048$ WCA fluid.

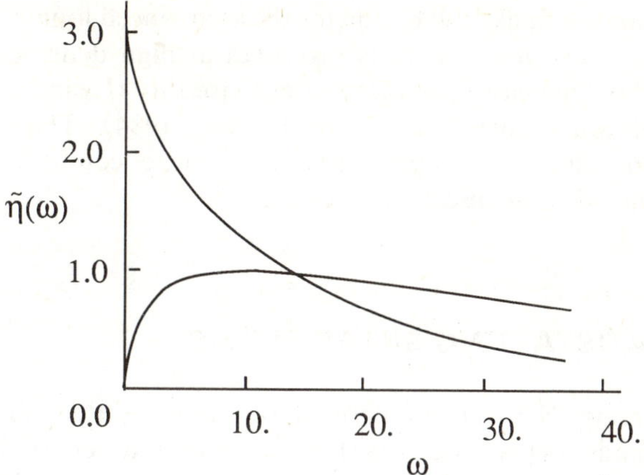

Fig. 6.10 Frequency-dependent shear viscosity at the Lennard–Jones triple point.

If one compares Fig. 6.10 with the Maxwell model for viscoelasticity (Fig. 2.4), one sees a qualitative similarity with the low-frequency response being viscous and the high-frequency response being elastic. The shape of the two sets of curves is, however, quite different, particularly at low frequencies. An analysis of the low-frequency data shows that this is consistent with a nonanalytic square-root dependence upon frequency.

$$\begin{aligned}\tilde{\eta}_R(\omega) &= \eta(0) - \eta_{\omega 1}\omega^{0.5} + O(\omega)\\ \tilde{\eta}_I(\omega) &= \eta_{\omega 1}\omega^{0.5} + O(\omega)\end{aligned} \tag{6.48}$$

where η_R and η_I, are the real and imaginary parts of the viscosity coefficient, respectively. Since the frequency-dependent viscosity is the Fourier–Laplace transform of the memory function (2.76), we can use the Tauberian theorems (Doetsch, 1961) to show that if (6.48) represents the asymptotic low-frequency behaviour of the frequency-dependent viscosity, then the memory function must have the form

$$\lim_{t\to\infty} \eta(t) = \frac{\eta_{\omega 1} t^{-\frac{3}{2}}}{\sqrt{2\pi}} \tag{6.49}$$

This time dependence is again consistent with the time dependence predicted by mode-coupling theory (Pomeau and Resibois, 1975). However, as was the case for the strain-rate dependence the amplitude of the effect shown in Fig. 6.10 is orders of magnitude larger than theoretical predictions. This matter is also the subject of much research and investigation.

Similar *enhanced* long-time tails have been observed subsequently in Green–Kubo calculations for the shear viscosity (Erpenbeck and Wood,

1981). Whatever the final explanation for these enhanced long-time tails, they are a ubiquitous feature of viscous processes at high densities. They have been observed in the wavevector-dependent viscosity (Evans, 1982a) and in shear flow of four-dimensional fluids (Evans, 1984). The situation for two-dimensional liquids is apparently even more complex (Evans and Morriss, 1983a; Morriss and Evans, 1989).

6.4 THERMOSTATTING SHEAR FLOWS

While performing NEMD simulations of thermostatted shear flow for hard-sphere fluids, Erpenbeck (1984) observed that at very high shear rates fluid particles organized themselves into strings. This was an early observation of a nonequilibrium phase transition. This organization of particles into strings reduces the rate at which entropy is produced in the system by the external field. This effect is in competition with the kink instability of the strings themselves. If the strings move too slowly across the simulation cell, thermal fluctuations in the curvature of the strings lead to their destruction. A snapshot of a string phase is shown in Fig. 6.11: the velocity gradient is

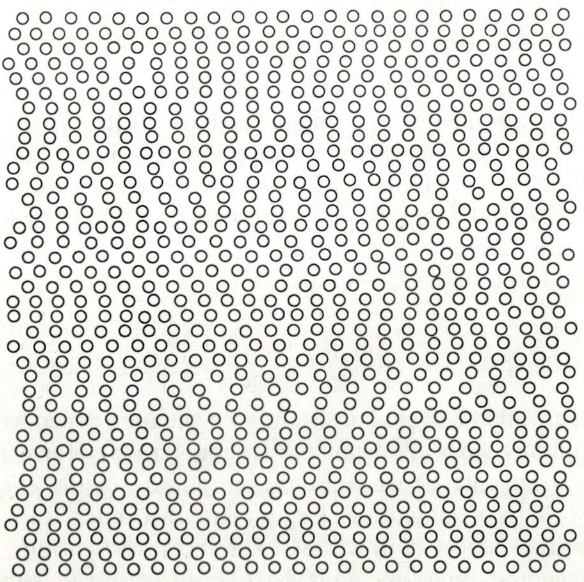

Fig. 6.11 High shear rate string phase in soft discs.

vertical and the streaming velocity is horizontal; the system comprizes 896 soft discs at a state point close to freezing and a reduced shear rate of 17.

The string phase is, in fact, stabilized by the use of a thermostat which *assumes* that a linear velocity profile (implicit in (6.44)) is stable. Thermostats which make some assumption about the form of the streaming velocity profile are called *profile-biased thermostats* (PBT). All the thermostats we have met so far are profile biased. At equilibrium there can be little cause for worry, the streaming velocity must be zero. Away from equilibrium we must be more careful.

Any kink instability that might develop in Erpenbeck's strings, leading to their breakup, would necessarily lead to the formation of large-scale *eddies* in the streaming velocity of the fluid. The profile-biased thermostat would interpret any incipient eddy motion as *heat*, and the thermostat would then try to *cool* the system by suppressing the eddy formation. This, in effect, stabilizes the string phase (Evans and Morriss, 1986).

Thermostats for secondary or convecting flows

Profile-biased thermostats (PBT) for shear flow assume that the kinetic temperature, T_B, for a system undergoing planar Couette flow can be defined by the equation

$$dNk_B T_B = \left\langle \sum_{i=1}^{N} m(\mathbf{v}_i - \mathbf{i}\gamma y_i)^2 \right\rangle \tag{6.50}$$

where d is the number of dimensions and N is the number of particles. The term $\mathbf{i}\gamma y_i$ is the *presumed* streaming velocity at the location of particle i. Once the form of the streaming velocity profile is established it is a simple matter to use Gaussian isokinetic or Nosé methods to thermostat the shearing system.

At low shear rates and low Reynolds number, the Lees–Edwards shearing periodic boundary conditions do indeed lead to a planar velocity profile of the form assumed in (6.50). In Erpenbeck's (1984) simulations the Reynolds numbers ($\mathrm{Re} = \rho m\gamma L^2/\eta$) were very large ($10^3$–$10^5$). The assumption of a linear streaming velocity profile under these conditions is extremely dubious. Suppose that at high Reynolds number the linear velocity profile assumed in (6.50) is not stable. In a freely shearing system with Lees–Edwards geometry, this might manifest itself in an S-shaped kink developing in the velocity profile. If (6.44) is used to maintain the temperature, the thermostat will interpret the development of this secondary flow as a component of the temperature. This increase in temperature will be continuously removed by the thermostat, leading to a damping of the secondary flow.

If we rewrite the SLLOD equations in terms of laboratory momenta

$$\dot{\mathbf{r}}_i = \frac{\mathbf{p}_i}{m}$$
$$\dot{\mathbf{p}}_i = \mathbf{F}_i - \alpha\left(\frac{\mathbf{p}_i}{m} - \mathbf{i}\gamma y_i\right) \tag{6.51}$$

then the momentum current, $\mathbf{J}$,

$$\mathbf{J}(\mathbf{r}, t) \equiv \rho(\mathbf{r}, t)\mathbf{u}(\mathbf{r}, t) = \sum \mathbf{p}_i \delta(\mathbf{r}_i(t) - \mathbf{r}) \tag{6.52}$$

satisfies the following continuity equation.

$$\frac{\mathrm{d}\mathbf{J}}{\mathrm{d}t} = -\nabla\cdot(\mathbf{P} + \rho\mathbf{uu}) - \alpha \sum_{i=1}^{N}\left(\frac{\mathbf{p}_i}{m} - \mathbf{i}\gamma y_i\right)\delta(\mathbf{r}_i - \mathbf{r})$$

$$= -\nabla\cdot(\mathbf{P} + \rho\mathbf{uu}) - \frac{\alpha}{m}(\mathbf{J}(\mathbf{r}, t) - \rho(\mathbf{r}, t)\mathbf{u}_{\text{linear}}(\mathbf{r}, t)) \tag{6.53}$$

The derivation of this equation is carried out by a simple supplementation to the Irving–Kirkwood procedure (Sections 3.7 and 3.8). We have to add the contribution of the thermostat to equations (3.115) to (3.123). Comparing (6.53) with the momentum-conservation equation (2.12), we see that the thermostat could exert a stress on the system. The expected divergence terms ($\rho\mathbf{uu} + \mathbf{P}$) are present on the right-hand side of (6.53). However, the term involving α, the thermostatting term, is new and represents the force exerted on the fluid by the thermostat. This term will only vanish if a linear velocity profile is *stable* and

$$\mathbf{J}(\mathbf{r}, t) = \rho(\mathbf{r}, t)\mathbf{u}(\mathbf{r}, t) = \mathbf{i}m\gamma y \qquad \forall\mathbf{r} \tag{6.54}$$

At high Reynolds numbers this condition might not be true. For simulations at high Reynolds numbers one needs a thermostat which makes no assumptions whatever about the form of the streaming velocity profile. The thermostat should not even assume that a stable profile exists. These ideas led to the development (Evans and Morriss, 1986) of profile-unbiased thermostats (PUT).

The PUT thermostat begins by letting the simulation itself define the local streaming velocity $\mathbf{u}(\mathbf{r}, t)$. This is easily done by replacing the delta functions in (6.52) by microscopically small cells in the simulation programme. The temperature of a particular cell at $\mathbf{r}$, $T(\mathbf{r}, t)$, can be determined from the equation,

$$\frac{dn(\mathbf{r}, t) - d}{2} k_{\mathrm{B}} T(\mathbf{r}, t) \equiv \sum \frac{m}{2}(\mathbf{v}_i - \mathbf{u}(\mathbf{r}, t))^2 \delta(\mathbf{r}_i(t) - \mathbf{r}) \tag{6.55}$$

where $n(\mathbf{r}, t)$ is the number density at $\mathbf{r}, t$ (the delta function has unit volume). The number of degrees of freedom in the cell is $dn(\mathbf{r}, t) - d$, because d degrees of freedom are used to determine the streaming velocity of the cell.

The profile-unbiased thermostatted SLLOD equations of motion can be written as

$$\frac{d\mathbf{r}_i}{dt} = \frac{\mathbf{p}_i}{m}$$
$$\frac{d\mathbf{p}_i}{dt} = \mathbf{F}_i - \alpha\left(\frac{\mathbf{p}_i}{m} - \mathbf{u}(\mathbf{r}, t)\right)\delta(\mathbf{r}_i - \mathbf{r}) \tag{6.56}$$

The streaming velocity, $\mathbf{u}(\mathbf{r}, t)$, is not known in advance but is computed as time progresses from its definition, (6.52). The thermostat multiplier, α, could be a Gaussian multiplier chosen to fix the peculiar kinetic energy (6.55). Equally well the multiplier could be a Nosé–Hoover multiplier. The momentum equation for the profile-unbiased thermostatted system reads

$$\frac{d\mathbf{J}}{dt} = -\nabla\cdot(\mathbf{P} + \rho\mathbf{u}\mathbf{u}) - \alpha\sum\left(\frac{\mathbf{p}_i}{m} - \mathbf{u}(\mathbf{r}, t)\right)\delta(\mathbf{r}_i - \mathbf{r})$$
$$= -\nabla\cdot(\mathbf{P} + \rho\mathbf{u}\mathbf{u}) - \frac{\alpha}{m}(\mathbf{J}(\mathbf{r}, t) - \rho(\mathbf{r}, t)\mathbf{u}(\mathbf{r}, t)) \tag{6.57}$$

From the definition of the streaming velocity of a cell we know that $n(\mathbf{r}, t)\mathbf{u}(\mathbf{r}, t) = \Sigma\,\mathbf{p}_i/m\delta(\mathbf{r}_i(t) - \mathbf{r})$. We also know that $n(\mathbf{r}, t)\mathbf{u}(\mathbf{r}, t) = \Sigma\,\mathbf{u}(\mathbf{r}, t)\delta(\mathbf{r}_i(t) - \mathbf{r}) = \mathbf{u}(\mathbf{r}, t)\,\Sigma\,\delta(\mathbf{r}_i(t) - \mathbf{r})$. Thus the thermostatting term in (6.57) vanishes for all values of $\mathbf{r}$.

In terms of practical implementation in computer programs, PUT can only be used in simulations involving large numbers of particles. Thus far their use has been restricted to simulations of two-dimensional systems. At low Reynolds numbers where no strings are observed in profile-biased simulations, it is found that profile-unbiased simulations yield results for all properties which are indistinguishable from those computed using PBT methods. However, at high strain rates the results obtained using the two different thermostatting methods are quite different. No one has observed a string phase while using a PUT.

6.5 THERMAL CONDUCTIVITY

Thermal conductivity has proven to be one of the most difficult transport coefficients to calculate. Green–Kubo calculations are notoriously difficult to perform. Natural nonequilibrium molecular dynamics (NEMD), where one might simulate heat flow between walls maintained at different temperatures (Tenenbaum *et al.*, 1982), is also fraught with major difficulties. Molecules stack against the walls leading to a major change in the microscopic fluid structure. This means that the results can be quite different from those characteristic of the bulk fluid. In order to measure a statistically significant heat flux, one must use enormously large temperature gradients. These gradients are so large that the absolute temperature of the system may change by 50% in a few tens of Angstroms. The thermal conductivity that one obtains from such simulations is an average over the wide range of temperatures and densities present in the simulation cell.

We will now describe the most efficient presently known algorithm for calculating the thermal conductivity (Evans, 1982b). This technique is synthetic in that a fictitious field replaces the temperature gradient as the force driving the heat flux. Unlike real heat flow, this technique is homogeneous with no temperature or density gradients. We start with the Green–Kubo expression for the thermal conductivity (Section 4.4),

$$\lambda = \frac{V}{k_B T^2} \int_0^\infty dt \, \langle J_{Q_z}(t) J_{Q_z}(0) \rangle \tag{6.58}$$

where J_{Q_z} is the z component of the heat-flux vector. It appears to be impossible to construct a Hamiltonian algorithm for the calculation of thermal conductivity. This is because the equations of motion so obtained are discontinuous when used in conjunction with periodic boundary conditions. We shall instead invent an external field and its coupling to the phase of the N-particle system so that the heat flux generated by this external field is trivially related to the magnitude of the heat flux induced by a real temperature gradient.

Aided by the realization that the heat-flux vector is the diffusive energy flux, computed in a co-moving coordinate frame (see equation (3.151)), we proposed the following equations of motion

$$\dot{\mathbf{q}}_i = \frac{\mathbf{p}_i}{m} \tag{6.59}$$

$$\dot{\mathbf{p}}_i = \mathbf{F}_i + (E_i - \bar{E})\mathbf{F}(t) - \tfrac{1}{2} \sum_{j=1}^{N} \mathbf{F}_{ij}(\mathbf{q}_{ij} \cdot \mathbf{F}(t)) + \frac{1}{2N} \sum_{j,k}^{N} \mathbf{F}_{jk}(\mathbf{q}_{jk} \cdot \mathbf{F}(t)) \tag{6.60}$$

where E_i is the energy of particle i, and

$$\bar{E} = \frac{1}{N} \left\{ \sum_{i=1}^{N} \frac{\mathbf{p}_i^2}{2m} + \tfrac{1}{2} \sum_{i \neq j} \phi_{ij} \right\} \tag{6.61}$$

the *instantaneous* average energy per particle.

There is no known Hamiltonian which generates these equations, but the equations do satisfy $AI\Gamma$. This means that linear response theory can be applied in a straightforward fashion. The equations of motion are momentum preserving, homogeneous and compatible with the usual periodic boundary conditions. It is clear from the term $(E_i - E^{\mathrm{bar}})\mathbf{F}(t)$ that these equations of motion will drive a heat current. A particle whose energy is greater than the average energy will experience a force in the direction of $\mathbf{F}$, while a particle whose energy is lower than average will experience a force in the $-\mathbf{F}$ direction. *Warmer* particles are driven with the field; *colder* particles are driven against the field.

If the total momentum is zero it will be conserved and the dissipation is

$$\dot{H}_0^{\mathrm{ad}} = \mathbf{F}(t)\cdot\left\{\sum_{i=1}^{N}\frac{\mathbf{p}_i E_i}{m} - \tfrac{1}{2}\sum_{i,j}^{N}\mathbf{q}_{ij}\left(\frac{\mathbf{p}_i\cdot\mathbf{F}_{ij}}{m}\right)\right\} = \mathbf{F}(t)\cdot\mathbf{J}_{\mathrm{Q}}V \tag{6.62}$$

Using linear response theory we have

$$\langle\mathbf{J}_{\mathrm{Q}}(t)\rangle = \beta V\int_0^t \mathrm{d}s\,\langle\mathbf{J}_{\mathrm{Q}}(t-s)\mathbf{J}_{\mathrm{Q}}\rangle\cdot\mathbf{F}(s) \tag{6.63}$$

Consider a field $\mathbf{F} = (0, 0, F_z)$, then taking the limit $t \to \infty$ we find that the ratio of the induced heat flux to the product of the absolute temperature and the magnitude of the external field is in fact the thermal conductivity.

$$\lambda = \frac{V}{k_{\mathrm{B}}T^2}\int_0^\infty \mathrm{d}t\,\langle J_{\mathrm{Q}_z}(t)J_{\mathrm{Q}_z}(0)\rangle = \lim_{F\to 0}\frac{\langle J_{\mathrm{Q}_z}(\infty)\rangle}{TF} \tag{6.64}$$

In the linear limit the effect that the heat field has on the system is identical to that of a logarithmic temperature gradient ($F = -\partial \ln T/\partial z$). The theoretical justification for this algorithm is tied to *linear* response theory. No meaning is known for the finite field susceptibility.

In 1983 Gillan and Dixon introduced a slightly different synthetic method for computing the thermal conductivity (Gillan and Dixon, 1983). Although their algorithm is considerably more complex to apply in computer simulations, their equations of motion look quite similar to those given above. Gillan's synthetic algorithm is of some theoretical interest since it is the only known algorithm which violates momentum conservation and $AI\Gamma$ (MacGowan and Evans, 1986b).

Figure 6.12 shows the thermal conductivity of the triple-point Lennard–Jones fluid computed as a function of the strength of the heat field. Also shown are the experimental data for argon, assuming that argon can be modelled by the standard Lennard–Jones model ($\varepsilon/k_{\mathrm{B}} = 119.8$ K, $\sigma = 3.405$ Å). The experimental uncertainties are so large that if an accurate potential function is used the thermal conductivity can be calculated more accurately than it can be measured.

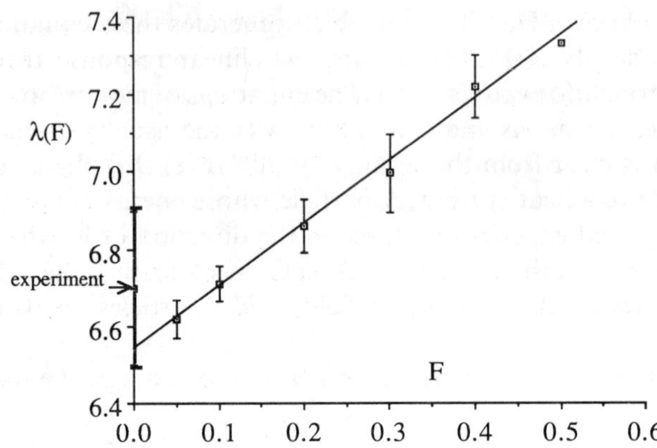

Fig. 6.12 Thermal conductivity: Lennard–Jones triple point.

6.6 NORTON ENSEMBLE METHODS

The Norton and Thévenin's theorems are of fundamental importance in electrical circuit theory (Brophy, 1966). They prove that any network of resistors and power supplies can be analysed in terms of equivalent circuits which include either ideal current or ideal voltage sources. These two theorems are an example of the macroscopic duality that exists between what are generally recognized as thermodynamic fluxes and thermodynamic forces—in the electrical circuit case, electrical currents and the electromotive force. Indeed, in our introduction to linear irreversible thermodynamics (Chapter 2), there was an apparent arbitrariness with respect to our definition of forces and fluxes. At no stage did we give a convincing macroscopic distinction between the two.

Microscopically one might think that there is a clear and unambiguous distinction which can be drawn. For an arbitrary mechanical system subject to a perturbing external field the dissipation can be written as $dH_0^{ad}/dt \equiv -J(\mathbf{\Gamma})F_e(t)$. The dissipative flux is the phase variable $J(\mathbf{\Gamma})$ and the force is the time-dependent independent variable, $F_e(t)$. This might seem to remove the arbitrariness. However, suppose that we complicate matters a little and regard the external field, $F_e(t)$, as a Gaussian multiplier in a feedback scheme designed to stop the flux, $J(\mathbf{\Gamma})$, from changing. We might wish to perform a constant current simulation. In this case the imposed *external* field, $F_e(t)$, is in fact a phase variable, $F_e(\mathbf{\Gamma})$. Even microscopically the distinction between forces and fluxes is more complex than is often thought.

In this section we will explore the statistical consequences of this duality. Until recently, the Green–Kubo relations were only known for the conventional Thévenin ensemble in which the forces are the independent state-defining variables. Here we will derive their Norton ensemble equivalents and will then show how these ideas have been applied to algorithms for isobaric molecular dynamics simulations. This work will provide the necessary background for the derivations of fluctuation expressions, for the derived properties of nonequilibrium steady states, including the nonlinear inverse Burnett coefficients (Chapter 9).

Gaussian constant-colour current algorithm

From the colour Hamiltonian (6.9) we see that the equations of motion for colour conductivity in the Thévenin ensemble are

$$\begin{aligned} \dot{\mathbf{q}}_i &= \frac{\mathbf{p}_i}{m} \\ \dot{\mathbf{p}}_i &= \mathbf{F}_i + c_i \mathbf{F}(t) \end{aligned} \tag{6.65}$$

These equations are the adiabatic version of (6.18) and (6.19). We will now treat the colour field as a Gaussian multiplier chosen to fix the colour current and introduce a thermostat.

Our first step is to redefine the momenta (Evans and Morriss, 1985), so that they are measured with respect to the species current of the particles. Consider the following equations of motion

$$\begin{aligned} \dot{\mathbf{q}}_i &= \frac{\mathbf{p}_i}{m} + \frac{c_i \mathbf{I}(t)}{\sum_{i=1}^{N} c_i^2} \\ \dot{\mathbf{p}}_i &= \mathbf{F}_i - c_i \boldsymbol{\lambda} - \alpha \mathbf{p}_i \end{aligned} \tag{6.66}$$

where α is the thermostatting multiplier and $\boldsymbol{\lambda}$ is the current multiplier. These equations are easily seen to be equivalent to (6.18) and (6.19). We distinguish two types of current, a canonical current $\mathbf{J}$ defined in terms of the canonical momenta,

$$\mathbf{J} \equiv \sum \frac{c_i \mathbf{p}_i}{m} \tag{6.67}$$

and a kinetic current $\mathbf{I}$, where

$$\mathbf{I} \equiv \sum c_i \dot{\mathbf{q}}_i \tag{6.68}$$

We choose $\boldsymbol{\lambda}$ so that the canonical current is always zero, and α so that the canonical (i.e. peculiar) kinetic energy is fixed. Our constraint equations are, therefore,

$$g_{\mathrm{d}} = \sum \left(\frac{c_i \mathbf{p}_i}{m} \right) - \mathbf{J} = 0 \tag{6.69}$$

and

$$g_{\mathrm{T}} = \frac{1}{m} \sum \mathbf{p}_i^2 - 3Nk_{\mathrm{B}}T = 0 \tag{6.70}$$

The Gaussian multipliers may be evaluated in the usual way by summing moments of the equations of motion and eliminating the accelerations using the differential forms of the constraints. We find that

$$\boldsymbol{\lambda} = \frac{\sum c_i \mathbf{F}_i}{\sum c_i^2} \tag{6.71}$$

and

$$\alpha = \frac{\sum \mathbf{F}_i \cdot \mathbf{p}_i}{\sum \mathbf{p}_i^2} \tag{6.72}$$

If we compare the Gaussian equations of motion with the corresponding Hamiltonian equations we see that the Gaussian multiplier $\boldsymbol{\lambda}$ can be identified as a fluctuating external colour field which maintains a constant-colour current. It is, however, a phase variable. Gauss' principle has enabled us to go from a constant-field nonequilibrium ensemble to the conjugate ensemble where the current is fixed. The Gaussian multiplier fluctuates in the precise manner required to fix the current. The distinction drawn between canonical and kinetic currents has allowed us to decouple the Lagrange multipliers appearing in the equations of motion. Furthermore, setting the canonical current to zero is equivalent to setting the kinetic current to the required value, **I**. This can be seen by taking the charge moment of (6.66). If the canonical current is zero then,

$$\sum c_i \dot{\mathbf{q}} = \frac{\sum c_i^2 \mathbf{I}(t)}{\sum c_i^2} = \mathbf{I}(t) \tag{6.73}$$

In this equation the current, which was formerly a phase variable has now become a possibly time-dependent external force.

In order to be able to interpret the response of this system to the external current field, we need to compare the system's equations of motion with a macroscopic constitutive relation. Under adiabatic conditions the second-order form of the equations of motion is

$$m\ddot{\mathbf{q}}_i = \mathbf{F}_i + m\dot{\mathbf{I}}(t)\frac{c_i}{\sum c_i^2} - \boldsymbol{\lambda} e_i \tag{6.74}$$

We see that to maintain a constant current $\mathbf{I}(t)$ we must apply a fluctuating colour field $\mathbf{E}_{\mathrm{eff}}$,

$$\mathbf{E}_{\mathrm{eff}}(t) = \frac{m\dot{\mathbf{I}}(t)}{\sum c_i^2} - \boldsymbol{\lambda} \tag{6.75}$$

The adiabatic rate of change of internal energy, H_0, is given by

$$\dot{H}_0^{\mathrm{ad}} = -\sum\left[\frac{c_i\mathbf{p}_i\cdot\boldsymbol{\lambda}}{m} + \frac{c_i\mathbf{F}_i\cdot\mathbf{I}}{\sum c_i^2}\right] = -\mathbf{J}\cdot\boldsymbol{\lambda} - \mathbf{I}\cdot\boldsymbol{\lambda} \tag{6.76}$$

As the current, $\mathbf{J} = \mathbf{J}(\boldsymbol{\Gamma})$, is fixed at a value of zero, the dissipation is $-\mathbf{I}(t)\cdot\boldsymbol{\lambda}(\boldsymbol{\Gamma})$. As expected the current is now an external time-dependent field, while the colour field is a phase variable. Using linear response theory we have

$$\langle\boldsymbol{\lambda}(t)\rangle = -\beta\int_0^t \mathrm{d}s\,\langle\boldsymbol{\lambda}(t-s)\boldsymbol{\lambda}\rangle\cdot\mathbf{I}(s) \tag{6.77}$$

which gives the linear response result for the phase variable component of the effective field. Combining (6.77) with (6.75) the effective field is, therefore,

$$\mathbf{E}_{\mathrm{eff}}(t) = \int_0^t \mathrm{d}s\,\chi(t-s)\mathbf{I}(s) + \frac{m\dot{\mathbf{I}}(t)}{\sum c_i^2} \tag{6.78}$$

where the susceptibility χ is the equilibrium λ autocorrelation function,

$$\chi(t) \equiv \beta\langle\lambda(t)\lambda(0)\rangle \tag{6.79}$$

By doing a Fourier–Laplace transform on (6.78) we obtain the frequency-dependent colour resistance, $\tilde{E} \equiv \tilde{R}\tilde{I}'$,

$$\tilde{R}(\omega) = \tilde{\chi}(\omega) + \frac{i\omega m}{\sum c_i^2} \tag{6.80}$$

To compare with the usual Green–Kubo relations which have always been derived for conductivities rather than resistances, we find

$$\tilde{\sigma}(\omega) = \frac{1}{V\left(\tilde{\chi}(\omega) + \dfrac{i\omega m}{\sum_{i=1}^{N} c_i^2}\right)} \tag{6.81}$$

This equation shows that the Fourier–Laplace transform of $\chi(t)$ is the memory function of the complex frequency-dependent conductivity. In the conjugate constant-force ensemble the frequency-dependent conductivity is related to the current autocorrelation function

$$\tilde{\sigma}(\omega) = \frac{1}{2Vk_BT}\int_0^\infty dt\, e^{-i\omega t}\langle \mathbf{J}(t)\cdot\mathbf{J}(0)\rangle_{\mathbf{E}=0} \tag{6.82}$$

From equations (6.79)–(6.82) we see that at *zero* frequency the colour conductivity is given by the integral of the Thévenin ensemble current correlation function, while the resistance, which is the reciprocal of the conductivity, is given by the integral of the colour field autocorrelation function computed in the Norton ensemble. Thus *at zero frequency* the integral of the Thévenin ensemble current correlation function is the reciprocal of the integral of the Norton ensemble field-correlation function. A comparison of the Norton and Thévenin algorithms for computing the colour conductivity is given in Fig. 6.2. The results obtained for the conductivity are ensemble independent—even in the nonlinear regime far from equilibrium.

In Fig. 6.12 we show the reduced colour conductivity plotted as a function of frequency (Evans and Morriss, 1985). The system is identical to the Lennard–Jones system shown in Fig. 6.2. The curves were calculated by taking the Laplace transforms of the appropriate equilibrium time-correlation functions computed in both the Thévenin and Norton ensembles. Within statistical uncertainties, the results are in agreement. The arrow shows the zero-frequency colour conductivity computed using nonequilibrium molecular dynamics (NEMD). The value is taken from Fig. 6.2.

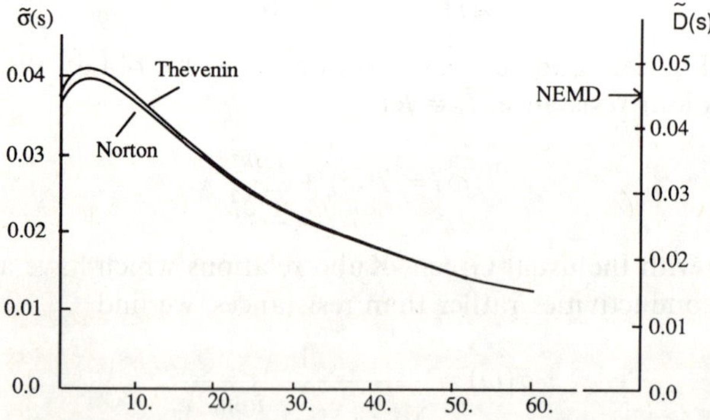

Fig. 6.13 The colour conductivity as a function of the Laplace transform variable s.

6.7 CONSTANT-PRESSURE ENSEMBLES

For its first 30 years, molecular dynamics was limited to the microcanonical ensemble. We have already seen how the development of thermostats has enabled simulations to be performed in the isochoric, canonical and isokinetic ensembles. We will now describe molecular dynamics algorithms for performing simulations at constant pressure or constant enthalpy. The technique used to make the pressure rather than the volume, the independent state-defining variable, uses essentially the same ideas as those employed in Section 6.6 to design Norton ensemble algorithms. The methods we describe now are of use for both equilibrium and nonequilibrium simulations.

It is often advantageous, particularly in studies of phase transitions, to work within the isobaric ensemble. It is possible to stabilize the pressure in a number of ways: we will describe the Gaussian method (Evans and Morriss, 1983b) since it was both the first deterministic isobaric technique to be developed and it is conceptually simpler than the corresponding Nosé–Hoover (Hoover, 1985) or Rahman–Parrinello (Parrinello and Rahman, 1980a,b, 1981) schemes. Although it may be slightly more difficult to write the computer programs for the Gaussian method, once written they are certainly easier to use. The Gaussian method has the distinct advantage that the pressure is a rigorous constant of the motion, whereas the Nosé based schemes (Nosé, 1984) and those of Parrinello and Rahman allow fluctuations in *both* the pressure and the volume.

If one makes a poor initial guess for the density, Nosé–Hoover isobaric algorithms induce sharp density changes when attempting to correct the density to that appropriate for the specified mean pressure. Because bulk oscillations damp quite slowly, Nosé–Hoover methods can easily result in the system exploding—a situation that cannot be reversed due to the finite range of the interaction potentials. Gaussian isobaric algorithms are free of these instabilities.

Isothermal–isobaric molecular dynamics

Consider the SLLOD equations of motion where the strain-rate tensor $\nabla\mathbf{u}$ is isotropic. The equations of motion become

$$\dot{\mathbf{q}}_i = \frac{\mathbf{p}_i}{m} + \dot{\varepsilon}\mathbf{q}_i \tag{6.83}$$

$$\dot{\mathbf{p}}_i = \mathbf{F}_i - \dot{\varepsilon}\mathbf{p}_i \tag{6.84}$$

Now if the system is *cold* ($\mathbf{p}_i = 0$ for all i), and non-interacting ($\phi_{ij} = 0$), these equations reduce to

$$\dot{\mathbf{q}}_i = \dot{\varepsilon}\mathbf{q}_i \tag{6.85}$$

Since this equation is true for all particles i, it describes a uniform dilation or contraction of the system. This dilation or contraction is the same in each coordinate direction, so if the system initially occupied a cube of volume V, then the volume would satisfy the following equation of motion.

$$\dot{V} = 3V\dot{\varepsilon} \tag{6.86}$$

For *warm*, interacting systems, the equation of motion for $\mathbf{q}_i$ shows that the canonical momentum $\mathbf{p}_i$ is in fact peculiar with respect to the streaming velocity $d\varepsilon/dt\, \mathbf{q}_i$. The dissipation for the system ((6.83) and (6.84)) is

$$\dot{H}_0 = -\dot{\varepsilon} \sum_{i=1}^{N} \left\{ \frac{1}{m} \mathbf{p}_i \cdot \mathbf{p}_i + \mathbf{F}_i \cdot \mathbf{q}_i \right\} = -3pV\dot{\varepsilon} \tag{6.87}$$

Since H_0 is the internal energy of the system we can combine (6.87) with the equation of motion for the volume to obtain the first law of thermodynamics for adiabatic compression,

$$\dot{H}_0 = -3pV\dot{\varepsilon} = -p\dot{V} \tag{6.88}$$

It is worth noting that these equations are true *instantaneously*. One does not need to employ any ensemble averaging to obtain (6.88). By choosing the dilation rate, $d\varepsilon/dt$, to be a sinusoidal function of time, these equations of motion can be used to calculate the bulk viscosity. Our purpose is, however, to use the dilation rate as a multiplier to maintain the system at a constant hydrostatic pressure. Before we do this we will introduce a Gaussian thermostat into the equations of motion:

$$\dot{\mathbf{q}}_i = \frac{\mathbf{p}_i}{m} + \dot{\varepsilon}\mathbf{q}_i \tag{6.89}$$

$$\dot{\mathbf{p}}_i = \mathbf{F}_i - \dot{\varepsilon}\mathbf{p}_i - \alpha\mathbf{p}_i \tag{6.90}$$

The form for the thermostat multiplier is determined by the fact that the momenta in (6.89) and (6.90) are peculiar with respect to the dilating coordinate frame. By taking the moment of (6.90) with respect to $\mathbf{p}_i$, and setting the time derivative of the peculiar kinetic energy to zero, we observe that

$$\alpha = -\dot{\varepsilon} + \frac{\sum_{i=1}^{N} \mathbf{F}_i \cdot \mathbf{p}_i}{\sum_{i=1}^{N} \mathbf{p}_i^2} \tag{6.91}$$

Differentiating the product pV (6.87) with respect to time gives

$$3\dot{p}V + 3p\dot{V} = \sum_{i=1}^{N} \left\{ \frac{2}{m} \dot{\mathbf{p}}_i \cdot \mathbf{p}_i + \dot{\mathbf{q}}_i \cdot \mathbf{F}_i + \mathbf{q}_i \cdot \frac{\partial \mathbf{F}_i}{\partial \mathbf{q}_i} \cdot \dot{\mathbf{q}}_i + \sum_{j \neq i} \mathbf{q}_i \cdot \frac{\partial \mathbf{F}_i}{\partial \mathbf{q}_j} \cdot \dot{\mathbf{q}}_j \right\} \tag{6.92}$$

The first term on the left-hand side is zero because the pressure is constant, and the first term on the right-hand side is zero because the peculiar kinetic energy is constant. Substituting the equations of motion for $d\mathbf{q}_i/dt$ and dV/dt, we can solve for the dilation rate.

$$\dot{\varepsilon} = -\frac{\dfrac{1}{2m} \sum_{i \neq j} \mathbf{q}_{ij} \cdot \mathbf{p}_{ij} \left(\phi''_{ij} + \dfrac{\phi'_{ij}}{q_{ij}} \right)}{\frac{1}{2} \sum_{i \neq j} q_{ij}^2 \left(\phi''_{ij} + \dfrac{\phi'_{ij}}{q_{ij}} \right) + 9pV} \tag{6.93}$$

Combining this equation with (6.91) gives a closed expression for the thermostat multiplier α.

In summary, our isothermal/isobaric molecular dynamics algorithm involves solving $6N + 1$ first-order equations of motion (equations (6.86), (6.89) and (6.90)). There are two subtleties to be aware of before implementing this method. Firstly, the pressure is sensitive to the long-range tail of the interaction potential. In order to obtain good pressure stability the long-range truncation of the potential needs to be handled carefully. Secondly, if a Gear predictor corrector scheme is used to integrate the equations of motion then some care must be taken in handling the higher-order derivatives of the coordinates and momenta under periodic boundary conditions. More details are given in Evans and Morriss (1983b, 1984a).

Isobaric–isoenthalpic molecular dynamics

For the adiabatic constant-pressure equations of motion we have already shown that the first law of thermodynamics for compression is satisfied

$$\dot{H}_0 = -p\dot{V} \tag{6.94}$$

It is now easy to construct equations of motion for which the enthalpy, $I = H_0 + pV$, is a constant of the motion. The constraint we wish to impose is that

$$\dot{I} = \dot{H}_0 + \dot{p}V + p\dot{V} = 0 \tag{6.95}$$

Combining these two equations we see that for our adiabatic constant-pressure equations of motion the rate of change of enthalpy is simply

$$\dot{I} = \dot{p}V \tag{6.96}$$

This equation says that if our adiabatic equations preserve the pressure then the enthalpy is automatically constant. The isobaric–isoenthalpic equations of motion are simply obtained from the isothermal–isobaric equations by dropping the constant-temperature constraint. The isoenthalpic dilation rate can be shown to be (Evans and Morriss, 1984a),

$$\dot{\varepsilon} = \frac{\dfrac{2}{m}\sum_{i=1}^{N} \mathbf{p}_i \cdot \mathbf{F}_i - \dfrac{1}{2m}\sum_{i \neq j} \mathbf{q}_{ij} \cdot \mathbf{p}_{ij}\left(\phi''_{ij} + \dfrac{\phi'_{ij}}{q_{ij}}\right)}{\dfrac{2}{m}\sum_{i=1}^{N} \mathbf{p}_i^2 + \tfrac{1}{2}\sum_{i \neq j} q_{ij}^2\left(\phi''_{ij} + \dfrac{\phi'_{ij}}{q_{ij}}\right) + 9pV} \tag{6.97}$$

6.8 CONSTANT-STRESS ENSEMBLE

We will now give another example of the usefulness of the Norton ensemble. Suppose we wish to calculate the yield stress of a Bingham plastic—a solid with a yield stress. If we use the SLLOD method outlined above a Bingham plastic will always yield, simply because the strain *rate* is an input into the simulation. If would not be easy to determine the yield stress from such a calculation. For simulating yield phenomena one would prefer the shear stress to be the input variable. If this were the case, simulations could be run for a series of incremented values of the shear stress. If the stress was less than the yield stress, the solid would strain elastically under the stress. Once the yield stress was exceeded, the material would shear.

Here we discuss a simple method for performing nonequilibrium molecular dynamics (NEMD) simulations in the stress ensemble. We will use this as an opportunity to illustrate the use of the Nosé–Hoover feedback mechanism. We will also derive linear response expressions for the viscosity within the context of the Norton ensemble. The equations of motion for shear flow, thermostatted using the Nosé–Hoover thermostat are

$$\dot{\mathbf{q}}_i = \frac{\mathbf{p}_i}{m} + \mathbf{i}\gamma y_i \tag{6.98}$$

$$\dot{\mathbf{p}}_i = \mathbf{F}_i - \mathbf{i}\gamma p_{yi} - \xi \mathbf{p}_i \tag{6.99}$$

$$\dot{\xi} = \frac{2(K(\mathbf{\Gamma}) - K_0)}{Q_\xi} = \frac{2}{\tau_\xi^2}\left(\frac{K(\mathbf{\Gamma})}{K_0} - 1\right) \tag{6.100}$$

Using the Nosé–Hoover feedback mechanism we relate the rate of change of the strain rate, γ, to the degree to which the instantaneous shear stress, $-P_{xy}(\mathbf{\Gamma})$, differs from a specified mean value, $-S_{xy}(t)$. We therefore determine the strain rate from the differential equation

$$\dot{\gamma} = \frac{(P_{xy}(\mathbf{\Gamma}) - S_{xy}(t))V}{Q_\gamma} \tag{6.101}$$

If the instantaneous stress is greater (i.e. more negative) than the specified value, the strain rate will decrease in an attempt to make the two stresses more nearly equal. The relaxation constant, Q_γ, should be chosen so that the time-scale for feedback fluctuations is roughly equal to the natural relaxation time of the system.

From the equations of motion, the time derivative of the internal energy $H_0 = \Sigma_i p_i^2/2m + \Phi$ is easily seen to be

$$\dot{H}_0 = -P_{xy}V\gamma - 2\xi K \tag{6.102}$$

The Nosé constant-stress, constant-temperature dynamics satisfy a Liouville equation in which phase space behaves as a compressible $6N + 2$-dimensional fluid. The equilibrium distribution function is a function of the $3N$-particle coordinates, the $3N$-particle momenta, the thermostatting multiplier ξ, and strain rate γ, $f_0 = f_0(\mathbf{\Gamma}, \xi, \gamma)$. The Liouville equation for this system is then

$$\frac{\mathrm{d}f_0}{\mathrm{d}t} = -f_0\left(\frac{\partial}{\partial\mathbf{\Gamma}}\cdot\dot{\mathbf{\Gamma}} + \frac{\partial}{\partial\xi}\dot{\xi} + \frac{\partial}{\partial\gamma}\dot{\gamma}\right) \tag{6.103}$$

Since

$$\dot{\xi} = \dot{\xi}(\mathbf{\Gamma})$$

and

$$\dot{\gamma} = \dot{\gamma}(\mathbf{\Gamma})$$

then

$$\frac{\partial}{\partial\xi}\dot{\xi} = \frac{\partial}{\partial\gamma}\dot{\gamma} = 0 \tag{6.104}$$

the phase-space compression factor, $\Lambda(\mathbf{\Gamma})$, is easily seen to be $-3N\xi$. If we consider the time derivative of the extended internal energy $H_0 + \frac{1}{2}Q_\xi\xi^2 + \frac{1}{2}Q_\gamma\gamma^2$ we find that

$$\begin{aligned}\frac{\mathrm{d}}{\mathrm{d}t}\left(H_0 + \tfrac{1}{2}Q_\gamma\gamma^2 + \tfrac{1}{2}Q_\xi\xi^2\right) &= \dot{H}_0 + Q_\gamma\gamma\dot{\gamma} + Q_\xi\xi\xi \\ &= -P_{xy}V\gamma - 2\xi K + 2\xi(K - K_0) \\ &\quad + (P_{xy} - S_{xy})V\gamma \\ &= -2\xi K_0 - S_{xy}V\gamma \end{aligned} \tag{6.105}$$

If we consider the situation at equilibrium when the set value of the shear stress, $-S_{xy}(t)$, is zero and $K_0 = 3N/2\beta$, the Liouville equation becomes

$$\frac{\mathrm{d}f_0}{\mathrm{d}t} = 2\beta\xi K_0 f_0 = -\beta f_0 \frac{\mathrm{d}(H_0 + \frac{1}{2}Q_\gamma\gamma^2 + \frac{1}{2}Q_\xi\xi^2)}{\mathrm{d}t} \tag{6.106}$$

Integrating both sides with respect to time gives the equilibrium distribution function for the constant-stress Norton ensemble.

$$f_0 = \frac{\exp[-\beta(H_0 + \frac{1}{2}Q_\gamma\gamma^2 + \frac{1}{2}Q_\xi\xi^2)]}{\int \mathrm{d}\mathbf{\Gamma} \int \mathrm{d}\gamma \int \mathrm{d}\xi \exp[-\beta(H_0 + \frac{1}{2}Q_\gamma\gamma^2 + \frac{1}{2}Q_\xi\xi^2)]} \tag{6.107}$$

The equilibrium distribution function is thus a generalized canonical distribution, permitting strain-rate fluctuations. Indeed the mean-square strain rate is

$$\langle\gamma^2\rangle_{S_{xy}=0} = \frac{1}{\beta Q_\gamma} \tag{6.108}$$

so the amplitude of the strain-rate fluctuations is controlled by the adjustable constant Q_γ.

We wish to calculate the linear response of an equilibrium ensemble of systems (characterized by the distribution f_0, at time $t = 0$) to an externally imposed time-dependent shear stress, $-S_{xy}(t)$. For the Nosé–Hoover feedback mechanism the external field is the mean shear stress, and it appears explicitly in the equations of motion (Hood *et al.*, 1987). This is in contrast to the more difficult Gaussian case (Brown and Clarke, 1986). For the Gaussian feedback mechanism the numerical value of the constraint variable does not usually appear explicitly in the equations of motion. This is a natural consequence of the differential nature of the Gaussian feedback scheme.

The linear response of an arbitrary phase variable $B(\mathbf{\Gamma})$ to an applied time-dependent external field is given by

$$\langle B(t)\rangle = \langle B(0)\rangle - \int_0^t \mathrm{d}s \int \mathrm{d}\mathbf{\Gamma}\, B(\mathbf{\Gamma}) \exp(-iL_0(t-s)) i\Delta L(s) f_0(\mathbf{\Gamma}) \tag{6.109}$$

where iL_0 is the equilibrium (Nosé–Hoover thermostatted) f-Liouvillean and $i\Delta L(s) = iL(s) - iL_0$ where $iL(s)$ is the full-field-dependent thermostatted f-Liouvillean. It only remains to calculate $i\Delta L(s) f_0$. Using the equations of motion and the equilibrium distribution function obtained previously, we see that

$$\begin{aligned} i\Delta L(s) f_0 &= \left[\dot{\mathbf{\Gamma}} \cdot \frac{\partial}{\partial \mathbf{\Gamma}} + \dot{\xi} \frac{\partial}{\partial \xi} + \dot{\gamma} \frac{\partial}{\partial \gamma}\right] f_0 + f_0 \left[\frac{\partial}{\partial \mathbf{\Gamma}} \cdot \dot{\mathbf{\Gamma}} + \frac{\partial}{\partial \xi} \dot{\xi} + \frac{\partial}{\partial \gamma} \dot{\gamma}\right] \\ &= -\beta(\dot{H}_0 + \dot{\gamma}\gamma Q_\gamma + \dot{\xi}\xi Q_\xi) f_0 - 3N\xi f_0 \\ &= \beta V S_{xy}(t) \gamma(\mathbf{\Gamma}) f_0 \end{aligned} \tag{6.110}$$

Here we make explicit reference to the phase dependence of γ, and the explicit time dependence of the external field $S_{xy}(t)$. The quantity $-S_{xy}(t) V \gamma(\mathbf{\Gamma})$ is the adiabatic derivative of the extended internal energy, $E = H_0 + \frac{1}{2} Q_\gamma \gamma^2$.

Combining these results the linear response of the phase variable B is

$$\langle B(t)\rangle = \langle B(0)\rangle - \beta V \int_0^t \mathrm{d}s\, \langle B(t-s)\gamma\rangle_0 S_{xy}(s) \tag{6.111}$$

In order to compute the shear viscosity of the system we need to calculate the time dependence of the thermodynamic force and flux which appear in the defining constitutive relation for shear viscosity. Because of the presence of the Nosé–Hoover relaxation time, which is controlled by the parameter Q_γ, the actual shear stress in the system, $-P_{xy}(\mathbf{\Gamma})$, does not match the externally imposed shear stress, $S_{xy}(t)$, instantaneously. To compute the shear viscosity we need to know the precise relation between P_{xy} and γ, not that between S_{xy} and the strain rate. The two quantities of interest are easily computed from (6.111).

$$\langle \gamma(t)\rangle = -\beta V \int_0^t \mathrm{d}s\, \langle \gamma(t-s)\gamma\rangle_0 S_{xy}(s) \tag{6.112}$$

$$\langle P_{xy}(t)\rangle = -\beta V \int_0^t \mathrm{d}s\, \langle P_{xy}(t-s)\gamma\rangle_0 S_{xy}(s) \tag{6.113}$$

Fourier–Laplace transforming we obtain the frequency-dependent linear response relations

$$\langle\tilde{\gamma}(\omega)\rangle = -\tilde{\chi}_{\gamma\gamma}(\omega)\tilde{S}_{xy}(\omega) \tag{6.114}$$

$$\langle\tilde{P}_{xy}(\omega)\rangle = -\tilde{\chi}_{P_{xy}\gamma}(\omega)\tilde{S}_{xy}(\omega) \tag{6.115}$$

where the Fourier–Laplace transform of $\chi(t)$ is defined as

$$\tilde{\chi}_{\mathbf{AB}}(\omega) = \int_0^\infty \mathrm{d}t\,\exp(-i\omega t)\chi_{\mathbf{AB}}(t) = \beta V\int_0^\infty \mathrm{d}t\,\exp(-i\omega t)\langle A(t)B\rangle_0 \tag{6.116}$$

The linear constitutive relation for the frequency-dependent shear viscosity is (Section 2.4)

$$\tilde{P}_{xy}(\omega) \equiv -\tilde{\eta}(\omega)\gamma(\omega) \tag{6.117}$$

so that the frequency-dependent viscosity is

$$\tilde{\eta}(\omega) = -\frac{\tilde{\chi}_{P_{xy}\gamma}(\omega)}{\tilde{\chi}_{\gamma\gamma}(\omega)} \tag{6.118}$$

This expression shows that the complex frequency-dependent shear viscosity is given by ratio of two susceptibilities. However, these two different time-correlation functions can be related by using the Nose´–Hoover equation of motion (6.101)

$$\begin{aligned}\dot{\chi}_{\gamma\gamma}(t) &= \beta V\langle\dot{\gamma}(t)\gamma\rangle_0\\ &= \frac{\beta V^2}{Q_\gamma}\langle P_{xy}(t)\gamma\rangle_0\\ &= \frac{V}{Q_\gamma}\chi_{P_{xy}\gamma}(t)\end{aligned} \tag{6.119}$$

In the frequency domain this relation becomes

$$\begin{aligned}\frac{V}{Q_\gamma}\tilde{\chi}_{P_{xy}\gamma}(\omega) &= -\chi_{\gamma\gamma}(t=0) + i\omega\tilde{\chi}_{\gamma\gamma}(\omega)\\ &= \frac{V}{Q_\gamma} + i\omega\tilde{\chi}_{\gamma\gamma}(\omega)\end{aligned} \tag{6.120}$$

The frequency-dependent shear viscosity in the constant-stress ensemble can be written as

$$\tilde{\eta}(\omega) = -\frac{1 + \dfrac{i\omega Q_\gamma\tilde{\chi}_{\gamma\gamma}(\omega)}{V}}{\tilde{\chi}_{\gamma\gamma}(\omega)} \tag{6.121}$$

In a similar way it is possible to write the frequency-dependent viscosity in terms of either the Norton ensemble stress autocorrelation function, or the Norton ensemble stress–strain cross-correlation function. Using (4.10) the stress autocorrelation function can be related to the strain autocorrelation function using the relation

$$\frac{d^2}{dt^2}\chi_{\gamma\gamma}(t) = -\frac{V^2}{Q_\gamma^2}\chi_{P_{xy}P_{xy}}(t) \tag{6.122}$$

In the frequency domain this becomes

$$\tilde{\chi}_{\gamma\gamma}(\omega) = -\frac{V}{i\omega Q_\gamma}\left(1 + \frac{V}{i\omega Q_\gamma}\tilde{\chi}_{P_{xy}P_{xy}}(\omega)\right) \tag{6.123}$$

Substituting this equation into (6.121) gives

$$\tilde{\eta}(\omega) = \frac{-\tilde{\chi}_{P_{xy}P_{xy}}(\omega)}{1 + \dfrac{V}{i\omega Q_\gamma}\tilde{\chi}_{P_{xy}P_{xy}}(\omega)} \tag{6.124}$$

In terms of the cross-correlation function, the frequency-dependent viscosity is

$$\tilde{\eta}(\omega) = -\frac{i\omega Q_\gamma}{V}\frac{\tilde{\chi}_{P_{xy}\gamma}(\omega)}{\tilde{\chi}_{P_{xy}\gamma}(\omega) - 1} \tag{6.125}$$

In Fig. 6.14 we show the results of a test of the theory given above. Hood *et al.* (1987) computed the strain-rate autocorrelation function in the Norton ensemble and the stress autocorrelation function in the Thévenin ensemble. They then used equation (6.121) to *predict* the strain-rate autocorrelation function on the basis of their Thévenin ensemble data. The system studied was the Lennard–Jones triple-point fluid. The smooth curves denote the autocorrelation function computed in the Norton ensemble and the points give the predictions from the Thévenin ensemble data. The two sets of data are in statistical agreement. This analysis shows that, in spite of the fact that the damping constant, Q_γ, has a profound influence on the time-dependent fluctuations in the system, the theory given above correctly relates the Q_γ-dependent fluctuations of strain rate and stress to the Q_γ-*independent*, frequency-dependent viscosity.

Fig. 6.15 shows the various Norton ensemble susceptibilities as a function of frequency for the same system.

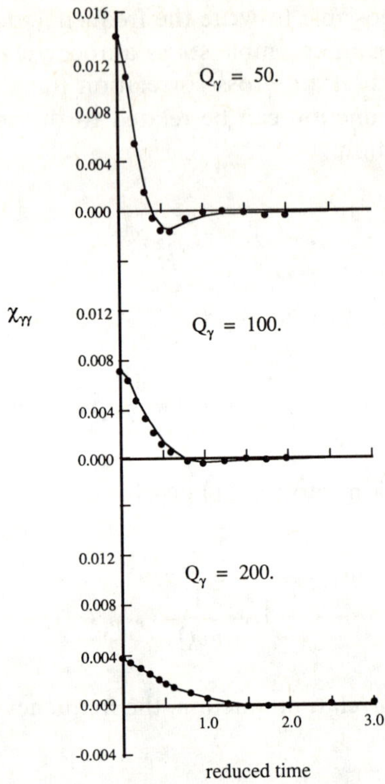

Fig. 6.14 A test of equation (6.121), for the Lennard–Jones triple-point fluid.

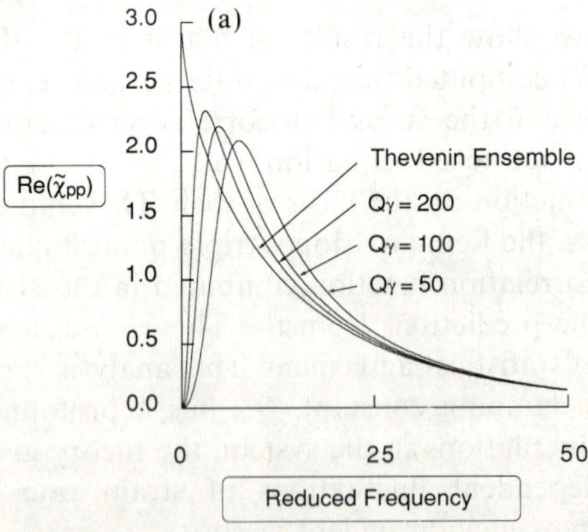

Fig. 6.15 The various Norton ensemble susceptibilities as a function of frequency. The system is the Lennard–Jones triple-point fluid.

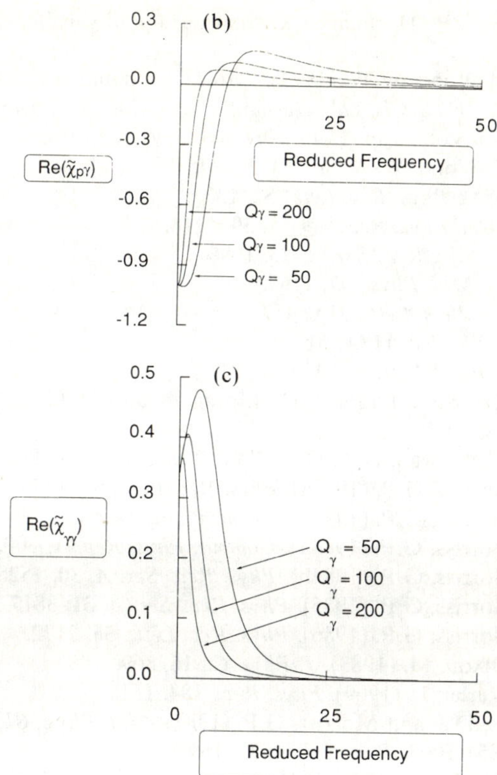

Fig. 6.15 *Continued.*

REFERENCES

Alder, B.J. and Wainwright, T.E. (1958). *Proceedings of the International Symposium on Statistical Mechanical Theory of Transport Processes* (Brussels 1956) (I. Prigogine, ed.), New York: Interscience.

Allen, M.P. and Tildesley, D.J. (1987). *Computer Simulation of Liquids*. Oxford: Clarendon Press.

Brophy, J.J. (1966). *Basic Electronics for Scientists*. New York: McGraw-Hill.

Brown, D. and Clarke, J.H.R. (1986). *Phys. Rev., Ser. A*, **34**, 2093.

de Schepper, I.M., Haffmans, A.F. and van Beijeren, H. (1986). *Phys. Rev. Lett.*, **57**, 1715.

Doetsch, G. (1961). *Guide to the Applications of Laplace Transforms*. New York: Van Nostrand.

Edberg, R.A., Morriss, G.P. and Evans, D.J. (1987). *J. Chem. Phys.*, **86**, 4555.

Ernst, M.H., Cichocki, B., Dorfman, J.R., Sharma, J. and van Beijeren, H. (1978). *J. Stat. Phys.*, **18**, 237.

Erpenbeck, J.J. and Wood, W.W. (1977). 'Molecular dynamics techniques for hard core particles'. In: *Statistical Mechanics, B: Modern Theoretical Chemistry* (B.J. Berne, ed.), Vol. 6, pp. 1–40. New York: Plenum Press.

Erpenbeck, J.J. and Wood, W.W. (1981). *J. Stat. Phys.*, **22**, 81.

Erpenbeck, J.J. (1984). *Phys. Rev. Lett.*, **52**, 1333.

Erpenbeck, J.J. (1989). *Phys. Rev., Ser. A*, **39**, 4718.

Evans, D.J. (1980). *Phys. Rev., Ser. A*, **23**, 1988.

Evans, D.J. (1982a). *Mol. Phys.*, **47**, 1165.

Evans, D.J. (1982b). *Phys. Lett.*, **91A**, 457.

Evans, D.J. (1983). *Physica*, **118A**, 51.

Evans, D.J. (1984). *Phys. Lett.*, **101A**, 100.

Evans, D.J., Hoover, W.G., Failor, B.H., Moran, B. and Ladd, A.J.C. (1983). *Phys. Rev., Ser. A*, **28**, 1016.

Evans, D.J. and MacGowan, D. (1987). *Phys. Rev., Ser. A*, **36**, 948.

Evans, D.J. and Morriss, G.P. (1983a). *Phys. Rev. Lett.*, **51**, 1776.

Evans, D.J. and Morriss, G.P. (1983b). *Chem. Phys.*, **77**, 63.

Evans, D.J. and Morriss, G.P. (1984a). *Comput. Phys. Rep.*, **1**, 300.

Evans, D.J. and Morriss, G.P. (1984b). *Phys. Rev., Ser. A*, **30**, 1528.

Evans, D.J. and Morriss, G.P. (1985). *Phys. Rev., Ser. A*, **31**, 3817.

Evans, D.J. and Morriss, G.P. (1986). *Phys. Rev. Lett.*, **56**, 2172.

Gillan, M.J. and Dixon, M. (1983). *J. Phys. C.*, **16**, 869.

Hansen, J.P. and Verlet, L. (1969). *Phys. Rev.*, **184**, 151.

Hood, L.M., Evans, D.J. and Morriss, G.P. (1987). *Mol. Phys.*, **62**, 419.

Hoover, W.G. (1985). *Phys. Rev., Ser. A*, **31**, 1695.

Hoover, W.G. and Ashurst, W.T. (1975). *Adv. Theor. Chem.*, **1**, 1.

Hoover, W.G., Evans, D.J., Hickman, R.B., Ladd, A.J.C., Ashurst, W.T. and Moran, B. (1980). *Phys. Rev., Ser. A*, **22**, 1690.

Kawasaki, K. and Gunton, J.D. (1973). *Phys. Rev., Ser. A*, **8**, 2048.

Kirkpatrick, T.R. (1984). *Phys. Rev. Lett.*, **53**, 1735.

Lees, A.W. and Edwards, S.F. (1972). *J. Phys., Ser. C*, **5**, 1921.

Levesque, D. and Verlet, L. (1970). *Phys. Rev., Ser. A*, **2**, 2514.

MacGowan, D. and Evans, D.J. (1986a). *Phys. Rev., Ser. A*, **34**, 2133.

MacGowan, D. and Evans, D.J. (1986b). *Phys. Lett.*, **A117**, 414.

Morriss, G.P. and Evans, D.J. (1989). *Phys. Rev., Ser. A*, **39**, 6335.

Nosé, S. (1984). *Mol. Phys.*, **52**, 255.

Parrinello, M. and Rahman, A. (1980a). *J. Chem. Phys.*, **76**, 2662.

Parrinello, M. and Rahman, A. (1980b). *Phys. Rev. Lett.*, **45**, 1196.

Parrinello, M. and Rahman, A. (1981). *J. Appl. Phys.*, **52**, 7182.

Pomeau, Y. and Resibois, P. (1975). *Phys. Rep.*, **19**, 63.

Rainwater, J.C. and Hanley, H.J.M. (1985). *Int. J. Thermophys.*, **6**, 595.

Rainwater, J.C., Hanley, H.J.M., Paskiewicz, T. and Petru, Z. (1985). *J. Chem. Phys.*, **83**, 339.

Tenenbaum, A., Ciccotti, G. and Gallico, R. (1982). *Phys. Rev., Ser. A*, **25**, 2778.

van Beijeren, H. (1984). *Phys. Lett.*, **105A**, 191.

Yamada, Y. and Kawasaki, K. (1973). *Prog. Theor. Phys.*, **53**, 1111.

7 Nonlinear Response Theory

7.1 KUBO'S FORM FOR THE NONLINEAR RESPONSE

In Chapter 6 we saw that nonequilibrium molecular dynamics leads inevitably to questions regarding the nonlinear response of systems. In this chapter we will begin a discussion of this subject.

It is not widely known that in his original paper Kubo (1957) did not only present results for adiabatic linear response theory, but that he also included a formal treatment of the adiabatic nonlinear response. The reason why this fact is not widely known is that, like many treatments of nonlinear response theory that followed, his formal results were exceedingly difficult to translate into useful, experimentally verifiable forms. This difficulty can be traced to three sources. Firstly, his results are not easily transformable into explicit representations that involve the evaluation of time correlation functions of explicit phase variables. Secondly, if one wants to study nonequilibrium steady states, the treatment of thermostats is mandatory. His theory did not include such effects. Thirdly, his treatment gave a power-series representation of the nonlinear response. We now believe that for most transport processes such expansions do not exist.

We will now give a presentation of Kubo's expansion for the nonequilibrium distribution function, $f(t)$. Consider an N-particle system evolving under the following dynamics,

$$\dot{\mathbf{q}}_i = \frac{\mathbf{p}_i}{m} + \mathbf{C}_i(\mathbf{\Gamma})F_e$$

$$\dot{\mathbf{p}}_i = \mathbf{F}_i + \mathbf{D}_i(\mathbf{\Gamma})F_e \tag{7.1}$$

The terms $\mathbf{C}_i(\mathbf{\Gamma})$ and $\mathbf{D}_i(\mathbf{\Gamma})$ describe the coupling of the external field, F_e, to the system. In this discussion we will limit ourselves to the case where the field is *switched on* at time zero, and thereafter remains at the same steady value. The f-Liouvillean is given by

$$iL = \dot{\mathbf{\Gamma}} \cdot \frac{\partial}{\partial \mathbf{\Gamma}} + \frac{\partial}{\partial \mathbf{\Gamma}} \cdot \dot{\mathbf{\Gamma}} = iL_0 + i\Delta L \tag{7.2}$$

where iL_0 is the equilibrium Liouvillean and $i\Delta L$ is the field-dependent perturbation which is a linear function of F_e. The Liouville equation is

$$\frac{\partial f(t)}{\partial t} = -iLf(t) \tag{7.3}$$

To go beyond the linear response treated in Section 5.1, Kubo assumed that $f(t)$ could be expanded as a power series in the external field, F_e,

$$f(t) = f_0 + f_1(t) + f_2(t) + f_3(t) + f_4(t) + \dots \tag{7.4}$$

where, $f_i(t)$ is the ith order in the external field F_e. The assumption that $f(t)$ can be expanded in a power series about $F_e = 0$ may seem innocent, but it is not. This assumption rules out any functional form containing a term of the form, F_e^α, where α is not an integer. Substituting (7.4) for $f(t)$, and equation (7.2) into the Liouville equation (7.3), and equating terms of the same order in F_e, we obtain an infinite sequence of partial differential equations to be solved

$$\frac{\partial f_i(t)}{\partial t} + iL_0 f_i(t) = -i\Delta L f_{i-1}(t) \tag{7.5}$$

where $i \geq 1$. The solution to this series of equations can be written as

$$f_i(t) = -\int_0^t ds \exp(-iL_0(t-s)) i\Delta L f_{i-1}(s) \tag{7.6}$$

To prove that this is correct, one differentiates both sides of the equation to obtain (7.5). Recursively substituting (7.6) into (7.4), we obtain a power-series representation of the distribution function

$$f(t) = f(0) + \sum_{i=1}^{\infty} (-1)^i \int_0^t ds_i \int_0^{s_{i-1}} ds_{i-1} \dots \int_0^{s_2} ds_1 \, e^{-iL_0(t-s)} \Delta L \dots e^{-iL_0(s_2-s_1)} \Delta L f(0) \tag{7.7}$$

Although this result is formally exact, there are a number of difficulties with this approach. The expression for $f(t)$ is a sum of convolutions of operators. In general, the operator $i\Delta L$ does not commute with the propagator $\exp(iL_0 t)$, and no further simplifications of the general result are possible. Furthermore, as we have seen in Chapter 6, there is a strong likelihood that fluxes associated with conserved quantities are nonanalytic functions of the thermodynamic force, F_e. This would mean that the average response of the shear stress, for example, cannot be expanded as a Taylor series about $F_e(=\gamma) = 0$. In Chapter 6 we saw evidence that the shear stress is of the form, $\langle P_{xy} \rangle = -\gamma(\eta_0 + \eta_1 \gamma^{1/2})$ (see Section 6.3). If this is true then $f_2(t)(\equiv \frac{1}{2}\gamma^2 \partial^2 f(\gamma)/\partial\gamma^2|_{\gamma=0})$ must be infinite.

7.2 KAWASAKI DISTRIBUTION FUNCTION

An alternative approach to nonlinear response theory was pioneered by Yamada and Kawasaki (1967). Rather than developing power-series expansions about $F_e = 0$, these authors derived a closed expression for the perturbed distribution function. The power of their method was demonstrated in a series of papers in which Kawasaki first predicted the nonanalytic nature of the shear viscosity with respect to strain rate (Kawasaki and Gunton, 1973; Yamada and Kawasaki, 1975). This work predates the first observation of these effects in computer simulations. The simplest application of the Kawasaki method is to consider the adiabatic response of a canonical ensemble of N-particle systems to a steady applied field F_e.

The Liouville equation for this system is

$$\frac{\partial f}{\partial t} = -iLf \tag{7.8}$$

The Liouvillean appearing in this equation is the field-dependent Liouvillean defined by the equations of motion (7.1). Equation (7.8) has the formal solution

$$f(t) = \exp[-iLt]f(0) \tag{7.9}$$

For simplicity we take the initial distribution function, $f(0)$, to be canonical, so that $f(t)$ becomes

$$f(t) = \mathrm{e}^{-iLt} \frac{\mathrm{e}^{-\beta H_0}}{\int d\mathbf{\Gamma}\, \mathrm{e}^{-\beta H_0}} \tag{7.10}$$

The *adiabatic* distribution function propagator is the Hermitian conjugate of the phase-variable propagator, so in this case $\exp(-iLt)$ is the negative-time phase-variable propagator, $\exp(iL(-t))$. This function operates on the phase variable in the numerator, moving time backwards in the presence of the applied field. This implies that

$$f(t) = \frac{\mathrm{e}^{-\beta H_0(-t)}}{\int d\mathbf{\Gamma}\, \mathrm{e}^{-\beta H_0}} \tag{7.11}$$

Formally, the f-propagator leaves the denominator invariant since it is not a phase variable. The phase dependence of the denominator has been integrated out. However, since the distribution function must be normalized, we can obviously also write,

$$f(t) = \frac{e^{-\beta H_0(-t)}}{\int d\mathbf{\Gamma}\, e^{-\beta H_0(-t)}} \tag{7.12}$$

This equation is an explicitly normalized version of (7.11) and we will have more to say concerning the relations between the so-called *bare* Kawasaki form, (7.11), and the renormalized Kawasaki form, (7.12), for the distribution function in Section 7.7. In Kawasaki's original papers he referred only to the *bare* Kawasaki form, (7.11).

Using the equations of motion (7.1) one can write the time derivative of H_0 as the product of a phase variable $J(\mathbf{\Gamma})$ and the magnitude of the perturbing external field, F_e.

$$\dot{H}_0^{ad} = -J(\mathbf{\Gamma})F_e \tag{7.13}$$

For the specific case of planar Couette flow, we saw in Section 6.2 that $(dH_0/dt)^{ad}$ is the product of the strain rate, the shear stress and the system volume, $-\gamma P_{xy}V$ and thus in the absence of a thermostat we can write,

$$H_0(-t) = H_0(0) - \int_0^t ds\, \dot{H}_0(-s) = H_0(0) + \gamma V \int_0^t ds\, P_{xy}(-s) \tag{7.14}$$

The *bare* form for the perturbed distribution function at time t is then

$$f(t) = \exp\left[-\beta\gamma V \int_0^t ds\, P_{xy}(-s)\right] f(0) \tag{7.15}$$

It is important to remember that the generation of $P_{xy}(-s)$ from $P_{xy}(0)$ is controlled by the field-dependent equations of motion.

A major problem with this approach is that in an adiabatic system the applied field will cause the system to heat up. This process continues indefinitely and a steady state can never be reached. What is surprising is that when the effects of a thermostat are included, the formal expression for the N-particle distribution function remains unaltered, the only difference being that thermostatted, field-dependent dynamics must be used to generate $H_0(-t)$ from $H_0(0)$. This is the next result we shall derive.

Consider an isokinetic ensemble of N-particle systems subject to an applied field. We will assume field dependent, Gaussian isokinetic equations of motion, (5.66). The f-Liouvillean therefore contains an extra thermostatting term. It is convenient to write the Liouville equation in operator form

$$\frac{\partial f(t)}{\partial t} = -iLf(t) = -i\text{L}f(t) - f(t)\Lambda = -\frac{\partial}{\partial \mathbf{\Gamma}}\cdot(\dot{\mathbf{\Gamma}} f) \tag{7.16}$$

The operator iL is the f-Liouvillean, and iL is the p-Liouvillean. The term Λ,

$$\Lambda = \frac{\partial}{\partial \mathbf{\Gamma}} \cdot \dot{\mathbf{\Gamma}} = -\frac{1}{f}\frac{\mathrm{d}f}{\mathrm{d}t} = -\frac{\mathrm{d}\ln f}{\mathrm{d}t} \tag{7.17}$$

is the phase-space compression factor (Section 3.3). The formal solution of the Liouville equation is given by

$$f(t) = \exp[-iLt]f(0) = \exp[-(\mathrm{iL} + \Lambda)t]f(0) \tag{7.18}$$

In the thermostatted case the p-propagator is no longer the Hermitian conjugate of the f-propagator.

We will use the Dyson decomposition derived in Section 3.6, to relate thermostatted p- and f-propagators. We assume that both the p-Liouvilleans have no explicit time dependence. We make a crucial observation, namely that the phase-space compression factor, Λ, is a phase variable rather than an operator. Taking the reference Liouvillean, iL_0, to be the adjoint of iL, we find

$$\exp[-\mathrm{iL}t - \Lambda t] = \exp[-\mathrm{iL}t] - \int_0^t \mathrm{d}s \exp[-\mathrm{iL}s - \Lambda s]\Lambda \exp[-\mathrm{iL}(t-s)] \tag{7.19}$$

Repeated application of the Dyson decomposition to $\exp[-\mathrm{iL}s - \Lambda s]$ on the right-hand side of (7.19) gives

$$\begin{aligned}
&\exp[-\mathrm{iL}t - \Delta] \\
&= \sum_{n=0}^{\infty} (-)^n \int_0^t \mathrm{d}s_1 \ldots \int_0^{s_{n-1}} \mathrm{d}s_n \exp[-\mathrm{iL}s_n]\Lambda \exp[-\mathrm{iL}(s_{n-1} - s_n)]\Lambda \ldots \exp[-\mathrm{iL}(t - s_1)] \\
&= \sum_{n=0}^{\infty} (-)^n \int_0^t \mathrm{d}s_1 \ldots \int_0^{s_{n-1}} \mathrm{d}s_n \Lambda(-s_n)\Lambda(-s_{n-1}) \ldots \Lambda(-s_1) \exp[-\mathrm{iL}t] \\
&= \exp\left[-\int_0^t \mathrm{d}s\, \Lambda(-s)\right] \exp[-\mathrm{iL}t]
\end{aligned} \tag{7.20}$$

In deriving the second line of this equation we use the fact for any phase variable, B, $\exp[-\mathrm{iL}s]B = B(-s)\exp[-\mathrm{iL}s]$. Substituting (7.20) into (7.18) and choosing $f(0) = f_T(0) = \delta(K - K_0)\exp(-\beta\Phi)/Z(\beta)$, we obtain

$$f(t) = \frac{\delta(K - K_0)\exp\left[-\int_0^t \mathrm{d}s\, \Lambda(-s)\right]\exp(-\beta\Phi(-t))}{Z(\beta)} \tag{7.21}$$

If we change variables in the integral of the phase-space-compression factor and calculate $\Phi(-t)$ from its value at time zero, we obtain

$$f(t) = \frac{\delta(K - K_0)\exp(-\beta\Phi(0))\exp\left[\int_0^{-t} \mathrm{d}s\, \Lambda(s) - \beta\dot{\Phi}(s)\right]}{Z(\beta)} \tag{7.22}$$

We know that for the isokinetic distribution, $\beta = 3N/2K$ (see Section 5.2). Since under the isokinetic equations of motion K is a constant of the motion, we can prove from (5.66) that

$$\Lambda - \beta\, \mathrm{d}\Phi/\mathrm{d}t = -\beta J F_e \tag{7.23}$$

If $AI\Gamma$ is satisfied, the dissipative flux, J, is defined by (7.13). Substituting (7.23) into (7.22) we find that the *bare* form of the thermostatted Kawasaki distribution function can be written as

$$\begin{aligned} f_T(t) &= f_T(0)\exp\left[\beta\int_0^{-t} \mathrm{d}s\, J(s)F_e\right] \\ &= f_T(0)\exp\left[-\beta\int_0^{t} \mathrm{d}s\, J(-s)F_e\right] \end{aligned} \tag{7.24}$$

Formally this equation is identical to the adiabatic response (7.15), This is despite the fact that the thermostat changes the equations of motion. The adiabatic and thermostatted forms are identical because the changes caused by the thermostat to the dissipation ($\mathrm{d}H_0/\mathrm{d}t$), are *exactly* cancelled by the changes caused by the thermostat to the form of the Liouville equation. This observation was first made by Morriss and Evans (1985). Clearly one can renormalize the thermostatted form of the Kawasaki distribution function giving (7.25), as the renormalized form of the isokinetic Kawasaki distribution function, $f_{Tr}(t)$.

$$f_{Tr}(t) = \frac{\exp\left[-\beta\int_0^{t} \mathrm{d}s\, J(-s)F_e\right] f_T(0)}{\left\langle \exp\left[-\beta\int_0^{t} \mathrm{d}s\, J(-s)F_e\right]\right\rangle} \tag{7.25}$$

As we will see, the renormalized Kawasaki distribution function is very useful for deriving relations between steady-state fluctuations and derivatives of steady-state phase deriving relations between steady-state fluctuations and derivatives of steady-state phase averages. However, it is not useful for computing nonequilibrium averages themselves. This is because it involves averaging exponentials of integrals which are extensive. We will now turn to an alternative approach to this problem.

7.3 THE TRANSIENT TIME CORRELATION FUNCTION FORMALISM

The transient time correlation function (TTCF) formalism provides perhaps the simplest nonlinear generalization of the Green–Kubo relations. A number of authors independently derived the TTCF expression for adiabatic phase averages (Visscher, 1974; Dufty and Lindenfeld, 1979; Cohen, 1983). We will illustrate the derivation for isokinetic planar Couette flow. However, the formalism is quite general and can easily be applied to other systems. The theory gives an exact relation between the nonlinear steady-state response and the so-called transient time-correlation functions. We will also describe the links between the TTCF approach and the Kawasaki methods outlined in Section 7.2. Finally, we will present some numerical results which were obtained as tests of the validity of the TTCF formalism.

Following Morriss and Evans (1987), we will give our derivations using the Heisenberg, rather than the customary Schrödinger picture. The average of a phase variable, $B(\boldsymbol{\Gamma})$, at time t is

$$\langle B(t) \rangle = \int \mathrm{d}\boldsymbol{\Gamma}\, B(\boldsymbol{\Gamma}) f(t) = \int \mathrm{d}\boldsymbol{\Gamma}\, f(0) B(\boldsymbol{\Gamma}; t) \tag{7.26}$$

where the second equality is a consequence of the Schrödinger–Heisenberg equivalence. For *time-independent* external fields, differentiating the Heisenberg form with respect to time yields

$$\frac{\mathrm{d}\langle B(t) \rangle}{\mathrm{d}t} = \int \mathrm{d}\boldsymbol{\Gamma}\, f(0) \dot{\boldsymbol{\Gamma}} \cdot \frac{\partial B(t)}{\partial \boldsymbol{\Gamma}} \tag{7.27}$$

In deriving (7.27) we have used the fact that, $\mathrm{d}B(t)/\mathrm{d}t = \mathrm{iL}\exp(\mathrm{iL}t)B = \exp(\mathrm{iL}t)\mathrm{iL}B$. This relies upon the time independence of the Liouvillean, L. The corresponding equation for the time-dependent case, is not true. Integrating (7.27) by parts, we see that

$$\frac{\mathrm{d}\langle B(t) \rangle}{\mathrm{d}t} = -\int \mathrm{d}\boldsymbol{\Gamma}\, B(t) \frac{\partial}{\partial \boldsymbol{\Gamma}} \cdot (\dot{\boldsymbol{\Gamma}} f(0)) \tag{7.28}$$

The boundary term vanishes because: the distribution function, $f(0)$, approaches zero when the magnitude of any component of any particle's momentum becomes infinite, and because the distribution function can be taken to be a periodic function of the particle coordinates. We are explicitly using the periodic boundary conditions used in computer simulations.

Integrating (7.28) with respect to time we see that the nonlinear nonequilibrium response can be written as

$$\langle B(t)\rangle = \langle B(0)\rangle - \int_0^t \mathrm{d}s \int \mathrm{d}\mathbf{\Gamma}\, B(s)\frac{\partial}{\partial \mathbf{\Gamma}}\cdot \dot{\mathbf{\Gamma}} f(0) \tag{7.29}$$

The dynamics implicit in $B(s)$ is of course driven by the full field-dependent, thermostatted equations of motion ((7.1) and (7.2)). For a system subject to the thermostatted shearing deformation, $\mathrm{d}\mathbf{\Gamma}/\mathrm{d}t$ is given by the thermostatted SLLOD equations of motion (6.44).

If the initial distribution is Gaussian isokinetic it is straightforward to show that $(\partial/\partial\mathbf{\Gamma})\cdot(f(0)\,\mathrm{d}\mathbf{\Gamma}/\mathrm{d}t) = \beta V P_{xy} f(0)$. If the initial ensemble is canonical then, to first order in the number of particles, $(\partial/\partial\mathbf{\Gamma})\cdot(f(0)\,\mathrm{d}\mathbf{\Gamma}/\mathrm{d}t)$ is $\beta V P_{xy} f(0)$. To show this one writes (following Section 5.3),

$$\langle B(t)\rangle_{\mathrm{C}} = \langle B(0)\rangle_{\mathrm{C}} - \beta\gamma V \int_0^t \mathrm{d}s \left\langle B(s)\left[P_{xy}(0) - P_{xy}^K(0)\frac{\Delta(K(0))}{\langle K\rangle_{\mathrm{C}}}\right]\right\rangle_{\mathrm{C}} \tag{7.30}$$

where $P_{xy}^K(0)$ is the kinetic part of the pressure tensor evaluated at time zero (compare this with the linear theory given in Section 5.3). Now we note that $\langle P_{xy}^K(0)\Delta(K(0))/\langle K\rangle\rangle_{\mathrm{C}} = 0$. This means that (7.30) can be written as

$$\langle B(t)\rangle_{\mathrm{C}} = \langle B(0)\rangle_{\mathrm{C}} - \beta\gamma V \int_0^t \mathrm{d}s \left\langle \Delta(B(s))\left[P_{xy}(0) - P_{xy}^K(0)\frac{\Delta(K(0))}{\langle K\rangle_{\mathrm{C}}}\right]\right\rangle_{\mathrm{C}} \tag{7.31}$$

As in the linear response case (Section 5.3), we assume, without loss of generality, that B is extensive. The kinetic fluctuation term involves the average of three zero mean, extensive quantities and because of the factor $1/\langle K(0)\rangle$, gives only an order-one contribution to the average. Thus, for both the isokinetic and canonical ensembles, we can write

$$\langle B(t)\rangle = \langle B(0)\rangle - \beta\gamma V \int_0^t \mathrm{d}s\, \langle \Delta(B(s)) P_{xy}(0)\rangle \tag{7.32}$$

This expression relates the non-equilibrium value of a phase variable B at time t to the integral of a TTCF (the correlation between P_{xy} in the equilibrium starting state, $P_{xy}(0)$, and B at time s after the field is turned on). The time-zero value of the transient correlation function is an equilibrium property of the system. For example, if $B = P_{xy}$, then the time-zero value is $\langle P_{xy}^2(0)\rangle$. Under some, but by no means all, circumstances the values of $B(s)$ and $P_{xy}(0)$ will become uncorrelated at long times. If this is the case the system is said to exhibit mixing. The transient correlation function will then approach $\langle B(t)\rangle\langle P_{xy}(0)\rangle$, which is zero because $\langle P_{xy}(0)\rangle = 0$.

The adiabatic systems treated by Visscher, Dufty, Lindenfeld and Cohen do not exhibit mixing because, in the absence of a thermostat, $\mathrm{d}\langle B(t)\rangle/\mathrm{d}t$ does not, in general, go to zero at large times. Thus the integral of the associated transient correlation function does not converge. This presumably means that the initial fluctuations in adiabatic systems are remembered forever. Other systems which are not expected to exhibit mixing are turbulent systems or systems which execute quasi-periodic oscillations.

If $AI\Gamma$ (Section 5.3) is satisfied, the result for the general case is,

$$\langle B(t)\rangle = \langle B(0)\rangle - \beta F_e \int_0^t \mathrm{d}s\, \langle \Delta B(s) J(0)\rangle \tag{7.33}$$

We can use recursive substitution to derive the Kawasaki form for the nonlinear response from the TTCF (7.33). The first step in the derivation of the Kawasaki representation is to rewrite the TTCF relation using iL to denote the phase variable Liouvillean, and $-iL$ to denote its adjoint, the f-Liouvillean. Thus $\mathrm{d}B/\mathrm{d}t \equiv \mathrm{iL}B$ and $\partial f/\partial t \equiv -iLf$. Using this notation, (7.32) can be written as

$$\langle B(t)\rangle = \int \mathrm{d}\boldsymbol{\Gamma}\, Bf(0) - \beta\gamma V \int_0^t \mathrm{d}s \int \mathrm{d}\boldsymbol{\Gamma}\, f(0)\mathrm{e}^{\mathrm{iL}s} B\, \mathrm{e}^{-\mathrm{iL}s} J \tag{7.34}$$

$$= \int \mathrm{d}\boldsymbol{\Gamma}\, Bf(0) - \beta\gamma V \int_0^t \mathrm{d}s \int \mathrm{d}\boldsymbol{\Gamma} (\mathrm{e}^{-iLs} f(0)) B\, \mathrm{e}^{-\mathrm{iL}s} J \tag{7.35}$$

where we have unrolled the first p-propagator onto the distribution function. Equation (7.35) can be written more simply as

$$\langle B(t)\rangle = \int \mathrm{d}\boldsymbol{\Gamma}\, Bf(0) - \beta\gamma V \int_0^t \mathrm{d}s \int \mathrm{d}\boldsymbol{\Gamma}\, B(0) J(-s) f(s) \tag{7.36}$$

Since this equation is true for all phase variables, B, the TTCF representation for the N-particle distribution function must be

$$f(t) = f(0) - \beta\gamma V \int_0^t \mathrm{d}s\, J(-s) f(s) \tag{7.37}$$

We can now successively substitute the transient correlation function expression for the nonequilibrium distribution function into the right-hand side of (7.37). This gives,

$$\langle B(t)\rangle = \int d\Gamma\, Bf(0) - \beta\gamma V \int_0^t ds_1 \int d\Gamma\, B(0)J(-s_1)f(0)$$
$$+ (\beta\gamma V)^2 \int_0^t ds_1 \int_0^{s_1} ds_2 \int d\Gamma\, B(0)J(-s_1)J(-s_2)f(0) + \ldots$$
$$= \int d\Gamma\, B(0) \exp\left[-\beta\gamma V \int_0^t ds\, J(-s) \right] f(0) \tag{7.38}$$

This is precisely the Kawasaki form of the thermostatted nonlinear response. This expression is valid for both the canonical and isokinetic ensembles. It is also valid for the canonical ensemble when the thermostatting is carried out using the Nosé–Hoover thermostat.

One can of course also derive the TTCF expression for phase averages from the Kawasaki expression. Following Morriss and Evans (1985), we simply differentiate (7.38) with respect to time, and then reintegrate.

$$\frac{d}{dt}\langle B(t)\rangle = -\beta\gamma V \int d\Gamma\, BJ(-t) \exp\left[-\beta\gamma V \int_0^t ds\, J(-s) \right] f(0)$$
$$= -\beta\gamma V \int d\Gamma\, BJ(-t)f(t)$$
$$= -\beta\gamma V \int d\Gamma\, B(t)J(0)f(0)$$
$$= -\beta\gamma V \langle B(t)J(0)\rangle \tag{7.39}$$

A simple integration of (7.39) with respect to time yields the TTCF relation (7.32). We have thus proved the formal equivalence of the TTCF and Kawasaki representations for the nonlinear thermostatted response.

Comparing the transient time-correlation expression for the nonlinear response with the Kawasaki representation, we see that the difference simply amounts to a time shift. In the transient time-correlation form, it is the dissipative flux, J, which is evaluated at time zero, whereas in the Kawasaki form, the response variable, B, is evaluated at time zero. For equilibrium or steady-state time-correlation functions the stationarity of averages means that such time shifts are essentially trivial. For transient response correlation functions there is, of course, no such invariance principle; consequently, the time-translation transformation is accordingly more complex.

The computation of the time-dependent response using the Kawasaki form directly, (7.38), is very difficult. The inevitable errors associated with the inaccuracy of the trajectory, as well as those associated with the finite grid size in the calculation of the extensive Kawasaki integrand, combine and are magnified by the exponential. This exponential is then multiplied by the phase variable $B(0)$, before the ensemble average is performed. In contrast, the calculation of the response using the transient correlation expression, (7.33), is, as we shall see, far easier.

It is trivial to see that in the linear regime both the TTCF and Kawasaki expressions reduce to the usual Green–Kubo expressions. The equilibrium time-correlation functions that appear in Green–Kubo relations are generated by the field-free thermostatted equations. In the TTCF formulae the field is 'turned on' at $t = 0$.

The coincidence at small fields, of the Green–Kubo and transient correlation formulae means that, unlike direct NEMD, the TTCF method can be used at small fields. This is impossible for direct NEMD because in the small field limit the signal-to-noise ratio goes to zero. The signal-to-noise ratio for the transient correlation function method becomes equal to that of the equilibrium Green–Kubo method. The transient correlation function method forms a bridge between the Green–Kubo method which can only be used at equilibrium, and direct NEMD which is the most efficient strong-field method. Because a field is required to generate TTCFs, their calculation using a molecular dynamics still requires a *nonequilibrium* computer simulation to be performed.

It is also easy to see that at short times there is no difference between the linear and nonlinear stress response. It takes time for the nonlinearities to develop. The way to see this is to expand the TTCF in a power series in γt. The coefficient of the first term in this series is just $V\langle P_{xy}^2\rangle/k_B T$, the infinite frequency shear modulus, G_∞. Since this is an equilibrium property its value is unaffected by the strain rate and is thus the same in both the linear and nonlinear cases. If we look at the response of a quantity like the pressure whose linear response is zero, the leading term in the short-time expansion is quadratic in the strain rate and in time. The linear response, is of course, the first to appear.

7.4 TRAJECTORY MAPPINGS

In calculations of transient time-correlation functions (TTCFs) it is convenient to generate the initial ensemble of starting states from a single field free, Gaussian isokinetic trajectory. As Gaussian isokinetic dynamics ergodically generate the isokinetic ensemble, a single field-free trajectory is sufficient to sample the ensemble. At equally spaced intervals along this single field-free trajectory (every N_e time-step), field-dependent simulations are started and followed for N_n time-steps. The number N_n should be greater than the characteristic time required for the system to relax to a steady state, and N_e should be large enough to ensure that the initial phases are uncorrelated. Each of these cycles gives one initial phase, $\mathbf{\Gamma}$, for the transient correlation function. This process can be made more efficient if we use this single equilibrium starting state to provide more than one initial phase for the nonequilibrium trajectories. To do this we use a group of phase-space mappings.

In this section we develop mappings of the phase, $\mathbf{\Gamma}$, which have useful properties for the theoretical interpretation and practical implementation of nonlinear response theory. For convenience we shall write the phase $\mathbf{\Gamma}$ as $(\mathbf{q}, \mathbf{p}) = (\mathbf{x}, \mathbf{y}, \mathbf{z}, \mathbf{p}_x, \mathbf{p}_y, \mathbf{p}_z)$, where each of the components $\mathbf{x}$, $\mathbf{y}$, $\mathbf{z}$, $\mathbf{p}_x$, $\mathbf{p}_y$ and $\mathbf{p}_z$ is itself an N-dimensional vector. The time evolution of an arbitrary phase variable $B(\mathbf{\Gamma})$ is governed by the phase variable propagator $\exp(\mathrm{i}Lt)$, so that $B(t) = B(\mathbf{\Gamma}(t)) = \exp(\mathrm{i}Lt)B(\mathbf{\Gamma})$. Note that the propagator is an operator which acts on the *initial* phases, $\mathbf{\Gamma}$; therefore in order to calculate the action of the propagator on a phase variable at a time other than zero, $B(t)$ must be expressed as a function of the initial phases $\mathbf{\Gamma}$ and not of the current phases $\mathbf{\Gamma}(t)$. We assume that the equations of motion have no explicit time dependence (by way of a time-dependent external field). The propagator is, therefore, a shift operator. In the time-dependent case, the propagator is not a simple shift operator and the results which follow will need to be generalized. We leave this generalization until Chapter 8.

The phase variable B at time t, $B(t)$, can be traced back to time zero by applying the negative-time phase-variable propagator $\exp(-\mathrm{i}Lt)$,

$$\exp(-\mathrm{i}Lt)B(t) = \exp(-\mathrm{i}Lt)\exp(\mathrm{i}Lt)B(\mathbf{\Gamma}) = B(\mathbf{\Gamma}) \tag{7.40}$$

Reversing the sign of the time in the propagator *retraces* the original trajectory. It is possible to *return* to the original phase point $\mathbf{\Gamma}(0)$ without changing the sign of the time. This is achieved by mapping the phase point $\mathbf{\Gamma}(t)$ so that a combination of positive time evolution and mapping takes $\mathbf{\Gamma}(t) \Rightarrow \mathbf{\Gamma}(0)$. This mapping is called the time reversal mapping $\mathbf{M}^{\mathrm{T}}$. For field-free equations of motion, this is straightforward as the mapping simply consists of reversing the signs of all the momenta.

$$\mathbf{M}^{\mathrm{T}}(\mathbf{q}, \mathbf{p}) = (\mathbf{q}^{\mathrm{T}}, \mathbf{p}^{\mathrm{T}}) = (\mathbf{q}, -\mathbf{p}) \tag{7.41}$$

It is important to realize that this process does not lead to a *retracing* of the original trajectory, as everywhere along the *return* path the momenta are of opposite sign to those of the *forward* path. Noting that $\mathrm{e}^{\mathrm{i}Lt} = \mathrm{e}^{\mathrm{i}L(\mathbf{\Gamma})t}$, this can be summarized as: $\mathbf{M}^{\mathrm{T}}\mathrm{e}^{\mathrm{i}Lt}\mathbf{M}^{\mathrm{T}}\mathrm{e}^{\mathrm{i}Lt}\mathbf{\Gamma}(0) = \mathbf{M}^{\mathrm{T}}\mathrm{e}^{\mathrm{i}Lt}\mathbf{M}^{\mathrm{T}}\mathbf{\Gamma}(t) = \mathrm{e}^{-\mathrm{i}Lt}\mathbf{\Gamma}(t) = \mathbf{\Gamma}(0)$. These results will be derived in more detail later.

Given an initial starting phase $\mathbf{\Gamma} = (\mathbf{x}, \mathbf{y}, \mathbf{p}_x, \mathbf{p}_y)$ then four starting phases, which occur within the equilibrium distribution with the same probability as $\mathbf{\Gamma}$, can be obtained using the mappings $\mathbf{M}^{\mathrm{I}}$, $\mathbf{M}^{\mathrm{T}}$, $\mathbf{M}^{\mathrm{Y}}$ and $\mathbf{M}^{\mathrm{K}}$;

$$\begin{aligned}
\mathbf{\Gamma}^{\mathrm{I}} &= \mathbf{M}^{\mathrm{I}}[\mathbf{\Gamma}] = (\mathbf{x}, \mathbf{y}, \mathbf{z}, \mathbf{p}_x, \mathbf{p}_y, \mathbf{p}_z) \\
\mathbf{\Gamma}^{\mathrm{T}} &= \mathbf{M}^{\mathrm{T}}[\mathbf{\Gamma}] = (\mathbf{x}, \mathbf{y}, \mathbf{z}, -\mathbf{p}_x, -\mathbf{p}_y, -\mathbf{p}_z) \\
\mathbf{\Gamma}^{\mathrm{Y}} &= \mathbf{M}^{\mathrm{Y}}[\mathbf{\Gamma}] = (\mathbf{x}, -\mathbf{y}, \mathbf{z}, \mathbf{p}_x, -\mathbf{p}_y, \mathbf{p}_z) \\
\mathbf{\Gamma}^{\mathrm{K}} &= \mathbf{M}^{\mathrm{K}}[\mathbf{\Gamma}] = (\mathbf{x}, -\mathbf{y}, \mathbf{z}, -\mathbf{p}_x, \mathbf{p}_y, -\mathbf{p}_z)
\end{aligned} \tag{7.42}$$

Here $\mathbf{M}^{\mathrm{I}}$ is the identity mapping; $\mathbf{M}^{\mathrm{T}}$ is the time-reversal mapping introduced above; $\mathbf{M}^{\mathrm{Y}}$ is termed the y-reflection mapping; and $\mathbf{M}^{\mathrm{K}}$ is called the Kawasaki mapping (it is the combined effect of time-reversal and y-reflection mapping $\mathbf{M}^{\mathrm{K}} = \mathbf{M}^{\mathrm{T}}\mathbf{M}^{\mathrm{Y}}$). For shear flow these four configurations give four different starting states, and lead to four different field-dependent trajectories from the single equilibrium phase point $\mathbf{\Gamma}$. Each of the mappings consists of a pair of reflections in a coordinate or momentum axis. In total there are 2^4 states that can be obtained using the reflections of a two-dimensional phase space; however, only 2^3 of these states will result in *at most* a sign change in the instantaneous shear stress $P_{xy}(\mathbf{\Gamma})$. Only 2^2 of the remaining mappings lead to different shearing trajectories. The shear stress obtained from trajectories starting from $\mathbf{\Gamma}_i$ and $-\mathbf{\Gamma}_i$, for example, are identical. The probability of each of these states occurring within the equilibrium distribution, is identical because the Hamiltonian, H_0, is invariant under these mappings.

There is a second, more important, advantage of this procedure. If we examine the transient response formula (7.32), we see that at long time the correlation function $\langle B(t)P_{xy}(0)\rangle$ approaches $\langle B(\infty)\rangle\langle P_{xy}(0)\rangle$. The steady-state average of B is usually non-zero (in contrast to equilibrium time-correlation functions). To minimize the statistical uncertainties in calculating the transient correlation integral, it is necessary to choose equilibrium starting states $\mathbf{\Gamma}$ in such a way that $\langle P_{xy}(0)\rangle \equiv 0$. The phase-mapping procedure described above achieves this. If the shear stress computed from the original starting phase is P_{xy}, then the shear stress from $\mathbf{\Gamma}^{\mathrm{T}}$ is also equal to P_{xy}, but the shear stresses from both $\mathbf{\Gamma}^{\mathrm{Y}}$ and $\mathbf{\Gamma}^{\mathrm{K}}$ are equal to $-P_{xy}$. This means that the sum of the shear stresses from these four starting phases is *exactly* zero, so if each chosen $\mathbf{\Gamma}$ is mapped in this way the average shear stress is exactly zero regardless of the number of samplings of $\mathbf{\Gamma}$. The statistical difficulties at long time, associated with a small non-zero value the average of $P_{xy}(0)$, are eliminated.

There are two further operations on these mappings which we need to complete the development of the mapping algebra. First, we need to know how each of the mappings affects the phase variables. Second, we must understand the effect of the mapping on the phase variable Liouvillean, $\mathrm{i}L(\mathbf{\Gamma})$, as it is also a function of the initial phase point $\mathbf{\Gamma}$. To do this we need to know how the equations of motion transform. First we will discuss the transformation of Hamiltonian equations of motion under the mapping, and then we will consider the transformation of the field-dependent dynamics. This will require an extension of the mappings to include the field itself.

To illustrate the development we will consider the time-reversal mapping $\mathbf{M}^{\mathrm{T}}$ in detail, and simply state the results for other mappings. In what follows the mapping operator $\mathbf{M}^{\mathrm{T}}$ operates on all functions and operators (which depend upon $\mathbf{\Gamma}$ and γ) to its right. A particular example is useful at this stage. Consider the shear stress P_{xy}

$$\begin{aligned}\mathbf{M}^{\mathrm{T}}[P_{xy}] &= \mathbf{M}^{\mathrm{T}}\left[\sum_{i=1}^{N}\frac{p_{xi}p_{yi}}{m} + \sum_{i=1}^{N} y_i F_{xi}\right] \\ &= \sum_{i=1}^{N}\frac{(-p_{xi})(-p_{yi})}{m} + \sum_{i=1}^{N} y_i F_{xi} \\ &= P_{xy} \end{aligned} \tag{7.43}$$

Here P_{xy} is mapped to the same value. For thermodynamically interesting phase variables the operation of the mappings involve simple parity changes

$$\mathbf{M}^{\mathrm{X}} B(\mathbf{\Gamma}) = p_B^{\mathrm{X}} B(\mathbf{\Gamma}) \tag{7.44}$$

where $p_B^{\mathrm{X}} = \pm 1$. In Table 7.1 we list the values of the parity operators for shear stress, pressure and energy for each of the mappings.

The operation of the mapping $\mathbf{M}^{\mathrm{T}}$ on the Hamiltonian equations of motion is

$$\mathbf{M}^{\mathrm{T}}\dot{\mathbf{q}}_i = \mathbf{M}^{\mathrm{T}}\left[\frac{\mathbf{p}_i}{m}\right] = \frac{\mathbf{p}_i^{\mathrm{T}}}{m} = \frac{-\mathbf{p}_i}{m} = -\dot{\mathbf{q}}_i \tag{7.45}$$

$$\mathbf{M}^{\mathrm{T}}\dot{\mathbf{p}}_i = \mathbf{M}^{\mathrm{T}}[\mathbf{F}_i(\mathbf{q})] = \mathbf{F}_i(\mathbf{q}^{\mathrm{T}}) = \mathbf{F}_i(\mathbf{q}) = \dot{\mathbf{p}}_i \tag{7.46}$$

where the transformed coordinate and momenta are denoted by the superscript T. The vector character of the force $\mathbf{F}$ is determined by the coordinate vector $\mathbf{q}$, so that under this mapping the force is invariant. Because

Table 7.1 Mapping parities.

Parity operator	Mapping	Shear stress	Pressure	Energy
$\mathbf{M}^{\mathrm{I}}$	Identity	1	1	1
$\mathbf{M}^{\mathrm{T}}$	Time reversal	1	1	1
$\mathbf{M}^{\mathrm{Y}}$	y-Reflection	−1	1	1
$\mathbf{M}^{\mathrm{K}}$	Kawasaki	−1	1	1

$d\mathbf{q}/dt$ and $\mathbf{p}$ change sign under the mapping $\mathbf{M}^{\mathrm{T}}$, the phase-variable Liouvillean becomes

$$\begin{aligned}\mathbf{M}^{\mathrm{T}}\mathrm{iL}(\boldsymbol{\Gamma}) &= \mathbf{M}^{\mathrm{T}}\left[\dot{\boldsymbol{\Gamma}}\cdot\frac{\partial}{\partial\boldsymbol{\Gamma}}\right] = \mathbf{M}^{\mathrm{T}}\left[\sum_{i=1}^{N}\left(\dot{\mathbf{q}}_i\cdot\frac{\partial}{\partial\mathbf{q}_i}+\dot{\mathbf{p}}_i\cdot\frac{\partial}{\partial\mathbf{p}_i}\right)\right]\\ &= \sum_{i=1}^{N}\left(\dot{\mathbf{q}}_i^{\mathrm{T}}\cdot\frac{\partial}{\partial\mathbf{q}_i^{\mathrm{T}}}+\dot{\mathbf{p}}_i^{\mathrm{T}}\cdot\frac{\partial}{\partial\mathbf{p}_i^{\mathrm{T}}}\right)\\ &= \sum_{i=1}^{N}\left(-\dot{\mathbf{q}}_i\cdot\frac{\partial}{\partial\mathbf{q}_i}+\dot{\mathbf{p}}_i\cdot\frac{\partial}{\partial(-\mathbf{p}_i)}\right) = -\mathrm{iL}(\boldsymbol{\Gamma})\end{aligned} \qquad (7.47)$$

It is straightforward to use this result and the series expansion of the propagator to show that

$$\mathbf{M}^{\mathrm{T}}\exp(\mathrm{iL}(\boldsymbol{\Gamma})t) = \exp(-\mathrm{iL}(\boldsymbol{\Gamma})t) \qquad (7.48)$$

To see exactly how this combination of the $\mathbf{M}^{\mathrm{T}}$ mapping and forward time propagation combine to give time reversal, we consider the time evolution of $\boldsymbol{\Gamma}$ itself.

$$\begin{aligned}\boldsymbol{\Gamma} &= \exp(-\mathrm{iL}(\boldsymbol{\Gamma})t)\exp(\mathrm{iL}(\boldsymbol{\Gamma})t)\boldsymbol{\Gamma}\\ &= \mathbf{M}^{\mathrm{T}}\mathbf{M}^{\mathrm{T}}\exp(-\mathrm{iL}(\boldsymbol{\Gamma})t)\exp(\mathrm{iL}(\boldsymbol{\Gamma})t)\boldsymbol{\Gamma}\\ &= \mathbf{M}^{\mathrm{T}}\exp(\mathbf{M}^{\mathrm{T}}[-\mathrm{iL}(\boldsymbol{\Gamma})]t)\mathbf{M}^{\mathrm{T}}\exp(\mathrm{iL}(\boldsymbol{\Gamma})t)\boldsymbol{\Gamma}\\ &= \mathbf{M}^{\mathrm{T}}\exp(\mathrm{iL}(\boldsymbol{\Gamma})t)\mathbf{M}^{\mathrm{T}}\exp(\mathrm{iL}(\boldsymbol{\Gamma})t)\boldsymbol{\Gamma}\end{aligned} \qquad (7.49)$$

This implies that

$$\mathbf{M}^{\mathrm{T}}\exp(\mathrm{iL}(\boldsymbol{\Gamma})t)\mathbf{M}^{\mathrm{T}}\exp(\mathrm{iL}(\boldsymbol{\Gamma})t) = 1 \qquad (7.50)$$

If we start with $\boldsymbol{\Gamma}(0)$, propagate forward to time t, map with $\mathbf{M}^{\mathrm{T}}$ (changing the signs of the momenta), propagate forward to time t, and map with $\mathbf{M}^{\mathrm{T}}$ (changing the signs of the momenta again), we return to $\boldsymbol{\Gamma}(0)$. An analogous identity can be constructed by considering $\boldsymbol{\Gamma}(0) = \exp(\mathrm{iL}(\boldsymbol{\Gamma})t)\boldsymbol{\Gamma}(-t)$, that is

$$\mathbf{M}^{\mathrm{T}}\exp(-\mathrm{iL}(\boldsymbol{\Gamma})t)\mathbf{M}^{\mathrm{T}}\exp(-\mathrm{iL}(\boldsymbol{\Gamma})t) = 1 \qquad (7.51)$$

This says that we can complete a similar cycle using the backward time propagator, $\exp(-\mathrm{iL}t)$, first. These results demonstrate the various uses of this time-reversal mapping.

When the equations of motion for the system involve an external field the time-reversal mapping can be generalized to include the field. This is necessary if we wish to determine whether a particular mapping leads to different field-dependent dynamics. Here we limit consideration to the isothermal

SLLOD algorithm for shear flow. It is clear that all the momenta must change sign, so a suitable choice for the mapping is

$$\mathbf{M}^{\mathrm{T}}(\mathbf{q},\mathbf{p},\gamma)=(\mathbf{q},-\mathbf{p},-\gamma) \tag{7.52}$$

As the field has units of inverse time, the field changes sign together with the momenta. The equations of motion for the mapped variables become

$$\begin{aligned}\mathbf{M}^{\mathrm{T}}\dot{\mathbf{q}}_i &= \mathbf{M}^{\mathrm{T}}\left[\frac{\mathbf{p}_i}{m}+\mathbf{n}_x\gamma y_i\right]\\ &=\frac{\mathbf{p}_i^{\mathrm{T}}}{m}+\mathbf{n}_x\gamma^{\mathrm{T}}y_i^{\mathrm{T}}\\ &=\frac{-\mathbf{p}_i}{m}+\mathbf{n}_x(-\gamma)y_i\\ &=-\dot{\mathbf{q}}_i\end{aligned} \tag{7.53}$$

and

$$\begin{aligned}\mathbf{M}^{\mathrm{T}}\dot{\mathbf{p}}_i &= \mathbf{M}^{\mathrm{T}}[\mathbf{F}_i(\mathbf{q})-\mathbf{n}_x\gamma p_{yi}-\alpha\mathbf{p}_i]\\ &=\mathbf{F}_i(\mathbf{q}^{\mathrm{T}})-\mathbf{n}_x\gamma^{\mathrm{T}}p_{yi}^{\mathrm{T}}-\alpha^{\mathrm{T}}\mathbf{p}_i^{\mathrm{T}}\\ &=\mathbf{F}_i(\mathbf{q})-\mathbf{n}_x(-\gamma)(-p_{yi})-(-\alpha)-\mathbf{p}_i)\\ &=\mathbf{F}_i(\mathbf{q})-\mathbf{n}_x\gamma p_{yi}-\alpha\mathbf{p}_i\\ &=\dot{\mathbf{p}}_i\end{aligned} \tag{7.54}$$

Note also that for the thermostatting variable α

$$\alpha^{\mathrm{T}}=\mathbf{M}^{\mathrm{T}}\alpha=\mathbf{M}^{\mathrm{T}}\left[\frac{\sum_{i=1}^{N}(\mathbf{F}_i\cdot\mathbf{p}_i-\gamma p_{xi}p_{yi})}{\sum_{i=1}^{N}\mathbf{p}_i^2}\right]=-\alpha \tag{7.55}$$

as the numerator changes sign and the denominator is invariant under the time-reversal mapping. The mapping of the Liouvillean is similar to the field-free case and it can be shown that

$$\mathbf{M}^{\mathrm{T}}\mathrm{i}L(\boldsymbol{\Gamma},\gamma)=-\mathrm{i}L(\boldsymbol{\Gamma},-\gamma) \tag{7.56}$$

In the field-dependent case the two operators, equations (7.50) and (7.51) generalize to

$$\mathbf{M}^{\mathrm{T}}\exp(\mathrm{i}L(\boldsymbol{\Gamma},-\gamma)t)\mathbf{M}^{\mathrm{T}}\exp(\mathrm{i}L(\boldsymbol{\Gamma},\gamma)t)=1 \tag{7.57}$$

$$\mathbf{M}^{\mathrm{T}}\exp(-\mathrm{i}L(\boldsymbol{\Gamma},-\gamma)t)\mathbf{M}^{\mathrm{T}}\exp(-\mathrm{i}L(\boldsymbol{\Gamma},\gamma)t)=1 \tag{7.58}$$

Because a phase variable is, by definition, not a function of the field, the parity operators associated with mapping phase variables are unchanged.

The second mapping we consider is the y-reflection mapping, $\mathbf{M}^{\mathrm{Y}}$, which acts to change the sign of the shear rate but not the time or the Liouvillean. This mapping is defined by

$$\mathbf{M}^{\mathrm{Y}}(\mathbf{x}, \mathbf{y}, \mathbf{z}, \mathbf{p}_x, \mathbf{p}_y, \mathbf{p}_z, \gamma) = (\mathbf{x}, -\mathbf{y}, \mathbf{z}, \mathbf{p}_x, -\mathbf{p}_y, \mathbf{p}_z, -\gamma) \tag{7.59}$$

This mapping consists of a coordinate reflection in the $\mathbf{x}, \mathbf{z}$-plane, and momenta reflection in the $\mathbf{p}_x, \mathbf{p}_z$-plane. Substituting this mapping into the SLLOD equations of motion shows that the time derivatives of both $\mathbf{y}$ and $\mathbf{p}_y$ change sign, while the thermostatting variable remains unchanged. The y-reflection Liouvillean is related to the standard Liouvillean by

$$\mathbf{M}^{\mathrm{Y}}\mathrm{iL}(\mathbf{\Gamma}, \gamma) = \mathrm{iL}(\mathbf{\Gamma}, -\gamma) \tag{7.60}$$

We now define the combination Kawasaki mapping, $\mathbf{M}^{\mathrm{K}}$, which consists of the time-reversal mapping followed by the y-reflection mapping, so that

$$\mathbf{M}^{\mathrm{K}}(\mathbf{x}, \mathbf{y}, \mathbf{z}, \mathbf{p}_x, \mathbf{p}_y, \mathbf{p}_z, \gamma) = (\mathbf{x}, -\mathbf{y}, \mathbf{z}, -\mathbf{p}_x, \mathbf{p}_y, -\mathbf{p}_z, \gamma) \tag{7.61}$$

Under the Kawasaki mapping, the Liouvillean is transformed to

$$\mathbf{M}^{\mathrm{K}}\mathrm{iL}(\mathbf{\Gamma}, \gamma) = -\mathrm{iL}(\mathbf{\Gamma}, \gamma) \tag{7.62}$$

Using the results obtained in this section which are summarized in Table 7.2, it is easy to show that the following four time evolutions of the phase variable B yield identical values. That is

$$\begin{aligned} \exp(\mathrm{iL}(\mathbf{\Gamma}, \gamma)t)B(\mathbf{\Gamma}) &= p_{\mathrm{B}}^{\mathrm{T}} \exp(-\mathrm{iL}(\mathbf{\Gamma}^{\mathrm{T}}, -\gamma^{\mathrm{T}})t)\mathrm{B}(\mathbf{\Gamma}^{\mathrm{T}}) \\ &= p_{\mathrm{B}}^{\mathrm{Y}} \exp(\mathrm{iL}(\mathbf{\Gamma}^{\mathrm{Y}}, -\gamma^{\mathrm{Y}})t)\mathrm{B}(\mathbf{\Gamma}^{\mathrm{Y}}) \\ &= p_{\mathrm{B}}^{\mathrm{K}} \exp(-\mathrm{iL}(\mathbf{\Gamma}^{\mathrm{K}}, \gamma^{\mathrm{K}})t)\mathrm{B}(\mathbf{\Gamma}^{\mathrm{K}}) \end{aligned} \tag{7.63}$$

Table 7.2 Summary of phase-space mappings.

Time reversal mapping.

$$\mathbf{M}^{\mathrm{T}}(\mathbf{q}, \mathbf{p}, \gamma) = (\mathbf{q}^{\mathrm{T}}, \mathbf{p}^{\mathrm{T}}, \gamma^{\mathrm{T}}) = (\mathbf{q}, -\mathbf{p}, -\gamma)$$

$$\mathbf{M}^{\mathrm{T}}\mathrm{iL}(\mathbf{\Gamma}, \gamma) = \mathrm{iL}(\mathbf{\Gamma}^{\mathrm{T}}, \gamma^{\mathrm{T}}) = -\mathrm{iL}(\mathbf{\Gamma}, -\gamma)$$

y-Reflection mapping

$$\mathbf{M}^{\mathrm{Y}}(\mathbf{x}, \mathbf{y}, \mathbf{z}, \mathbf{p}_x, \mathbf{p}_y, \mathbf{p}_z, \gamma) = (\mathbf{x}, -\mathbf{y}, \mathbf{z}, \mathbf{p}_x, -\mathbf{p}_y, \mathbf{p}_z, -\gamma)$$

$$\mathbf{M}^{\mathrm{Y}}\mathrm{iL}(\mathbf{\Gamma}, \gamma) = \mathrm{iL}(\mathbf{\Gamma}^{\mathrm{Y}}, \gamma^{\mathrm{Y}}) = \mathrm{iL}(\mathbf{\Gamma}, -\gamma)$$

Kawasaki mapping

$$\mathbf{M}^{\mathrm{K}}(\mathbf{x}, \mathbf{y}, \mathbf{z}, \mathbf{p}_x, \mathbf{p}_y, \mathbf{p}_z, \gamma) = (\mathbf{x}, -\mathbf{y}, \mathbf{z}, -\mathbf{p}_x, \mathbf{p}_y, -\mathbf{p}_z, \gamma)$$

$$\mathbf{M}^{\mathrm{K}}\mathrm{iL}(\mathbf{\Gamma}, \gamma) = \mathrm{iL}(\mathbf{\Gamma}^{\mathrm{K}}, \gamma^{\mathrm{K}}) = -\mathrm{iL}(\mathbf{\Gamma}, \gamma)$$

Note that these four time evolutions involve changing the sign of the time and/or the sign of the field. If we consider the phase variable $P_{xy}(\mathbf{\Gamma})$, the time evolution leads to a negative average value at long time, and where a single sign change is made in the propagator, the parity operator is -1. The third equality has been used to interpret the propagation of the dissipative flux in the Kawasaki exponent; negative time evolution with a positive external field from $\mathbf{\Gamma}$, is equivalent to positive time evolution with a positive field from $\mathbf{\Gamma}^{\mathrm{K}}$. As each of the time evolutions in (7.63) represent different mathematical forms for the same trajectory, the stabilities are also the same.

The Kawasaki mapping is useful as an aid to understanding the formal expression for the Kawasaki distribution function. The particular term we consider is the time integral of the dissipative flux in the Kawasaki exponent

$$-\beta \int_0^t \mathrm{d}s\, \gamma P_{xy}(-s, \gamma, \mathbf{\Gamma}) V$$

using the Kawasaki mapping the negative time evolution can be transformed to an equivalent positive time evolution. To do this consider

$$\begin{aligned} P_{xy}(-s, \mathbf{\Gamma}, \gamma) &= \exp(-\mathrm{i}L(\mathbf{\Gamma}, \gamma)s) P_{xy}(\mathbf{\Gamma}) \\ &= \exp(\mathrm{i}L(\mathbf{\Gamma}^{\mathrm{K}}, \gamma^{\mathrm{K}})s) p^{\mathrm{K}}_{P_{xy}} P_{xy}(\mathbf{\Gamma}^{\mathrm{K}}) \\ &= -\exp(\mathrm{i}L(\mathbf{\Gamma}^{\mathrm{K}}, \gamma^{\mathrm{K}})s) P_{xy}(\mathbf{\Gamma}^{\mathrm{K}}) \\ &= -P_{xy}(s, \mathbf{\Gamma}^{\mathrm{K}}, \gamma^{\mathrm{K}}) \\ &= -P_{xy}(s, \mathbf{\Gamma}^{\mathrm{K}}, \gamma) \end{aligned} \tag{7.64}$$

The last equality follows from the fact that $\gamma^{\mathrm{K}} = \gamma$. So we may think of $P_{xy}(-s, \gamma, \mathbf{\Gamma})$ as equivalent (apart from the sign of the parity operator) to the propagation of P_{xy} forward in time, with the same γ, but starting from a different phase point $\mathbf{\Gamma}^{\mathrm{K}}$. The probability of this new phase point $\mathbf{\Gamma}^{\mathrm{K}}$ in the canonical (or isothermal) distribution is the same as the original $\mathbf{\Gamma}$, as the equilibrium Hamiltonian, H_0, is invariant under time reversal and reflection. Therefore, the Kawasaki distribution function can be written as

$$\begin{aligned} f(t, \mathbf{\Gamma}, \gamma) &= \exp\left[\beta\gamma V \int_0^t \mathrm{d}s\, P_{xy}(-s, \mathbf{\Gamma}, \gamma) \right] f(0, \mathbf{\Gamma}, 0) \\ &= \exp\left[-\beta\gamma V \int_0^t \mathrm{d}s\, P_{xy}(s, \mathbf{\Gamma}^{\mathrm{K}}, \gamma) \right] f(0, \mathbf{\Gamma}^{\mathrm{K}}, 0) \end{aligned} \tag{7.65}$$

In this form the sign of the exponent itself changes, as does the sign of the time evolution. At sufficiently large time $P_{xy}(s, \mathbf{\Gamma}^{\mathrm{K}}, \gamma)$ approaches the steady-state value $\langle P_{xy}(s, \gamma) \rangle$, regardless of the initial phase point $\mathbf{\Gamma}^{\mathrm{K}}$.

7.5 NUMERICAL RESULTS FOR THE TRANSIENT TIME-CORRELATION FUNCTION

Computer simulations have been carried out for two different systems (Evans and Morriss, 1988). Two statistical mechanical systems were studied. The first was a system of 72 soft discs, ($\phi = 4\varepsilon(\sigma/r)^{1/2}$), in two dimensions at a reduced density, $\rho^* = \rho\sigma^2 = 0.6928$, a reduced temperature, $T^* = k_b T/\varepsilon = 1$, and for a range of reduced strain rates, $\gamma^* = \gamma(m/\varepsilon)^{1/2}\sigma = \partial u_x/\partial y\,(m/\varepsilon)^{1/2}\sigma$. The second system was studied more extensively. It consisted of 256 WCA particles. The system was three-dimensional and the density was set to $\rho^* = \rho\sigma^3 = 0.8442$, while the temperature was $T^* = k_b T/\varepsilon = 0.722$ (i.e. the Lennard–Jones triple-point state).

In each simulation the direct nonequilibrium molecular dynamics (NEMD) average of the shear stress, pressure, normal stress difference and thermostat multiplier, α, were calculated along with their associated transient correlation functions using, typically, 60 000 nonequilibrium starting states. For the three-dimensional system each nonequilibrium trajectory was run for a reduced time of 1.5 (600 time-steps). Each 60 000-starting-state simulation consisted of a total of 54 million time-steps made up of $2 \times 15\,000 \times 600$ time-steps at equilibrium and $4 \times 15\,000 \times 600$ perturbed nonequilibrium time-steps. The trajectory mappings described in Section 7.4 were used to generate the four starting states for the nonequilibrium trajectories.

In Fig. 7.1 we present the results obtained for the reduced shear stress $P^*_{xy} = P_{xy}(\sigma^2/\varepsilon)$, in the two-dimensional soft-disc system. The imposed reduced strain rate is unity. The values of the shear stress calculated from the transient correlation function expression (P_{xy}(T)) agree within error bars, with those calculated directly (P_{xy}(D)). The errors associated with the direct average are less than the size of the plotting symbols, whereas the error in the integral of the transient correlation function is approximately $\pm 2.5\%$ at the longest times. Although the agreement between the direct-simulation results and the transient time-correlation function (TTCF) prediction is very good, it must be remembered that the total response for the shear stress is the sum of a large linear effect which could be correctly predicted by using the Green–Kubo formula and a smaller (ca. 25%) nonlinear effect. The statistical agreement regarding the TTCF prediction of the intrinsically nonlinear component of the total response is, therefore, approximately 10%.

The shear-induced increase in pressure with increasing strain rate (shear dilatancy) is an intrinsically nonlinear effect and is not observed in Newtonian fluids. The Green–Kubo formulae predict that there is no coupling of the pressure and the shear stress because the equilibrium correlation function, $\langle \Delta p(t) P_{xy}(0) \rangle$, is exactly zero at all times. In Fig. 7.2 we present the direct and transient correlation function values of the difference between the pressure

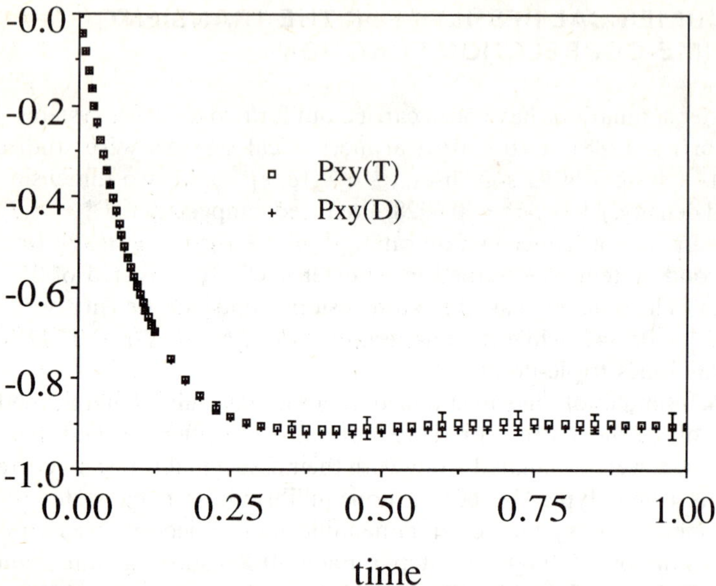

Fig. 7.1 Reduced shear stress, $P^*_{xy} = P_{xy}(\sigma^2/\varepsilon)$, in the two-dimensional soft-disc system. $P_{xy}(\mathrm{T})$, calculated from the transient correlation function; $P_{xy}(\mathrm{D})$, calculated directly.

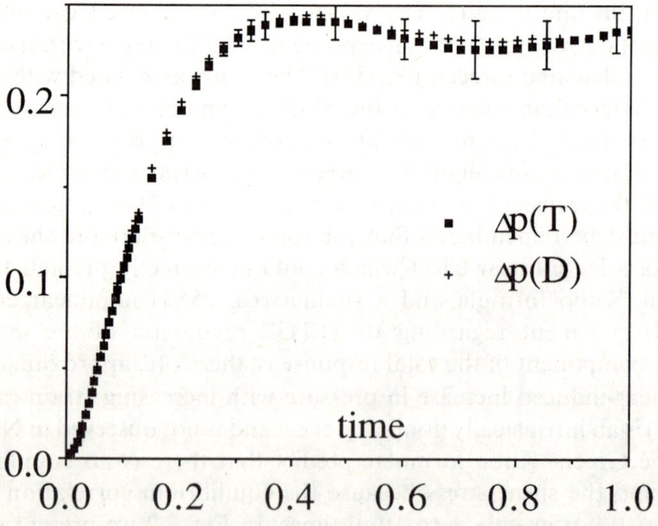

Fig. 7.2 Direct (D) and transient (T) correlation function values of $\Delta p^* = p^* - p^*_0$.

$p^* = p(\sigma^2/\varepsilon)$ and its equilibrium value, p_0^*, ($\Delta\rho^* = p^* - p_0^*$). The agreement between the direct average and the value obtained from the transient correlation function expression at $\gamma^* = 1.0$ is impressive. It is important to note that the agreement between theory and simulation shown in Fig. 7.2 is a test of the predictions of the theory for an *entirely* nonlinear effect. It is a more convincing check on the validity of the TTCF formalism than are the results for the shear stress because there is no underlying linear effect.

The results for the x–y element of the pressure tensor in the three-dimensional WCA system are given in Fig. 7.3. Again the agreement between the TTCF prediction and the direct simulation is excellent. We also show the long-time steady-state stress computed using conventional NEMD. It is clear that our time limit for the integration of the TTCF is sufficient to obtain convergence of the integrals (i.e. to ensure relaxation to the nonequilibrium steady state). We also show the Green–Kubo prediction for the stress. A comparison of the linear and nonlinear responses shows that the intrinsically nonlinear response is only generated at comparatively late times. The response is essentially linear until the stress overshoot time ($t^* \simeq 0.3$). The figure also

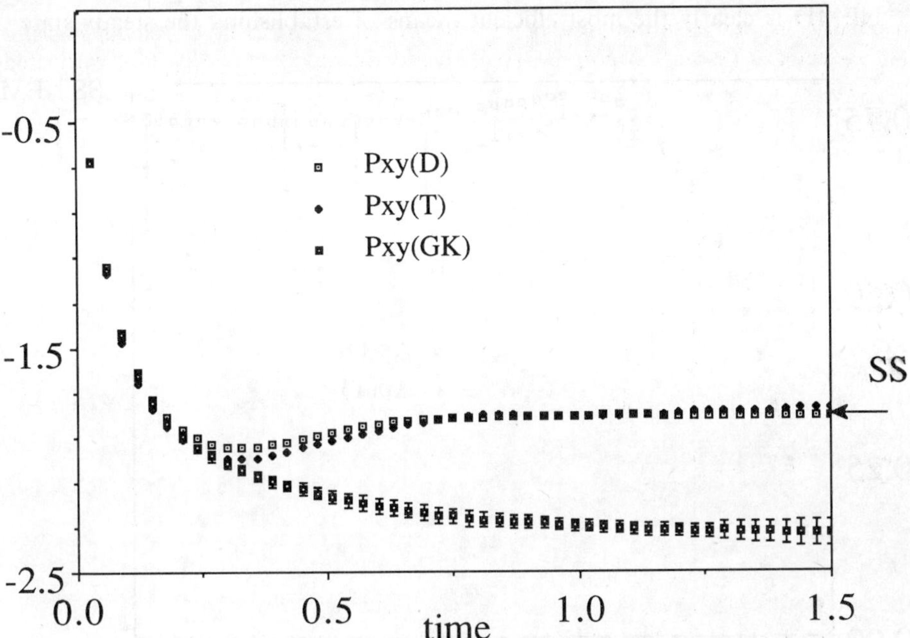

Fig. 7.3 The x–y element of the pressure tensor in the three-dimensional WCA system. T, TTCF prediction; D, direct simulation; GK, Green–Kubo prediction. SS, long-time steady-state stress computed using conventional NEMD.

shows that the total nonlinear response converges far more rapidly than does the linear Green–Kubo response. The linear Green–Kubo response has obviously not relaxed to its steady state limiting value at a t^* value of 1.5. This is presumably due to long-time tail effects which predict that the linear response relaxes very slowly as $t^{-1/2}$, at long times.

In Fig. 7.4 we show the corresponding results for shear dilatancy in three dimensions. Again the TTCF predictions are in statistical agreement with the results from direct simulation. We also show the steady-state pressure shift obtained using conventional NEMD. Again it is apparent that $t^* = 1.5$ is sufficient to obtain convergence of the TTCF integral. Although it is not easy to see in the figure, the initial slope of the pressure response is zero. This contrasts with the initial slope of the shear stress response which is G_∞. This is in agreement with the predictions of the TTCF made in Section 7.3. Figures 7.1 and 7.3 clearly show that at short time the stress is controlled by linear-response mechanisms. It takes time for the nonlinearities to develop but, paradoxically perhaps, convergence to the steady-state asymptotic values is ultimately much faster in the nonlinear, large field regime.

Comparing the statistical uncertainties of the transient correlation and direct NEMD results shows that at reduced strain rates of unity conventional NEMD is clearly the most efficient means of establishing the steady-state

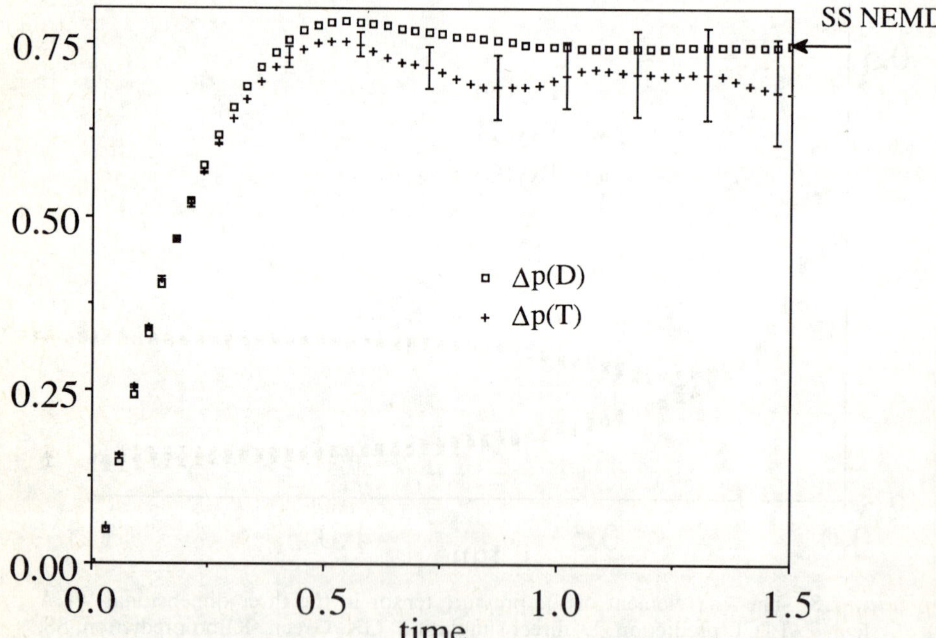

Fig. 7.4 Shear dilatancy in three dimensions. For abbreviations see Fig. 7.3.

response. For example, under precisely the same conditions: after 54 million time-steps the TTCF expression for P_{xy} is accurate to $\pm 0.05\%$, but the directly averaged transient response is accurate to $\pm 0.001\%$. Because time is not wasted in establishing the steady state from each of the 60000 time origins, conventional steady-state NEMD needs only 120000 time-steps to obtain an uncertainty of $\pm 0.0017\%$. If we assume that errors are inversely proportional to the square root of the run length, then the relative uncertainties for a 54-million time-step run would be $\pm 0.05\%$, $\pm 0.001\%$ and 0.00008% for the TTCF, the directly averaged transient response and for conventional NEMD, respectively. Steady-state NEMD is about 600 times more accurate than TTCF for the same number of time-steps. On the other hand, the transient correlation method has a computational efficiency which is similar to that of the equilibrium Green–Kubo method. For TTCFs time origins cannot be taken more frequently than the time interval over which the TTCFs are calculated. An advantage of the TTCF formalism is that it models the rheological problem of stress growth (Bird *et al.*, 1977), not simply steady shear flow, and we can observe the associated effects such as stress overshoot, and the time development of normal stress differences.

Figure 7.5 shows the transient responses for the normal stress differences, $P_{yy} - P_{zz}$ and $P_{xx} - P_{yy}$, for the three-dimensional WCA system at a reduced

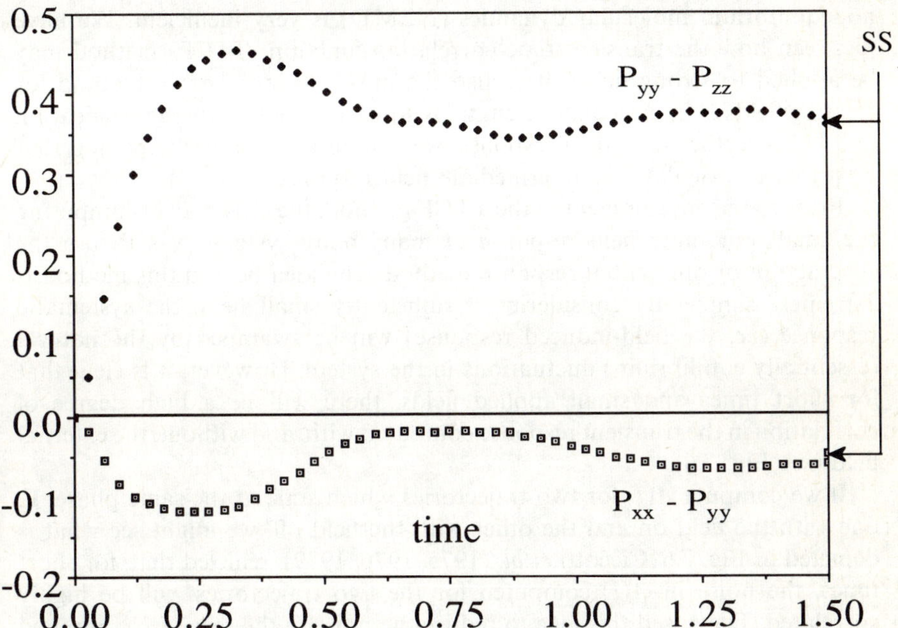

Fig. 7.5 Transient responses for the normal stress differences $P_{yy} - P_{zz}$ and $P_{xx} - P_{yy}$ for the three-dimensional WCA system at a reduced strain rate of unity.

strain rate of unity. The normal stress differences are clearly more subtle than either the shear stress or the hydrostatic pressure. Whereas the latter two functions seem to exhibit a simple overshoot before relaxing to their final steady-state values, the normal stress differences show two maxima before achieving their steady-state values. As before it is apparent that $t^* = 1.5$ is sufficient time for an essentially complete relaxation to the steady state.

Over the years a number of numerical comparisons have been made between the Green–Kubo expressions and the results of NEMD simulations. The work we have just described takes this comparison one step further. It compares NEMD simulation results with the thermostatted, nonlinear generalization of the Green–Kubo formulae. It provides convincing numerical evidence for the usefulness and correctness of the TTCF formalism. The TTCF formalism is the natural thermostatted, nonlinear generalization of the Green–Kubo relations.

7.6 DIFFERENTIAL RESPONSE FUNCTIONS

Surprisingly often we are interested in the intermediate regime where the Green–Kubo method cannot be applied and where, because of noise, direct nonequilibrium molecular dynamics (NEMD) is very inefficient. We have just seen how the transient time-correlation function (TTCF) method may be applied to strong fields. It is also the most efficient known method for treating fields of intermediate strength. Before we demonstrate the application of TTCFs to the small-field response, we will describe an early method that was used to calculate the intermediate field response.

Prior to the development of the TTCF method, the only way of computing the small but finite field response of many-body systems was to use the subtraction or differential response method. The idea behind this method is extremely simple. By considering a sufficiently small field, the systematic response (i.e. the field-induced response) will be swamped by the natural (essentially equilibrium) fluctuations in the system. However, it is clear that for short times and small applied fields, there will be a high degree of correlation in the transient response computed with, and without, the external field (see Fig. 7.6).

If we compute $A(t)$ for two trajectories which start at the same phase, $\boldsymbol{\Gamma}$, one with the field on and the other with the field off, we might see what is depicted in Fig. 7.6. Ciccotti *et al.* (1975, 1976, 1979), realized that, for short times, the noise in $A(t)$ computed for the two trajectories, will be highly correlated. They used this idea to reduce the noise in the response computed at small applied fields.

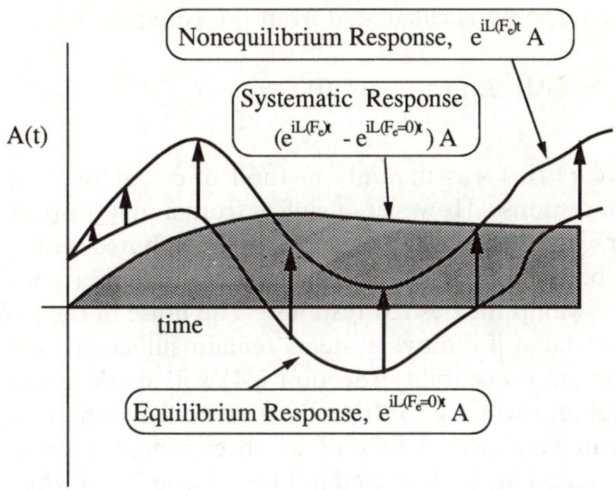

Fig. 7.6 We depict the systematic nonequilibrium response (the shaded curve) as the difference of the nonequilibrium response from the equilibrium response. By taking this difference we can dramatically reduce the noise in the computed systematic nonequilibrium response. To complete this calculation one averages this difference over an ensemble of starting states.

To use their *subtraction method* one performs an equilibrium simulation ($F_e = 0$) from which one periodically initiates nonequilibrium calculations ($F_e \neq 0$). The general idea is shown in Fig. 7.7. The phases $\{\mathbf{\Gamma}_i\}$, are taken as time origins from which one calculates the differences of the response in a phase variable with and without the applied field. The *systematic* or

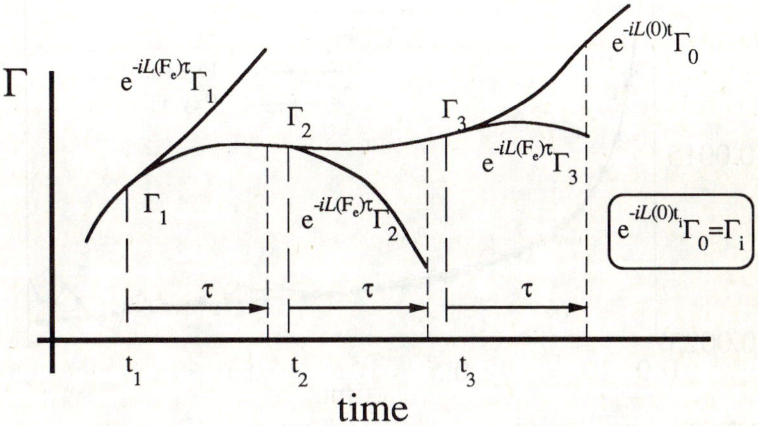

Fig. 7.7 Illustration of the subtraction method.

nonequilibrium response is calculated from the equation

$$\langle A(t;F_e)\rangle = \langle A(t;F_e)\rangle - \langle A(t;0)\rangle = \frac{1}{N}\sum_{i=1}^{N}[e^{iL(F_e)t} - e^{iL(0)t}]A(\Gamma_i) \tag{7.66}$$

For many years this was the only method of calculating the small-field nonequilibrium response. However, it suffers from a major problem. For the method to work, the noise in the value of $A(t)$ computed with and without the field must be highly correlated, otherwise the equilibrium fluctuations will completely swamp the desired response. The noise in the two responses will only be correlated if the two systems remain sufficiently close in phase space. The Lyapunov instability (Section 3.4) will work against this. The Lyapunov instability will try to drive the two systems apart exponentially quickly. This can be expected to lead to an exponential growth of noise with respect to time. This is illustrated in Figs 7.8 and 7.9 in which the TTCF and subtraction techniques are compared for the 256-particle WCA system considered in Section 7.5.

Figure 7.8 shows the shear stress for the three-dimensional WCA system at the comparatively small strain rate of $\gamma^* = 10^{-3}$. At this field strength conventional steady-state NEMD is swamped by noise. However, the subtraction technique can be used to improve the statistics substantially. It

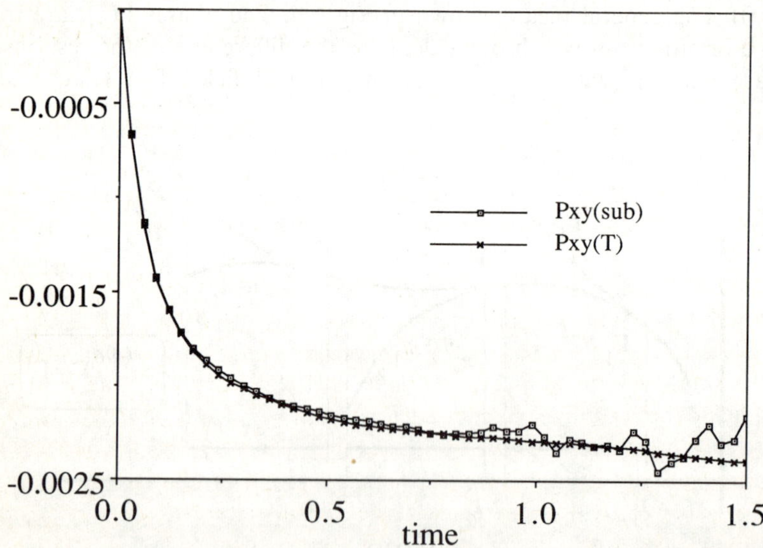

Fig. 7.8 Shear stress for the three-dimensional WCA system at a strain rate of $\gamma^* = 10^{-3}$. sub, Subtraction technique; T, TTCF.

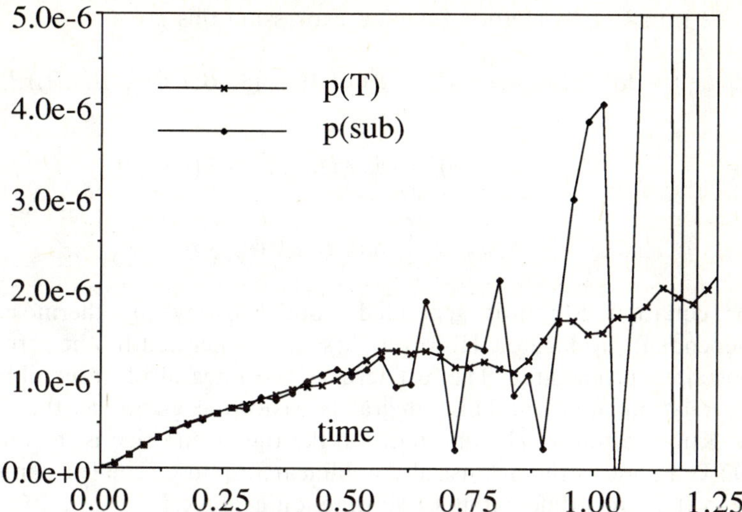

Fig. 7.9 Shear dilatancy for the three-dimensional WCA system at a strain rate of $\gamma^* = 10^{-3}$, sub, Subtraction technique; T, TTCF.

is important to note that both the subtraction and TTCF techniques are based on an analysis of the transient response of systems. The results compared in Fig. 7.8 were computed for *exactly* the same system using *exactly* the same data. The only difference between the two sets of results is the way in which the data were analysed. Lyapunov noise is clearly evident in the subtraction results. For longer times, during which we expect the slow nonlinearities to complete the relaxation to the steady state, the subtraction technique becomes very noisy.

Figure 7.9 shows the corresponding results for shear dilatancy. Here the subtraction technique is essentially useless. Even the TTCF method becomes somewhat noisy at long times. The TTCF results clearly show the existence of a measurable, intrinsically nonlinear effect even at this small strain rate.

Although the TTCF method allows us to compute the response of systems to fields of arbitrary, even zero, strength, we often require more information about the small-field response than it is capable of providing. For example, at small fields the response is essentially linear. Nonlinear effects that we may be interested in are completely swamped by the linear response terms. The *differential transient time-correlation function* (DTTCF) is an attempt to provide an answer to this problem. It uses a subtraction technique on the TTCFs themselves formally to subtract the linear response.

In the DTTCF method we consider the difference between $B(s)$ evaluated with and without the external field, starting from the same initial phase point.

From the transient correlation function expression this gives

$$\begin{aligned}\langle (B(t,\gamma) - B(0,\gamma))\rangle &= -\beta\gamma V \int_0^t \mathrm{d}s\, \langle (B(s,\gamma) - B(s,0) + B(s,0))P_{xy}\rangle \\ &= -\beta\gamma V \int_0^t \mathrm{d}s\, \langle (B(s,\gamma) - B(s,0))P_{xy}\rangle \\ &\quad + \beta\gamma V \int_0^t \mathrm{d}s\, \langle B(s,0)P_{xy}\rangle \end{aligned} \tag{7.67}$$

In this equation $B(s,\gamma)$ is generated from $B(0)$ by the thermostatted field-dependent propagator. However, $B(s,0)$ is generated by the zero-field thermostatted propagator. The last term is the integral of an equilibrium time-correlation function. This integral is easily recognized as the *linear*, Green–Kubo response. The first term on the right-hand side is the integral of a DTTCF, and is the intrinsically nonlinear response. The left-hand side is termed the direct differential, or subtraction average.

There are two possible cases: (i) B has a non-zero linear response term; and (ii) the linear response is identically zero. If B is chosen to be P_{xy} the third term in (7.67) is the Green–Kubo expression for the response of the shear stress $\eta(0)\gamma$, where $\eta(0)$ is the zero-shear-rate shear viscosity. The definition of the shear-rate dependent viscosity, $\eta(\gamma) \equiv -\langle P_{xy}\rangle/\gamma$ gives

$$\eta(\gamma) - \eta(0) = \beta V \int_0^\infty \mathrm{d}s\, \langle (P_{xy}(s,\gamma) - P_{xy}(s,0))P_{xy}\rangle \tag{7.68}$$

as the intrinsically nonlinear part of the shear viscosity. As $s \to \infty$ the DTTCF (using the mixing assumption) becomes $\langle (P_{xy}(s,\gamma) - P_{xy}(s,0))\rangle\langle P_{xy}(0)\rangle = \langle P_{xy}(s,\gamma)\rangle\langle P_{xy}(0)\rangle$. This is zero because $\langle P_{xy}(0)\rangle$ is zero. However, $\langle (P_{xy}(s,\gamma)\rangle$ is clearly non-zero which means that the use of our trajectory mappings will improve the statistics as $s \to \infty$.

To apply the phase-space mappings in the differential response method we consider the identity (7.63). We can obtain four different time evolutions of $B(\mathbf{\Gamma})$ by simply removing the minus signs and parity operators from each of the equivalent forms in equation (7.63). If we use the index α to denote the four mappings $\{\mathrm{I}, \mathrm{T}, \mathrm{Y}, \mathrm{K}\}$, then

$$\begin{aligned}\sum_{\alpha=\{\mathrm{I,T,Y,K}\}} B(t,\mathbf{\Gamma}^\alpha,\gamma^\alpha) &= \sum_\alpha \exp(\mathrm{i}L(\mathbf{\Gamma}^\alpha,\gamma^\alpha)t)B(\mathbf{\Gamma}^\alpha) \\ &= \sum_\alpha \exp(\mathbf{M}^\alpha[\mathrm{i}L(\mathbf{\Gamma},\gamma)]t)p_{\mathrm{B}}^\alpha B(\mathbf{\Gamma}) \\ &= \{\exp(\mathrm{i}L(\mathbf{\Gamma},\gamma)t)p_{\mathrm{B}}^{\mathrm{I}} + \exp(-\mathrm{i}L(\mathbf{\Gamma},-\gamma)t)p_{\mathrm{B}}^{\mathrm{T}} \\ &\quad + \exp(\mathrm{i}L(\mathbf{\Gamma},-\gamma)t)p_{\mathrm{B}}^{\mathrm{Y}} + \exp(-\mathrm{i}L(\mathbf{\Gamma},\gamma)t)p_{\mathrm{B}}^{\mathrm{K}}\}B(\mathbf{\Gamma})\end{aligned} \tag{7.69}$$

This is the direct response of the phase variable $B(\mathbf{\Gamma})$ from one sampling of $\mathbf{\Gamma}$, where the mappings are used to generate four starting phase points. To calculate the differential response of B we need to subtract the field-free time evolution of $B(\mathbf{\Gamma})$ from each of these four starting states. The four field-free time evolutions are found by setting $\gamma^\alpha = 0$ in (7.69); that is

$$\sum_{\alpha=\{\mathrm{I,T,Y,K}\}} B(t, \mathbf{\Gamma}^\alpha, \gamma^\alpha = 0) = \{p_{\mathrm{B}}^{\mathrm{I}} \exp(\mathrm{i}\mathrm{L}t) + p_{\mathrm{B}}^{\mathrm{T}} \exp(-\mathrm{i}\mathrm{L}t) + p_{\mathrm{B}}^{\mathrm{Y}} \exp(\mathrm{i}\mathrm{L}t) + p_{\mathrm{B}}^{\mathrm{K}} \exp(-\mathrm{i}\mathrm{L}t)\} B(\mathbf{\Gamma}) \tag{7.70}$$

Clearly there are only two different field-free time evolutions; the remaining two can be obtained from these by the sign changes of the parity operators. In practice, a single cycle of the numerical evaluation of a DTTCF will involve the calculation of four field dependent trajectories and two field-free trajectories, yielding four starting states.

The use of the symmetry mappings implies some redundancies in the various methods of calculating the response. In particular, the *direct* response of $P_{xy}(t)$ is exactly equal to the *subtraction* response for all values of the time. This means that the contribution from the field-free time evolutions is exactly equal to zero. This is easy to see from (7.69) as there are only two different time evolutions; those corresponding to $\exp(\mathrm{i}\mathrm{L}t)$ and $\exp(-\mathrm{i}\mathrm{L}t)$, respectively, and for P_{xy} each comes with a positive and negative parity operator. Therefore, these two responses exactly cancel for all values of time.

The second redundancy of interest is that the *transient* response of the pressure, $p(t)$, is exactly equal to the *differential transient* response for all values of time. This implies that the contribution to the *equilibrium* time-correlation function, $\langle p(t)P_{xy}\rangle$, from a single sampling of $\mathbf{\Gamma}$ is exactly zero. Clearly this equilibrium time correlation is zero when the full ensemble average is taken, but the result we prove here is that the mappings ensure that $\Sigma p(t)P_{xy}$ is zero for each starting state $\mathbf{\Gamma}$ for all values of t. The contribution from the field-free trajectories is

$$\sum_{\alpha=\{\mathrm{I,T,Y,K}\}} p(t, \mathbf{\Gamma}^\alpha, \gamma^\alpha = 0) P_{xy}(\mathbf{\Gamma}^\alpha) = P_{xy}(\mathbf{\Gamma})\{p_{P_{xy}}^{\mathrm{I}} p_p^{\mathrm{I}} \exp(\mathrm{i}\mathrm{L}t) + p_{P_{xy}}^{\mathrm{T}} p_p^{\mathrm{T}} \exp(-\mathrm{i}\mathrm{L}t) + p_{P_{xy}}^{\mathrm{Y}} p_p^{\mathrm{Y}} \exp(\mathrm{i}\mathrm{L}t) + p_{P_{xy}}^{\mathrm{K}} p_p^{\mathrm{K}} \exp(-\mathrm{i}\mathrm{L}t)\} p(\mathbf{\Gamma}) = 0 \tag{7.71}$$

Again the product of parities ensures that the two field-free time evolutions $\exp(\mathrm{i}\mathrm{L}t)$, and $\exp(-\mathrm{i}\mathrm{L}t)$ occur in cancelling pairs. Therefore, the field-free contribution to the DTTCF is exactly zero and the differential transient results are identical to the transient correlation function results.

The DTTCF method suffers from the same Lyapunov noise characteristic of all differential or subtraction methods. In spite of this problem, Evans and

Morriss (1987) were able to show, using the DTTCF method, that the intrinsically nonlinear response of three-dimensional fluids undergoing shear flow is given by the classical Burnett form (see Section 9.5). This is at variance with the nonclassical behaviour predicted by mode-coupling theory. However, nonclassical behaviour can only be expected in the large-system limit. The classical behaviour observed by Morriss and Evans (1987) is presumably the small strain rate, asymptotic response for *finite* periodic systems.

A much better approach to calculating and analysing the asymptotic nonlinear response will be discussed in Section 9.5.

7.7 NUMERICAL RESULTS FOR THE KAWASAKI REPRESENTATION

We now present results of a direct numerical test of the Kawasaki representation for the nonlinear isothermal response. We show that phase averages calculated using the explicitly normalized Kawasaki distribution function agree with those calculated directly from computer simulation.

The system we consider is the thermostatted nonequilibrium molecular dynamics (NEMD) simulation of planar Couette flow using the isothermal SLLOD algorithm (Section 6.3). As remarked earlier, the primary difficulty in using the Kawasaki expression in numerical calculations arises because it involves calculating an extensive exponential. For a 100-particle Lennard–Jones triple-point system we would have to average quantities of the order of e^{200} to determine the viscosity. Larger system sizes would involve proportionately larger exponents! The simulations presented here attempt to reduce these difficulties by using two strategies: (i) using a comparatively small number, $N = 18$, of soft discs in two dimensions; and (ii) using a low density, $\rho^* = 0.1$, where the viscosity is ca. 50 times smaller than its triple-point value. For small systems it is necessary to take into consideration terms of order $1/N$ in the definition of the temperature, $T = (\Sigma_i p_i^2/m)/(dN - d - 1)$, and the shear stress, $P_{xy}V = (dN - d)/(dN - d - 1)\Sigma_i p_{xi} p_{yi}/m - (1/2)\Sigma_{ij} y_{ij} F_{xij}$.

The first-order equations of motion were solved using the fourth-order Runge–Kutta method with a reduced time-step of 0.005. The reduced shear rate, γ^*, and the reduced temperature were both set to unity.

The simulation consisted of a single equilibrium trajectory. At regular intervals (every 399 time-steps) the current configuration was used to construct four different configurations using the trajectory mappings described in Section 7.4. Each of these configurations was used as an initial starting point for a non-equilibrium simulation of 400 time-steps, with a negative time-step and reduced shear rate, $\gamma = 1$. Time-dependent averages were calculated, with the time being measured from the preceding equilibrium starting state. The aim was to produce the Kawasaki averages by exactly programming the dynamics in the Kawasaki distribution function (equation 7.24).

The phase-space integral of the *bare* Kawasaki distribution function, $f(t)$, equation (7.24), is

$$Z(t) = \int \mathrm{d}\mathbf{\Gamma}\, f(\mathbf{\Gamma}, t) = \int \mathrm{d}\mathbf{\Gamma}\, f(\mathbf{\Gamma}, 0) \exp\left[-\beta F_{\mathrm{e}} \int_0^t \mathrm{d}s\, J(-s) \right] \quad (7.72)$$

$Z(0)$ is the phase integral of the equilibrium distribution function which is equal to unity since $f(0)$ is the normalized equilibrium distribution function. It is interesting to consider the rate of change in $Z(t)$ after the external field is switched on. Using manipulations based on the reversible Liouville equation we can show that

$$\begin{aligned}\frac{\mathrm{d}Z(t)}{\mathrm{d}t} &= \int \mathrm{d}\mathbf{\Gamma}\, f(0) \frac{\partial}{\partial t} \exp\left[-\beta F_{\mathrm{e}} \int_0^t \mathrm{d}s\, J(-s) \right] \\ &= -\beta F_{\mathrm{e}} \int \mathrm{d}\mathbf{\Gamma}\, f(t) J(-t) \\ &= -\beta F_{\mathrm{e}} \int \mathrm{d}\mathbf{\Gamma}\, f(0) J(0) = 0 \end{aligned} \quad (7.73)$$

The last equality is a consequence of the Schrödinger–Heisenberg equivalence (Section 3.3). This implies that the bare Kawasaki distribution function is normalized for all times t. This is a direct result of the reversibility of the classical equations of motion. In Fig. 7.10 we present the numerical results for $Z(t)$. Figure 7.10 shows that (7.73) is clearly false· The normalization is unity only for a time proportional to the Lyapunov time for the system. After this time the normalization decreases rapidly. The explanation of this apparent paradox is that the analysis used to establish (7.73) is based on the reversible Liouville equation. The simulation used to generate the results shown in Fig. 7.10 is, however, not time reversible. Trajectories which imply a long period (compared with the Lyapunov time) of entropy *decrease* are mechanically unstable both in nature and in computer simulations. Because it is impossible to integrate the equations of motion exactly, these entropy-decreasing trajectories are not observed for times longer than the Lyapunov time which characterizes the irreversible instability of the equations of motion.

The form of the function, $Z(t)$, shown in Fig. 7.10, is determined by the accuracy with which the calculations are carried out. In principle, by using ever more powerful computers one could, by increasing the word length and by decreasing the integration time-step, ensure that the computed $Z(t)$ stayed close to unity for longer and longer times. The exact result is that $Z(t) = 1$.

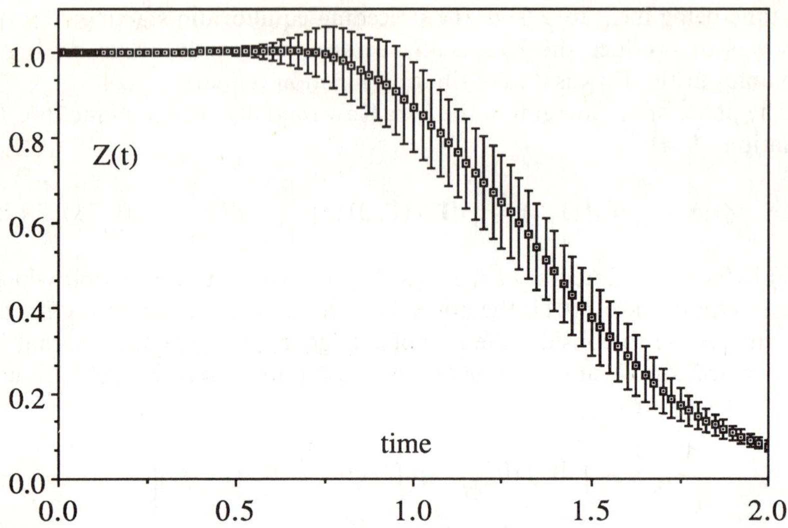

Fig. 7.10 We show computer simulation results for the Kawasaki normalization, $Z(t)$. According to the Liouville equation this function should be unity for all times, t. This is clearly not the case (see Evans, 1990, for details).

For a hard-sphere system, the time over which the trajectory accuracy is better than a set tolerance only increases as $-\ln(\varepsilon^{1/\lambda})$, where λ is the largest Lyapunov exponent for the system and ε is the magnitude of the least-significant digit representable on the computer. However, our ability to extend numerically the times over which $Z(t) \simeq 1$, is much worse than this analysis implies. As we compute (7.72) for longer times, the variance in $\langle \exp[-\beta F_e \int_0^t ds\, J(-s)] \rangle$ grows exponentially with time, regardless of the accuracy with which the trajectories are computed!

We have discussed the Kawasaki normalization in terms of numerical procedures. However, exactly the same arguments apply to the experimental observation of the normalization. In nature, the problems in observing $Z(t) \simeq 1$ for long times result from uncontrolled external perturbations on the system rather than from numerical errors. However, numerical error can be regarded as a particular form of external perturbation (ε above, would then be a measure of the background-noise level). Of course the act of observation itself is a source of 'external' noise.

The results shown in Fig. 7.10 show that the *computed* bare Kawasaki distribution function is not properly normalized. Thus we should not be surprised to see that the bare Kawasaki expression for the average shear stress is inconsistent with the results of direct calculation, as shown in Fig. 7.11.

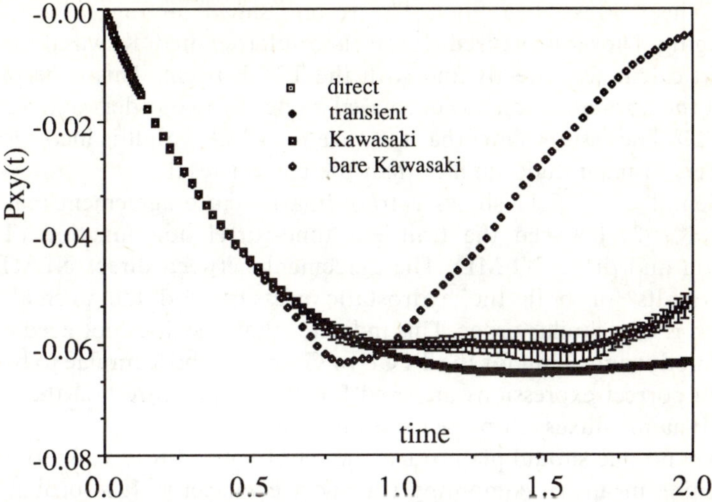

Fig. 7.11 We compare four different methods of computing the nonlinear non-equilibrium response of a system of 18 soft discs to a suddenly imposed shear flow. The agreement between the Transient Time Correlation Function method and direct nonequilibrium molecular dynamics is better than 2 parts in 10^3 over the entire range of times studied. This is the most convincing numerical verification yet made of the correctness of the Transient Time Correlation Function method. The renormalized Kawasai method (denoted Kawasaki) is in statistical agreement with the direct calculations but the bare Kawasaki method is clearly incorrect (see Evans, 1990, for details).

The obvious way around this problem is to *explicitly normalize* the distribution function (Morriss and Evans, 1987). The explicitly normalized form is

$$f(t) = \frac{f(0)\exp\left[-\beta F_e \int_0^t ds\, J(-s)\right]}{\int d\Gamma\, f(0)\exp\left[-\beta F_e \int_0^t ds\, J(-s)\right]} \tag{7.74}$$

The renormalized average of the shear stress is then

$$\langle P_{xy}(t)\rangle = \frac{\int d\Gamma\, P_{xy}(\Gamma) f(0)\exp\left[-\beta F_e \int_0^t ds\, J(-s)\right]}{\int d\Gamma\, f(0)\exp\left[-\beta F_e \int_0^t ds\, J(-s)\right]} \tag{7.75}$$

We used computer simulation to compare the direct NEMD averages and the bare and renormalized Kawasaki expressions for the time-dependent average shear stress in a fluid. The results shown in Fig. 7.11 are very encouraging. The renormalized Kawasaki result (denoted 'Kawasaki') agrees with that calculated directly and with the TTCF result. This is despite the fact that the normalization has decreased by nearly two orders of magnitude at $t^* = 2.0$. The results show that the *bare* Kawasaki result is incorrect; it is two orders of magnitude smaller than the correct results.

Incidentally, Fig. 7.11 shows extraordinarily close agreement (ca. 0.2% for $0 < t^* < 2$) between the transient time-correlation function (TTCF) prediction and direct NEMD. The agreement between direct NEMD and TTCF results for both the hydrostatic pressure and the normal stress difference is of a similar order. This indicates that one does not need to take the thermodynamic limit for the TTCF or Green–Kubo formulae to be valid. Provided correct expressions are used for the temperature and the various thermodynamic fluxes, 18 particles seems sufficient.

Clearly no one should plan to use the renormalized Kawasaki formalism as a routine means of computing transport coefficients. It is probably the least efficient known method for computing nonequilibrium averages. The Kawasaki formalism is, however, a very important theoretical tool. It was of crucial importance in the development of nonlinear response theory and it provided an extremely useful basis for subsequent theoretical derivations. As we will see in Chapter 9, the renormalized Kawasaki formalism, in contrast to the TTCF formalism, is very useful in providing an understanding of fluctuations in nonequilibrium steady states.

7.8 THE VAN KAMPEN OBJECTION TO LINEAR RESPONSE THEORY

Having explored some of the fundamentals of nonlinear response theory, we are now in a better position to comment on one of the early criticisms of linear response theory. In an oft-cited paper, van Kampen (1971) criticized linear response theory on the basis that microscopic linearity which is assumed in linear response theory is quite different from the macroscopic linearity manifest in linear constitutive relations. Van Kampen correctly noted that to observe linear *microscopic* response (i.e. of individual particle trajectories) over macroscopic time (seconds, minutes, or even hours) requires external fields which are orders of magnitude smaller than those for which linear macroscopic behaviour is actually observed. Therefore, so the argument goes, the theoretical justification of the Green–Kubo relations for linear transport coefficients is suspect.

In order to explain his assertion that *linearity of microscopic motion is entirely different from macroscopic linearity*, van Kampen considered a system composed of electrons which move, apart from occasional collisions with impurities, freely through a conductor. An imposed external field, F_e, accelerates the particles between collisions. The distance which an electron moves in a time t, under the influence of the field, is $1/2t^2(eF_e/m)$. In order for the induced current to be linear it is required that $t^2(eF_e/2m) \ll d$, the mean spacing of the impurities. Taking $d \simeq 100$ Å and t to be a macroscopic time, say 1 s, we see that the field must be less than ca. 10^{-18} V cm^{-1}!

As a criticism of the derivation of linear response theory, this calculation implies that, for linear response theory to be valid, trajectories must be subject to a linear perturbation over macroscopic times—the time taken for experimentalists to make sensible measurements of the conductivity. This however is incorrect.

The linear response theory expression for the conductivity, $\sigma(\equiv J/F_e)$ is

$$\sigma = \beta V \int_0^\infty \mathrm{d}t \, \langle J(t)J(0) \rangle_{\mathrm{eq}} \tag{7.76}$$

Now it happens that in three-dimensional systems the integral of the equilibrium current autocorrelation function converges rapidly. (In two-dimensional systems this is not so.) The integral in fact converges in microscopic time, a few collision times in the above example. Indeed if this were not so one could never use equilibrium molecular dynamics to compute transport coefficients from the Green–Kubo formulae. Molecular dynamics is based on the assumption that transport coefficients for simple fluids can be computed from simulations which only follow the evolution systems for ca. 10^{-10} s. These times are sufficient to ensure convergence of the Green–Kubo correlation functions for all the Navier–Stokes transport coefficients. If we require microscopic linearity over 10^{-10} s (rather than van Kampen's 1 s), then we see that the microscopic response will be linear for fields of less than about 100 V cm^{-1}, not an unreasonable number. It simply does not matter that for times longer than those characterizing the relaxation of the relevant Green–Kubo correlation function, the motion is perturbed in a nonlinear fashion. In order for linear response theory to yield correct results for linear transport coefficients, linearity is only required for times characteristic of the decay of the relevant correlation functions. These times are microscopic.

We used a nonequilibrium molecular dynamics (NEMD) simulation of shear flow in an atomic system to explore the matter in more detail (Morriss *et al.*, 1989). We performed a series of simulations with and without an imposed strain rate, γ $(\equiv \partial u_x/\partial y)$, to measure the actual separation, d, of phase-space trajectories as a function of the imposed strain rate. The phase-space separation is defined as

$$d(t,\gamma) \equiv \sqrt{(\mathbf{\Gamma}(t,\gamma) - \mathbf{\Gamma}(t,0))^2} \tag{7.77}$$

where $\mathbf{\Gamma} \equiv (\mathbf{q}_1, \mathbf{q}_2, \ldots \mathbf{q}_N, \mathbf{p}_1, \mathbf{p}_2, \ldots \mathbf{p}_N)$ is the $6N$-dimensional phase-space position for the system. In measuring the separation of phase-space trajectories we imposed the initial condition that at time zero the equilibrium and nonequilibrium trajectories start from *exactly* the same point in phase space, $d(0, \gamma) = 0, \forall \gamma$. We used the 'infinite checker board' convention to define the Cartesian coordinates of a particle in a periodic system; this eliminates trivial discontinuities in these coordinates. We also reported the ensemble average of the phase-space separation, averaged over an equilibrium ensemble of initial phases, $\mathbf{\Gamma}(0, 0)$.

The equations of motion employed were the SLLOD equations. As we have seen the linear response computed from these equations is given precisely by the Green–Kubo expression for the shear viscosity. The system studied in these simulations was the Lennard–Jones fluid at its triple point ($\rho^* = \rho\sigma^3 = 0.8442$, $T^* = k_B T/\varepsilon = 0.722$, $t^* = t(\varepsilon/m)^{1/2}\sigma^{-1}$). A Lees–Edwards periodic system of 256 particles with a potential truncated at $r^* = r/\sigma = 2.5$ was employed.

Before we begin to analyse the phase-separation data we need to review some of the relevant features of Lennard–Jones triple-point rheology. Firstly, as we have seen (Section 6.3), this fluid is shear thinning. The strain-rate dependent shear viscosities of the Lennard–Jones triple-point fluid are set out in Table 7.3. The most important relevant fact that should be noted from these results is that for reduced strain rates, $\gamma^* < \sim 10^{-2}$, the fluid is effectively Newtonian with a viscosity which varies at most, by less than ca. 4% of its zero shear value. (Because of the uncertainty surrounding the zero shear viscosity, we cannot be more certain about the degree of nonlinearity present at $\gamma^* = 0.01$.)

The second relevant fact that we should remember is that the Green–Kubo equilibrium time-correlation function, whose integral gives the shear viscosity, has decayed to less than 1% of its zero-time value at a reduced time $t^* = 2.0$. The values are given in Table 7.4. Of course the viscosity which is the time integral of this correlation function, converges, relatively slowly due to the

Table 7.3 Strain-rate dependent shear viscosities for the triple-point Lennard–Jones fluid.

Reduced strain rate	Reduced viscosity	Nonlinearity (%)
1.0	2.17 ± 0.03	37
0.1	3.04 ± 0.03	12
0.01	3.31 ± 0.08	~4
0.0	3.44 ± 0.2	0‡ NEMD est

‡ NEMD estimated.

Table 7.4 Green–Kubo equilibrium stress correlation function for shear viscosity.

t^*	Correlation function	Percentage of $t = 0$ value
0.0	24.00	100
0.1	7.17	29
1.0	0.26	1
2.0	0.09	0.3

presence of the slowly decaying $t^{-3/2}$ long time tail. Here again there is some uncertainty. If one believes that *enhanced long-time tail* phenomena (Section 6.3) are truly asymptotic and persist indefinitely, then one finds that the viscosity converges to within ca. 13% of its asymptotic value at $t^* = 1.0$ and to within ca. 5% of the asymptotic value at $t^* = 10.0$. (If we map our simulation onto the standard Lennard–Jones representation of argon, $t^* = 1.0$ corresponds to a time of 21.6 ps.) If *enhanced long-time tails* are not asymptotic then the Green–Kubo integrand for the shear viscosity converges to within ca. 5% of its infinite time value by $t^* = 2$.

The only important observation that concerns us here is that the Green–Kubo estimate for the shear viscosity is determined in *microscopic* time, a few hundreds of picoseconds at the very most, for argon. This observation was omitted from van Kampen's argument. We call the range of times required to ensure say 5% convergence of the Green–Kubo expression for the viscosity, the *Green–Kubo time window.*

Figure 7.12 shows the common logarithm of the ensemble average of the phase-space separation plotted as a function of reduced time for various values of the imposed shear rate. The shear rates employed were: $\gamma^* = 1.0$, 10^{-1}, 10^{-2}, 10^{-3}, 10^{-5} and 10^{-7}. Note that for the standard Lennard–Jones argon representation, these strain rates correspond to shear rates of 4.6×10^{11} to 4.6×10^{5} Hz. It is clear from the present results that no new phenomena would be revealed at strain rates of less than $\gamma^* \simeq 10^{-4}$.

One can see from the figure that at a shear rate of 10^{-7} the phase-space separation increases very rapidly initially and then slows to an exponential increase with time. The same pattern is followed at a strain rate of 10^{-5} except that the initial rise is even more rapid than for a strain rate of 10^{-7}. Remember that at $t = 0$ the phase-space separations start from zero and, therefore, the logarithm of the $t = 0$ separations is $-\infty$ for all strain rates.

For strain rates of $> 10^{-5}$, we notice that at long times the phase separation is a constant independent of time. We see an extremely rapid initial rise, followed by an exponential increase with a slope *which is independent of strain rate*, followed at later times by a plateau. The plateau is easily understood.

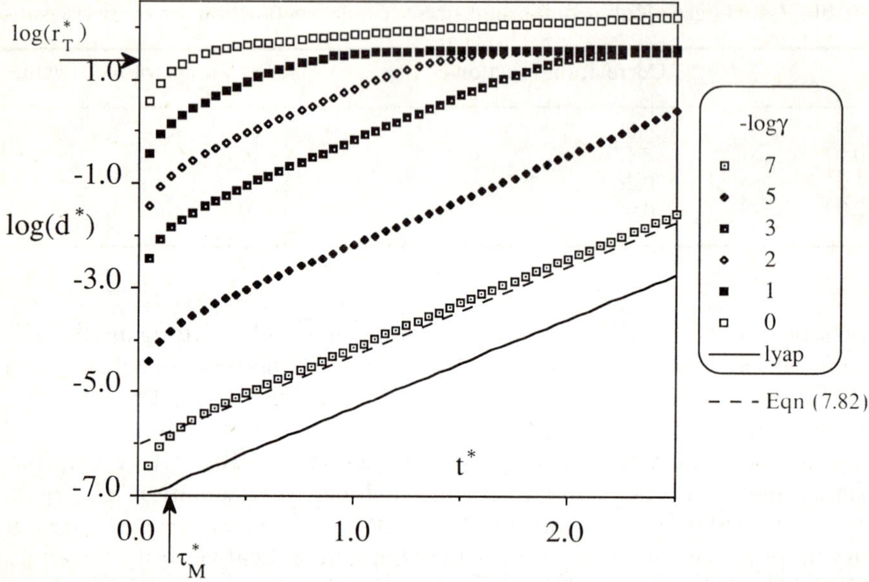

Fig. 7.12 Logarithm of the ensemble average of the phase-space separation plotted as a function of reduced time for various values of the imposed shear rate, γ^* (see text).

The simulations shown in Figs 7.12 and 7.13 are carried out at constant peculiar kinetic energy $\Sigma p_i^2/2m = 3Nk_B T$. The $3N$ components of the phase-space momenta therefore lie on the surface of a $3N$-dimensional sphere of radius, $r_T = \sqrt{(3Nmk_B T)}$. Once the phase-space separation exceeds this radius, the curved nature of accessible momentum space will be apparent in our phase-space-separation plots. The arrow marked on Fig. 7.12 shows when the logarithm of the separation is equal to this radius. The maximum separation of phase points within the momentum sub-space is of course $2r$. It is clear, therefore, that the exponential separation must end at approximately $d(t, \boldsymbol{\Gamma}) = r_T$; this is exactly what is observed in Fig. 7.12.

Between the plateau and the initial (almost vertical) rise is an exponential region. As can be seen from the graph the slope of this rise is virtually independent of strain rate. The slope is related to the largest positive Lyapunov exponent for the system at equilibrium. The Lyapunov exponent measures the rate of separation of phase trajectories that start a small distance apart, but which are governed by identical dynamics. After initially being separated by the external field, the rate of phase-space separation thereafter is governed by the usual Lyapunov instability. The fact that the two trajectories employ slightly different dynamics is a secondary consideration.

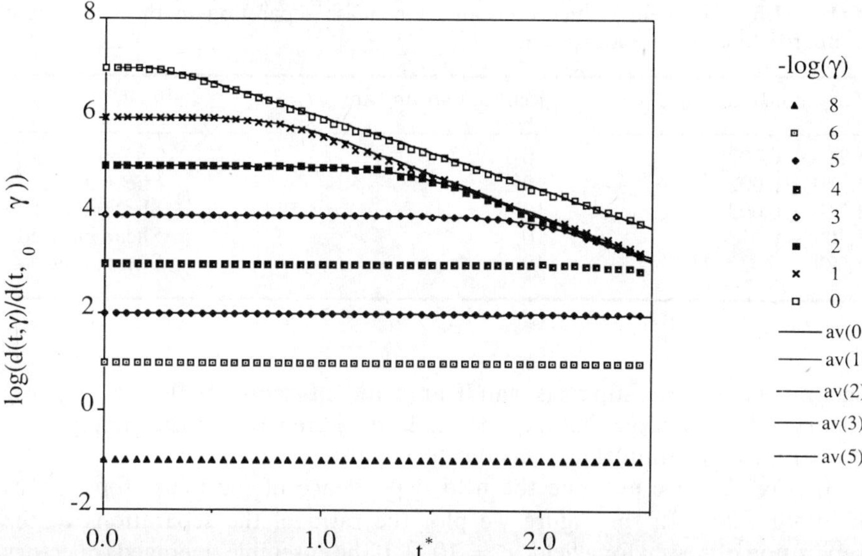

Fig. 7.13 We plot the ratio of the phase space separations as a function of strain rate and time. The ratios are computed relative to the separation at a reduced strain rate of 10^{-7}. Curves denoted by 'av' are ensemble averages. Those not so denoted give the results for individual phase trajectories. Since the integrals of the Green–Kubo correlation functions converge to within a few percent by a reduced time of ~ 1.5, we see that the trajectory separation is varying linearly with respect to strain rate for reduced strain rates less than ~ 2. This is precisely the strain rate at which direct nonequilibrium molecular dynamics shows a departure of the computed shear viscosity from linear behaviour.

The Lyapunov exponents are known to be insensitive to the magnitude of the perturbing external field for field strengths of less than 10^{-2}.

This conjecture regarding the role played by the Lyapunov exponent in the separation of equilibrium and nonequilibrium trajectories which start from a common phase origin is easily verified numerically. Instead of measuring the separation, d, induced by the strain rate, we ran a simulation in which two trajectories started at slightly different phases and which evolved under (identical) zero strain-rate equations of motion. The resulting displacement is shown in Fig. 7.12 and labelled as 'lyap' in the legend. One can see that the slope of this Lyapunov curve has identical shape to the exponential portions of the strain-rate driven curves. The time constants for the exponential portions of the curves are given in Table 7.5.

At this stage we see that even at our smallest strain rate, the trajectory separation is exponential in time. It may be thought that this exponential

Table 7.5 Exponential time constants for phase separation in the triple-point Lennard–Jones fluid under shear.

Time constant	Reduced strain rate	
1.715 ± 0.002	0.0	Lyapunov
1.730 ± 0.002	10^{-7}	Shear induced
1.717 ± 0.002	10^{-5}	Shear induced
1.708 ± 0.012	10^{-3}	Shear induced
1.689 ± 0.03	10^{-2}	Shear induced

separation in time supports van Kampen's objection to linear response theory. The assertion that exponentially diverging trajectories imply non-linearity in the strain rate turns out to be false.

In Fig. 7.13 we examine the field dependence of the phase separations in more detail. In this figure we plot the ratio of the separations to the separation observed for a field, $\gamma^* = 10^{-7}$. If the ensemble-averaged trajectory response is linear then each of the curves in Fig. 7.13 will be equispaced horizontal lines. The curves denoted 'av' refer to the ensemble averaged separations shown in Fig. 7.12. One can see immediately that within the Green–Kubo time window, $t^* < 2.0$, all the separations are linear in the field except for the largest two strain rates, $\gamma^* = 1.0$ and 0.1. We should expect that all strain rates exhibiting linearity within the Green–Kubo time window should correspond to those systems exhibiting macroscopic linear behaviour (i.e. those which are Newtonian). Those exhibiting microscopic nonlinearity within the Green–Kubo time window should display non-Newtonian macroscopic behaviour. Comparing Table 7.3 with Fig. 7.12, this is exactly what is seen.

Although systems at shear rates $\gamma^* = 10^{-2}$ and 10^{-4} do exhibit a nonlinear growth in the phase-space separation, it occurs at times which are so late that it cannot possibly affect the numerical values of the shear viscosity. These nonlinearities occur outside the Green–Kubo time window.

A possible objection to these conclusions might be that, since we are computing ensemble *averages* of the phase-space separations, it might be the averaging process which ensures the observed microscopic linearity. Individual trajectories might still be perturbed nonlinearly with respect to the strain rate. This, however, is not the case. In Fig. 7.13 the symbols plotted represent the phase-space separation induced in *single* trajectories. For all strain rates a common phase origin is used. We did not average over the time zero phase origins of the systems.

What we see is a slightly noisier version of the ensemble-averaged results. Detailed analysis of the unaveraged results reveals that:

(1) for $\gamma^* < 10^{-2}$ linearity in strain rate is observed for individual trajectories; and
(2) the exponential behaviour in time is only observed when $d(\gamma, t)$ is averaged over some finite but small, time interval.

The exponential Lyapunov separation is of course only expected to be observed 'on average' either by employing time or ensemble averages. The main point we make here is that even for individual trajectories where phase separation is not exactly exponential in time, trajectory separation is, to four significant figures of accuracy, linear in the field. The linearity of the response is not produced by ensemble averaging.

We conclude from these studies that within the Green–Kubo time window, macroscopic and microscopic linearity are observed for identical ranges of strain rates. For times shorter than those required for convergence of the linear response theory expressions for transport coefficients, the individual phase-space trajectories are perturbed linearly with respect to the strain rate for those values of the strain rate over which the fluid exhibits linear macroscopic behaviour. This is in spite of the fact that within this domain the strain rate induces an exponential separation of trajectories with respect to time. We believe that many people have assumed an exponential trajectory separation in time implies an exponential separation with respect to the magnitude of the external field. This work shows that within the Green–Kubo time window, the dominant microscopic behaviour in fluids which exhibit linear macroscopic behaviour is linear in the external field but exponential in time.

It is shown in Fig. 7.12 that for intermediate times the phase separation takes the form

$$d = A\gamma \exp[t/\tau_L] \tag{7.78}$$

where the *Lyapunov time*, τ_L, is the inverse of the largest Lyapunov exponent for the system at equilibrium. We can explain why the phase separation exhibits this functional form and, moreover, we can make a rough calculation of the absolute magnitude of the coefficient, A. We know that the exponential separation of trajectories only begins after a time which is roughly the Maxwell relaxation time, τ_M, for the fluid. Before the particles sense their mutual interactions, the particles are freely streaming with trajectories determined by the initial values of $(\mathrm{d}\mathbf{q}/\mathrm{d}t, \mathrm{d}\mathbf{p}/\mathrm{d}t)$. After this initial motion the particles will have coordinates and momenta as follows:

$$\mathbf{q}_i(t) = \mathbf{q}_i(0) + \left[\frac{\mathbf{p}_i(0)}{m} + \mathbf{i}\gamma y_i(0)\right]t$$
$$\mathbf{p}_i(t) = \mathbf{p}_i(0) + [\mathbf{F}_i(0) - \mathbf{i}\gamma p_{yi}(0)]t \tag{7.79}$$

When this approximation breaks down, approximately at the Maxwell relaxation time, $\tau_M \equiv \eta/G$, the phase separation $d(\tau_M, \gamma)$ is given by

$$d(\tau_M, \gamma) = \gamma\tau_M \sqrt{\sum_{i=1}^{N} [y_i^2(0) + p_{yi}^2(0)]} \tag{7.80}$$

For our system this distance is

$$d(\tau_M, \gamma) = \gamma\tau_M \left\{\frac{N^{5/3}}{3n^{2/3}} + NT\right\}^{1/2} \simeq 8.7\gamma \tag{7.81}$$

We have used the fact that the reduced Maxwell time is 0.137. After this time the phase separation can be expected to grow as

$$d(\gamma, t) \simeq d(\gamma, \tau_M) \exp\left[\frac{t}{\tau_L + O(\gamma^2)}\right] \tag{7.82}$$

where, as before, τ_L is the inverse of the largest zero-strain-rate Lyapunov exponent. For fields of less than $\gamma^* = 10^{-2}$, the equilibrium Lyapunov time dominates the denominator of the above expression. This explains why the slopes of the curves in Fig. 7.12 are independent of strain rate. Furthermore, by combining equations (7.70), (7.81) and (7.82) we see that in the regime where the strain-rate corrections to the Lyapunov exponents are small, the phase separation takes the form given by equation (7.78) with the coefficient $A \simeq 8.7$. Equation (7.82) is plotted, for a reduced strain rate of 10^{-7}, as a dashed line in Fig. 7.12; it is in reasonable agreement with the results. The results for other strain rates are similar. The greatest uncertainty in the prediction is the estimation of the precise time at which Lyapunov behaviour begins.

REFERENCES

Bird, R.B., Armstrong, R.C. and Hassager, O. (1977). *Dynamics of Polymeric Liquids*, Vol. 1. New York: Wiley.
Ciccotti, G. and Jacucci, G. (1975). *Phys. Rev. Lett.*, **35**, 789.
Ciccotti, G., Jacucci, G. and McDonald, I.R. (1976). *Phys. Rev., Ser. A*, **13**, 426.
Ciccotti, G., Jacucci, G. and McDonald, I.R. (1979). *J. Stat. Phys.*, **21**, 1.
Cohen, E.G.D. (1983). *Physica*, **118A**, 17.

Dufty, J.W. and Lindenfeld, M.J. (1979). *J. Stat. Phys.*, **20**, 259.
Evans, D.J. (1990). *Phys. Rev.*, to appear.
Evans, D.J. and Morriss, G.P. (1987). *Mol. Phys.*, **61**, 1151.
Evans, D.J. and Morriss, G.P. (1988). *Phys. Rev., Ser. A*, **38**, 4142.
Kawasaki, K. and Gunton, J.D. (1973). *Phys. Rev., Ser. A*, **8**, 2048.
Kubo, R. (1957). *J. Phys. Soc. Jpn.*, **12**, 570.
Morriss, G.P. and Evans, D.J. (1985). *Mol. Phys.*, **54**, 629.
Morriss, G.P. and Evans, D.J. (1987). *Phys. Rev., Ser. A*, **35**, 792.
Morriss, G.P., Evans, D.J., Cohen, E.G.D. and van Beijeren, H. (1989). *Phys. Rev. Lett.*, **62**, 1579.
van Kampen, N.G. (1971). *Phys. Norvegica*, **5**, 279.
Visscher, W.M. (1974). *Phys. Rev., Ser. A*, **10**, 2461.
Yamada, T. and Kawasaki, K. (1967). *Prog. Theor. Phys.*, **38**, 1031.
Yamada, T. and Kawasaki, K. (1975). *Prog. Theor. Phys.*, **53**, 111.

8 Time Dependent Response Theory

8.1 INTRODUCTION

In this chapter we extend the nonlinear response theory discussed in Chapter 7 to describe the response of classical, many-body systems to time-dependent external fields. The resulting formalism is applicable to both adiabatic and thermostatted systems. The results are then related to a number of known special cases: time-dependent linear response theory, and time independent nonlinear response theory as described by the transient time-correlation approach and the Kawasaki response formula.

We begin by developing a formal operator algebra for manipulating distribution functions and mechanical phase variables in a thermostatted system subject to a time-dependent applied field. The analysis parallels perturbation treatments of quantum field theory (Raimes, 1972; Parry, 1973). The mathematical techniques required for the time-dependent case are sufficiently different from, and more complex than, those required in the time-independent case that we have reserved their discussion until now. One of the main differences between the two types of nonequilibrium system is that time-ordered exponentials are required for the definition of propagators in the time-dependent case. New commutivity constraints, which have no counterparts in the time-independent case, place severe limitations on the mathematical forms allowed to express the nonlinear time-dependent response. In the time-independent case two forms have already been met in Chapter 7: the Kawasaki and the transient time correlation function (TTCF) forms; in this chapter we will meet yet another. Of these three forms only *one* is applicable in the time-dependent case.

8.2 TIME EVOLUTION OF PHASE VARIABLES

When a system is subject to time-dependent external fields the equations of motion for both the distribution function and phase variables become quite complex. There are two time dependences in such a system. One is associated with the time at which you wish to know the phase position $\mathbf{\Gamma}(t)$, and the other is associated with the explicit time dependence of the field, $F_e(t)$. In order to deal with this complexity in a convenient way we introduce a more compact notation for the propagator. Apart from some important notational differences the initial development parallels that of Holian and Evans (1985). We define the p-propagator, $U_R(0,t)$, to be the operator which advances a function of $\mathbf{\Gamma}$ only, forward in time from 0 to t (the meaning of the subscript will emerge later). That is

$$\mathbf{\Gamma}(t) = U_R(0,t)\mathbf{\Gamma}(0) \tag{8.1}$$

The operator $U_R(0,t)$ operates on all functions of phase located to its right. The equations of motion for the system at time t, which are themselves a function of phase $\mathbf{\Gamma}$, are given by

$$\dot{\mathbf{\Gamma}}(\mathbf{\Gamma}(t),t) = U_R(0,t)\dot{\mathbf{\Gamma}}(\mathbf{\Gamma}(0),t) \tag{8.2}$$

The notation $\dot{\mathbf{\Gamma}}(\mathbf{\Gamma}(t),t)$ implies that the derivative should be calculated on the current phase $\mathbf{\Gamma}(t)$, using the current field $F_e(t)$. However, $\dot{\mathbf{\Gamma}}(\mathbf{\Gamma}(0),t)$ implies that the derivative should be calculated on the initial phase $\mathbf{\Gamma}(0)$, using the current field $F_e(t)$. The p-propagator $U_R(0,t)$ has no effect on explicit time. Its only action is to advance the implicit time dependence of the phase $\mathbf{\Gamma}$.

The total time derivative of a phase function $B(\mathbf{\Gamma})$ with no explicit time dependence (by definition a phase function cannot have an explicit time dependence) is

$$\begin{aligned}
\frac{\mathrm{d}}{\mathrm{d}t}B(\mathbf{\Gamma}(t)) &= \dot{\mathbf{\Gamma}}[\mathbf{\Gamma}(t),t]\cdot\left.\frac{\partial B(\mathbf{\Gamma})}{\partial \mathbf{\Gamma}}\right|_{\mathbf{\Gamma}=\mathbf{\Gamma}(t)} \\
&= U_R(0,t)\dot{\mathbf{\Gamma}}[\mathbf{\Gamma},t]\cdot\left.\frac{\partial}{\partial \mathbf{\Gamma}}B(\mathbf{\Gamma})\right|_{\mathbf{\Gamma}} \\
&= U_R(0,t)\mathrm{i}L(t)B(\mathbf{\Gamma}) \\
&= \frac{\partial}{\partial t}U_R(0,t)B(\mathbf{\Gamma})
\end{aligned} \tag{8.3}$$

where we have introduced the time-dependent p-Liouvillean, $\mathrm{i}L(t) \equiv \mathrm{i}L(\mathbf{\Gamma},t)$ which acts on functions of the *initial* phase $\mathbf{\Gamma}$, but contains the external field at the *current* time. The partial derivative of B with respect to initial phase $\mathbf{\Gamma}$ is simply another phase function, so that the propagator $U_R(0,t)$ advances this phase function to time t (that is the partial derivative of B with respect to phase evaluated at time t). In writing the last line of (8.3) we have used the fact that the p-propagator is an explicit function of time (as well as phase), and that when written in terms of the p-propagator, $\mathrm{d}B(\Gamma(t))/\mathrm{d}t$, must *only* involve the partial time derivative of the p-propagator. Equation (8.3) implies that the p-propagator $U_{\mathrm{R}}(0, t)$ satisfies an operator equation of the form

$$\frac{\partial}{\partial t} U_{\mathrm{R}}(0, t) = U_{\mathrm{R}}(0, t)\mathrm{iL}(t) \tag{8.4}$$

where the order of the two operators on the right-hand side is crucial. As we shall see shortly, $U_{\mathrm{R}}(0, t)$ and $\mathrm{iL}(t)$ do not commute since the propagator $U_{\mathrm{R}}(0, t)$ contains sums of products of $\mathrm{iL}(s_i)$ at different times s_i, and $\mathrm{iL}(s_i)$ and $\mathrm{iL}(s_j)$, do not commute unless $s_i = s_j$. The formal solution of this operator equation is

$$U_{\mathrm{R}}(0, t) = \sum_{n=0}^{\infty} \int_0^t \mathrm{d}s_1 \int_0^{s_1} \mathrm{d}s_2 \ldots \int_0^{s_{n-1}} \mathrm{d}s_n\, \mathrm{iL}(s_n) \ldots \mathrm{iL}(s_2)\mathrm{iL}(s_1) \tag{8.5}$$

Notice that the p-Liouvilleans are *right ordered* in time (latest time to the right). As Liouvilleans do not commute, this time ordering is fixed. The integration limits imply that $t > s_1 > s_2 > \ldots > s_n$, so that the time arguments of the p-Liouvilleans in the expression for $U_{\mathrm{R}}(0, t)$ increase as we move from the left to the right. It is very important to remember that in generating $B(t)$ from $B(0)$ using (8.5), if we write the integrals as, say, a trapezoidal approximation it is the Liouvillean at the *latest* time $\mathrm{iL}(t)$, which attacks $B(0)$ first. The Liouvilleans attack B in an *anticausal* order. This issue is discussed further in Section 8.4.

We can check that (8.5) is the solution to (8.4) by differentiating with respect to time. We see that, $\int_0^\infty \mathrm{d}s_1$ disappears and the argument $\mathrm{iL}(s_1)$, changes to $\mathrm{iL}(t)$. This term appears on the right-hand side, as it must to satisfy the differential operator equation. It is easy to derive an equation for the incremental p-propagator $U_{\mathrm{R}}(\tau, t)$ which advances a phase function from time τ to t,

$$U_{\mathrm{R}}(\tau, t) = \sum_{n=0}^{\infty} \int_\tau^t \mathrm{d}s_1 \int_\tau^{s_1} \mathrm{d}s_2 \ldots \int_\tau^{s_{n-1}} \mathrm{d}s_n\, \mathrm{iL}(s_n) \ldots \mathrm{iL}(s_2)\mathrm{iL}(s_1) \tag{8.6}$$

Our convention for the time arguments of the U-propagators is that the first argument (in this case τ) is the lower limit of all the integrals. The second argument (in this case t) is the upper limit of the first integral.

8.3 THE INVERSE THEOREM

We will assume that $t > 0$. Intuitively it is obvious that the inverse of $U_{\mathrm{R}}(0, t)$, which we write as $U_{\mathrm{R}}(0, t)^{-1}$, should be the propagator that propagates backwards in time from t to 0. From (8.6) we can write

$$U_R(0,t)^{-1} = \sum_{n=0}^{\infty} \int_t^0 ds_1 \int_t^{s_1} ds_2 \ldots \int_t^{s_{n-1}} ds_n \, iL(s_n) \ldots iL(s_2) iL(s_1) \quad (8.7)$$

Before proceeding further we will introduce an identity which is useful for manipulating these types of integrals. Often we will have a pair of integrals which we want to exchange. The limits of the innermost integral depend on the integration variable for the outer integral. The result we shall use is

$$\int_{t_0}^{t} ds_1 \int_{t_0}^{s_1} ds_2 = \int_{t_0}^{t} ds_2 \int_{s_2}^{t} ds_1 \quad (8.8)$$

As can be seen from Fig. 8.1, the range of integration for both integrals is the same. If we approximate the integral as a sum we see that the difference is in the order in which the contributions are summed. As long as the original integral is absolutely convergent, the result is true. We will assume that all integrals are absolute convergent.

It is illustrative to develop other representations of $U_R(0,t)^{-1}$ so we consider expression (8.7) term by term,

$$U_R(0,t)^{-1} = 1 + \int_t^0 ds \, iL(s) + \int_t^0 ds_1 \int_t^{s_1} ds_2 \, iL(s_2) iL(s_1)$$
$$+ \int_t^0 ds_1 \int_t^{s_1} ds_2 \int_t^{s_2} ds_3 \, iL(s_3) iL(s_2) iL(s_1) + \ldots \quad (8.9)$$

Interchanging the integration limits in every integral gives a factor of -1 for each interchange.

$$U_R(0,t)^{-1} = 1 - \int_0^t ds \, iL(s) + \int_0^t ds_1 \int_{s_1}^t ds_2 \, iL(s_2) iL(s_1)$$
$$- \int_0^t ds_1 \int_{s_1}^t ds_2 \int_{s_2}^t ds_3 \, iL(s_3) iL(s_2) iL(s_1) + \ldots \quad (8.10)$$

We can use the integral interchange result (8.8) on the third term on the right-hand side (note that the integrand is unchanged by this operation). In the fourth term we can use the interchange result three times to completely reverse the order of the integrations, giving

$$U_R(0,t)^{-1} = 1 - \int_0^t ds \, iL(s) + \int_0^t ds_2 \int_0^{s_2} ds_1 \, iL(s_2) iL(s_1)$$
$$- \int_0^t ds_3 \int_0^{s_3} ds_2 \int_0^{s_2} ds_1 \, iL(s_3) iL(s_2) iL(s_1) + \ldots \quad (8.11)$$

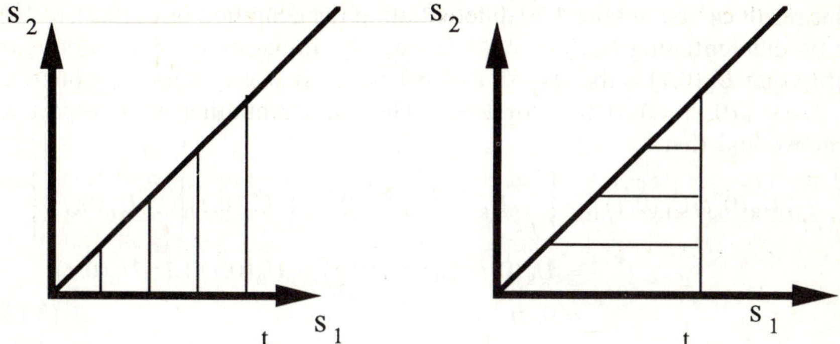

Fig. 8.1 We give a diagrammatic representation of the exchange of order of integrations in equation (8.8).

The final step is to relabel the dummy integration variables to give

$$U_R(0,t)^{-1} = 1 - \int_0^t ds\, iL(s) + \int_0^t ds_1 \int_0^{s_1} ds_2\, iL(s_1)iL(s_2)$$

$$- \int_0^t ds_1 \int_0^{s_1} ds_2 \int_0^{s_2} ds_3\, iL(s_1)iL(s_2)iL(s_3) + \ldots$$

$$U_R(0,t)^{-1} = \sum_{n=0}^{\infty} (-)^n \int_0^t ds_1 \int_0^{s_1} ds_2 \ldots \int_0^{s_{n-1}} ds_n\, iL(s_1)iL(s_2)\ldots iL(s_n) \tag{8.12}$$

As $t > 0$, an examination of the integration limits reveals that the Liouvilleans in this expression are left-ordered. Comparing this expression with the definition of $U_R(0,t)$ there are two differences: the time ordering *and* the factor of $(-)^n$. We now define the operator $U_L(0,t)$ to be equal to the right-hand side of (8.12), so we have

$$U_L(0,t) \equiv \sum_{n=0}^{\infty} (-)^n \int_0^t ds_1 \int_0^{s_1} ds_2 \ldots \int_0^{s_{n-1}} ds_n\, iL(s_1)iL(s_2)\ldots iL(s_n) \tag{8.13}$$

and

$$U_R(0,t)^{-1} = U_L(0,t) \tag{8.14}$$

From this definition of $U_L(0,t)$, it can be shown that $U_L(0,t)$ satisfies the operator equation

$$\frac{\partial}{\partial t} U_L(0,t) = -iL(t)U_L(0,t) \tag{8.15}$$

This result can be obtained by differentiating the definition of $U_L(0,t)$, (8.13), or by differentiating $U_R(0,t)^{-1}$, (8.7), directly. Equation (8.15) allows us to verify that $U_L(0,t)$ is the inverse of $U_R(0,t)$ in a new way. First we note that $U_L(0,t)U_R(0,t) = 1$ is true for $t = 0$. Then, differentiating with respect to time we find that

$$\begin{aligned}\frac{\partial}{\partial t}[U_R(0,t)U_L(0,t)] &= \left[\frac{\partial}{\partial t}U_R(0,t)\right]U_L(0,t) + U_R(0,t)\left[\frac{\partial}{\partial t}U_L(0,t)\right] \\ &= U_R(0,t)\mathrm{i}L(t)U_L(0,t) - U_R(0,t)\mathrm{i}L(t)U_L(0,t) \\ &= 0, \forall t \qquad (8.16)\end{aligned}$$

As the result is true at $t = 0$, and the time derivative of each side of the equation is true for all time, the result is true for all time.

8.4 THE ASSOCIATIVE LAW AND COMPOSITION THEOREM

The action of the p-propagator $U_R(0,t)$ is to advance the phase $\mathbf{\Gamma}$, or a phase variable, forward in time from 0 to t. This must be equivalent to advancing time from 0 to s, then advancing time from s to t, whenever $0 < s < t$. This implies that

$$U_R(0,t)B(\mathbf{\Gamma}) = U_R(s,t)[U_R(0,s)B(\mathbf{\Gamma})] = U_R(s,t)B(s) \qquad (8.17)$$

The right-hand side of (8.17) is a physical rather than mathematical statement. It is a statement of causality. If we wish to understand how we can *generate* $B(t)$ from $B(0)$ through an intermediate time s, we find that we will have to attack B first with the operator $U_R(s,t)$ and then attack the resultant expression with $U_R(0,s)$. The operator expression $U_R(s,t)U_R(0,s)B$ cannot be equal to $U_R(0,t)$, because its time arguments are not ordered from left to right. The correct operator equation is

$$U_R(0,t) = U_R(0,s)U_R(s,t) \qquad (8.18)$$

To prove (8.18) we consider the product on the right-hand side and show that it is equal to $U_R(0,t)$.

$$\begin{aligned}&U_R(0,s)U_R(s,t)\\ &= \left(\sum_{m=0}^{\infty}\int_0^s \mathrm{d}s_1 \ldots \int_0^{s_{m-1}} \mathrm{d}s_m\, \mathrm{i}L(s_m)\ldots\mathrm{i}L(s_1)\right)\left(\sum_{n=0}^{\infty}\int_s^t \mathrm{d}s_1' \ldots \int_s^{s_{n-1}'} \mathrm{d}s_n'\, \mathrm{i}L(s_n')\ldots\mathrm{i}L(s_1')\right)\\ &= 1 + \int_0^t \mathrm{d}s_1\, \mathrm{i}L(s_1) + \int_0^s \mathrm{d}s_1 \int_0^{s_1} \mathrm{d}s_2\, \mathrm{i}L(s_2)\mathrm{i}L(s_1)\\ &\quad + \int_0^s \mathrm{d}s_1\, \mathrm{i}L(s_1)\int_s^t \mathrm{d}s_1'\, \mathrm{i}L(s_1') + \int_s^t \mathrm{d}s_1' \int_s^{s_1'} \mathrm{d}s_2'\, \mathrm{i}L(s_2')\mathrm{i}L(s_1') + \ldots \qquad (8.19)\end{aligned}$$

The first two terms are straightforward so we will consider in detail the three second-order terms. In the second of these three terms the integration limits imply that

$$0 < s_1 < s < s_1' < t$$

so that the time arguments of the operator product are correctly ordered, and we relabel them as follows:

$$s_1' \to s_1 \qquad \text{and} \qquad s_1 \to s_2$$

The integration limits are independent, so we can interchange the order of integration, (8.8). After dropping the primes in the third term, all three terms have the same integrand, so we need only consider the integration limits. The three second-order terms are

$$\int_0^s ds_1 \int_0^{s_1} ds_2 + \int_s^t ds_1 \int_0^s ds_2 + \int_s^t ds_1 \int_s^{s_1} ds_2$$

In the second and third terms, the s_1 integrals are the same and the s_2 integrals add together to give

$$\int_0^s ds_1 \int_0^{s_1} ds_2 + \int_s^t ds_1 \int_0^{s_1} ds_2$$

Now the s_2 integrals are identical and the s_1 integrals add together to give the required result

$$\int_0^t ds_1 \int_0^{s_1} ds_2$$

This is exactly the second-order term in $U_R(0, t)$. It may seem that we have laboured through the detail of the second-order term, but it is now straightforward to apply the same steps to all the higher-order terms and see that the result is true to all orders. Indeed, it is a useful exercise for the reader to examine the third-order term, as there are four integrals to consider, and after the same relabelling process is applied to the second and third terms, the four integrals obtained collapse from the right-hand side.

Combining equations (8.17) and (8.18) we see that the p-propagator, U_R, obeys an anticausal associative law, (8.17). The fundamental reason for its anticausal form is implicit in the form of the p-propagator itself, U_R. In applying the p-propagator to a phase variable it is, as we have seen, the latest times that attack the phase variable first.

Apart from the present discussion we will always write operators in a form which reflects the mathematical rather than the causal ordering. As we will see, any confusion that arises from the anticausal ordering of p-propagators can always be removed by considering the f-propagator form and then unrolling the operators in sequence to attack the phase variables. The f-propagators are causally ordered.

8.5 TIME EVOLUTION OF THE DISTRIBUTION FUNCTION

The Liouville equation for a system subject to a time-dependent external field is given by

$$\frac{\partial}{\partial t} f(t) = -\frac{\partial}{\partial \mathbf{\Gamma}} \cdot [\dot{\mathbf{\Gamma}}(\mathbf{\Gamma}, t) f(\mathbf{\Gamma}, t)] = -iL(t) f(t) \tag{8.20}$$

where we have defined the time-dependent f-Liouvillean, $iL(t)$. This equation tells you that if you sit at a *fixed* point in phase space denoted by the dummy variable $\mathbf{\Gamma}$, the density of phase points near $\mathbf{\Gamma}$ changes with time according to (8.20). In the derivation of this equation we related the partial derivative of $f(t)$ to various fluxes in phase space at the *same* value of the explicit time.

We define the distribution function propagator $U_{\mathrm{R}}^{\dagger}(0, t)$ which advances the time dependence of the distribution function from 0 to t, by

$$f(\mathbf{\Gamma}, t) = U_{\mathrm{R}}^{\dagger}(0, t) f(\mathbf{\Gamma}, 0) \tag{8.21}$$

In this equation $U_{\mathrm{R}}^{\dagger}(0, t)$ is the adjoint to $U_{\mathrm{R}}(0, t)$. It is therefore closely related to $U_{\mathrm{L}}(0, t)$, except that the Liouvilleans appearing in (8.13) are replaced by their adjoints $iL(s_i)$. Combining (8.21) with the Liouville equation (8.20) we find that $U_{\mathrm{R}}^{\dagger}(0, t)$ satisfies the following equation of motion

$$\frac{\partial}{\partial t} U_{\mathrm{R}}^{\dagger}(0, t) = -iL(\mathbf{\Gamma}, t) U_{\mathrm{R}}^{\dagger}(0, t) \tag{8.22}$$

The formal solution to this operator equation is

$$U_{\mathrm{R}}^{\dagger}(0, t) = \sum_{n=0}^{\infty} (-)^n \int_0^t \mathrm{d}s_1 \int_0^{s_1} \mathrm{d}s_2 \ldots \int_0^{s_{n-1}} \mathrm{d}s_n \, iL(s_1) iL(s_2) \ldots iL(s_n) \tag{8.23}$$

In distinction to the propagator for phase variables, the integration limits imply that $t > s_1 > s_2 > \ldots > s_n$, so that the f-Liouvilleans are *left time ordered.* The time arguments increase as we go from the right to the left. This is opposite to the time ordering in the p-propagator, $U_{\mathrm{R}}(0, t)$, but the definition of $U_{\mathrm{R}}^{\dagger}(0, t)$ is consistent with the definition of $U_{\mathrm{L}}(0, t)$.

For the f-propagator, $U_R^\dagger(0,t)$, the usual associative law is satisfied as the time arguments are ordered right to left,

$$U_R^\dagger(0,t)f(0) = [U_R^\dagger(s,t)U_R^\dagger(0,s)]f(0) = U_R^\dagger(s,t)[U_R^\dagger(0,s)f(0)] \quad (8.24)$$

This equation can be verified directly using similar arguments to those used in Section 8.4.

8.6 TIME-ORDERED EXPONENTIALS

A notation which is common in quantum mechanics is to refer to the phase and distribution function propagators as *right-* and *left-ordered exponentials* ($\exp_R$ and $\exp_L$), respectively. To exploit this notational simplification we introduce the time-ordering operators T_R and T_L. The operator T_R simply reorders a product of operators so that the time arguments increase from left to right. In this notation we write the p-propagator $U_R(0,t)$ as

$$U_R(0,t) = \exp_R\left(\int_0^t ds\, iL(s)\right) = T_R \exp\left(\int_0^t ds\, iL(s)\right) \quad (8.25)$$

Using the series expansion for the exponential this becomes

$$U_R(0,t) = T_R \sum_{n=0}^{\infty} \frac{1}{n!} \int_0^t ds_1 \int_0^t ds_2 \ldots \int_0^t ds_n\, iL(s_n)\ldots iL(s_2)iL(s_1) \quad (8.26)$$

Taking this series term by term the first two terms are trivial. We will consider the second-order term in some detail.

$$T_R \frac{1}{2!} \int_0^t ds_1 \int_0^t ds_2\, iL(s_2)iL(s_1)$$
$$= \frac{1}{2!} T_R \left\{ \int_0^t ds_1 \int_0^{s_1} ds_2\, iL(s_2)iL(s_1) + \int_0^t ds_1 \int_{s_1}^t ds_2\, iL(s_2)iL(s_1) \right\} \quad (8.27)$$

The time arguments in the first integral are time ordered from left to right so the operator will have no effect. In the second integral the order of the integrations can be interchanged to give

$$\int_0^t ds_2 \int_0^{s_2} ds_1\, iL(s_2)iL(s_1) = \int_0^t ds_1 \int_0^{s_1} ds_2\, iL(s_1)iL(s_2) \quad (8.28)$$

The second form is obtained by relabelling the dummy variables s_1 and s_2. Now both integrals have the same integration limits, and after the operation of T_R both integrands are the same, so the second-order term is

$$\int_0^t ds_2 \int_0^{s_2} ds_1\, iL(s_2)iL(s_1)$$

Using exactly the same steps we can show that each of the higher-order terms are the same as those in the original representation of $U_R(0,t)$. After manipulating the integrals to obtain the same range of integration for each term of a particular order, the integrand is the sum of all permutations of the time arguments. At the nth order there are $n!$ permutations, which after the operation of T_R are all identical. This $n!$ then cancels the $(n!)^{-1}$ from the expansion of the exponential, and the result follows. Using the same arguments, the f-propagator, $U_R^\dagger(0,t)$, can also be written in this form

$$U_R^\dagger(0,t) = \exp_L\left(-\int_0^t ds\, iL(s)\right) = T_L \exp\left(-\int_0^t ds\, iL(s)\right) \quad (8.29)$$

The use of the time-ordering operator can realize considerable simplifications in many of the proofs that we have given.

Using time-ordered exponentials, the composition theorem can be derived quite easily.

$$\begin{aligned} B(t) &= T_R \exp\left[\int_0^t d\tau\, iL(\tau)\right] B \\ &= T_R \exp\left[\int_0^s d\tau\, iL(\tau)\right] \exp\left[\int_s^t d\tau\, iL(\tau)\right] B \end{aligned} \quad (8.30)$$

Because the exponentials are already right ordered, we can write them as

$$\begin{aligned} B(t) &= T_R\left\{\exp\left[\int_0^s d\tau\, iL(\tau)\right]\right\} T_R\left\{\exp\left[\int_s^t d\tau\, iL(\tau)\right]\right\} B \\ &= \exp_R\left[\int_0^s d\tau\, iL(\tau)\right] \exp_R\left[\int_s^t d\tau\, iL(\tau)\right] B \end{aligned} \quad (8.31)$$

8.7 SCHRÖDINGER AND HEISENBERG REPRESENTATIONS

In this section we will derive some of the more important properties of the Liouville operators. These will lead us naturally to the discussion of the various representations for the properties of classical systems. The first property we shall discuss relates the p-Liouvillean to the f-Liouvillean as follows:

$$\int d\Gamma\, f(0)iL(t)B(\Gamma) = -\int d\Gamma\, B(\Gamma)iL(t)f(0) \quad (8.32)$$

The proof is a straightforward application of integration by parts.

$$\int d\mathbf{\Gamma}\, f(0)\left[\dot{\mathbf{\Gamma}}(\mathbf{\Gamma},t)\cdot\frac{\partial}{\partial\mathbf{\Gamma}}B\right|_{\mathbf{\Gamma}}$$

$$= f(0)\dot{\mathbf{\Gamma}}(\mathbf{\Gamma},t)B(\mathbf{\Gamma})\Big]_S - \int d\mathbf{\Gamma}\, B(\mathbf{\Gamma})\frac{\partial}{\partial\mathbf{\Gamma}}\cdot(f(0)\dot{\mathbf{\Gamma}}(\mathbf{\Gamma},t))$$

$$= -\int d\mathbf{\Gamma}\, B(\mathbf{\Gamma})\left\{\dot{\mathbf{\Gamma}}(\mathbf{\Gamma},t)\cdot\frac{\partial}{\partial\mathbf{\Gamma}}+\frac{\partial}{\partial\mathbf{\Gamma}}\cdot\dot{\mathbf{\Gamma}}(\mathbf{\Gamma},t)\right\}f(0)$$

$$= -\int d\mathbf{\Gamma}\, B(\mathbf{\Gamma})iL(t)f(0) \tag{8.33}$$

Equation (8.32) shows that $\mathrm{iL}(t)$ and $-iL(t)$ are adjoints.

We can compute the average of a phase variable B at time t by following the value of $B(t)$ as it changes along single trajectories in phase space. The average is taken by summing over the values of B for trajectories starting from each possible initial phase point $\mathbf{\Gamma}$, but weighting each $B(t)$ with the probability of that *starting* phase. These probabilities are chosen from an initial distribution function $f(\mathbf{\Gamma},0)$. This is the so-called Heisenberg picture.

$$\langle B(t)\rangle = \int d\mathbf{\Gamma}\, B(\mathbf{\Gamma}(t))f(\mathbf{\Gamma},0) = \int d\mathbf{\Gamma}\, f(\mathbf{\Gamma},0)U_{\mathrm{R}}(0,t)B(\mathbf{\Gamma}) \tag{8.34}$$

The Heisenberg picture is exactly analogous to the Lagrangian formulation of fluid mechanics, we can imagine that the phase space *mass point* has a differential box $d\mathbf{\Gamma}$ surrounding it which changes shape (and volume for a compressible fluid) with time as the phase point follows its trajectory. The probability of the differential element, or mass $f(\mathbf{\Gamma})\,d\mathbf{\Gamma}$ remains constant, but the value of the observable changes implicitly in time.

The second view is the Schrödinger, or distribution-based picture, where $\mathbf{\Gamma}$ refers not to the initial value of the phase joint, but to a stationary point (fixed for all time) inside a stationary differential box $d\mathbf{\Gamma}$. Just as in the Eulerian formulation of fluid mechanics, the observable takes on a fixed value for all time, $B(\mathbf{\Gamma})$, while mass points with different probability flow through the box.

$$\langle B(t)\rangle = \int d\mathbf{\Gamma}\, B(\mathbf{\Gamma})f(\mathbf{\Gamma},t) = \int d\mathbf{\Gamma}\, B(\mathbf{\Gamma})U_{\mathrm{R}}^{\dagger}(0,t)f(\mathbf{\Gamma},0) \tag{8.35}$$

The average value of B changes with time as the distribution function changes. The average of B is computed by multiplying the value of B at $\mathbf{\Gamma}$, by the probability to find the phase point $\mathbf{\Gamma}$ at time t, that is $f(\mathbf{\Gamma},t)$.

The average value of a phase variable B at time t can be evaluated in two ways. The mathematical proof of the equivalence of the Schrödinger and

Heisenberg pictures can be obtained by successive integrations by parts. Consider

$$\int \mathrm{d}\mathbf{\Gamma}\, f(0)B(\mathbf{\Gamma}(t)) = \int \mathrm{d}\mathbf{\Gamma}\, f(0)U_{\mathrm{R}}(0,t)B(\mathbf{\Gamma})$$
$$= \sum_{n=0}^{\infty}\int_0^t \mathrm{d}s_1 \ldots \int_0^{s_{n-1}} \mathrm{d}s_n \int \mathrm{d}\mathbf{\Gamma}\, f(0)\mathrm{i}\mathrm{L}(s_n)\ldots \mathrm{i}\mathrm{L}(s_1)B(\mathbf{\Gamma}) \tag{8.36}$$

One can unroll each Liouvillean in turn from the phase variable onto the distribution function using equation (8.32). For the first transfer we consider $\mathrm{iL}(s_{n-1})\ldots \mathrm{iL}(s_1)B$ to be the composite phase variable, so that the right-hand side becomes

$$= \sum_{n=0}^{\infty}\int_0^t \mathrm{d}s_1 \ldots \int_0^{s_{n-1}} \mathrm{d}s_n(-) \int \mathrm{d}\mathbf{\Gamma}\, (iL(s_n)f(0))\mathrm{iL}(s_{n-1})\ldots \mathrm{iL}(s_1)B(\mathbf{\Gamma})$$

We can then repeat this operator, unrolling

$$= \sum_{n=0}^{\infty}\int_0^t \mathrm{d}s_1 \ldots \int_0^{s_{n-1}} \mathrm{d}s_n(-)^2 \int \mathrm{d}\mathbf{\Gamma}\, (iL(s_{n-1})iL(s_n)f(0))\mathrm{iL}(s_{n-2})\ldots \mathrm{iL}(s_1)B(\mathbf{\Gamma})$$

Repeated unrolling leads to

$$= \sum_{n=0}^{\infty}\int_0^t \mathrm{d}s_1 \ldots \int_0^{s_{n-1}} \mathrm{d}s_n(-)^n \int \mathrm{d}\mathbf{\Gamma}\, B(\mathbf{\Gamma})iL(s_1)iL(s_2)\ldots iL(s_n)f(0)$$
$$= \int \mathrm{d}\mathbf{\Gamma}\, B(\mathbf{\Gamma}) \sum_{n=0}^{\infty}(-)^n \int_0^t \mathrm{d}s_1 \ldots \int_0^{s_{n-1}} \mathrm{d}s_n\, iL(s_1)\ldots iL(s_n)f(0)$$
$$= \int \mathrm{d}\mathbf{\Gamma}\, B(\mathbf{\Gamma})U_{\mathrm{R}}^{\dagger}(0,t)f(0)$$
$$= \int \mathrm{d}\mathbf{\Gamma}\, B(\mathbf{\Gamma})f(t) \tag{8.37}$$

We have obtained this result where the Liouvilleans explicitly depend on time. The derivation we have used has not made any reference to the details of either the initial distribution function or the first-order equations of motion of the system. This means that these results are valid for arbitrary equations of motion, in particular the equations of motion can contain a time-dependent external field. The initial distribution function is also arbitrary, the only constraint is that the distribution function at time t must have evolved from the initial distribution function under the influence of the perturbed equations of motion. They are also valid regardless of whether the equations of motion can be derived from a Hamiltonian or whether they satisfy $AI\Gamma$ (Section 5.3).

8.8 THE DYSON EQUATION

The Dyson equation is useful for deriving relationships between propagators. We first met a restricted form of this equation in Section 3.6 when we were dealing with time-independent propagators. We will now give a general derivation of the Dyson equation.

For two arbitrary p-Liouvilleans, the most general form of the Dyson equation is

$$U_{\mathrm{R}}(0,t) = U_{\mathrm{R0}}(0,t) + \int_0^t \mathrm{d}s\, U_{\mathrm{R}}(0,s)(\mathrm{iL}(s) - \mathrm{iL}_0(s))U_{\mathrm{R0}}(s,t) \quad (8.38)$$

and

$$U_{\mathrm{R}}(0,t) = U_{\mathrm{R0}}(0,t) + \int_0^t \mathrm{d}s\, U_{\mathrm{R0}}(0,s)(\mathrm{iL}(s) - \mathrm{iL}_0(s))U_{\mathrm{R}}(s,t) \quad (8.39)$$

Both Liouvilleans $\mathrm{iL}(t)$ and $\mathrm{iL}_0(t)$ may be time dependent. One can prove the correctness of these equations by showing that the left- and right-hand sides of (8.38) and (8.39) satisfy the same differential equations with identical initial conditions. The corresponding equations for left-ordered propagators are:

$$U_{\mathrm{R}}^\dagger(0,t) = U_{\mathrm{R0}}^\dagger(0,t) - \int_0^t \mathrm{d}s\, U_{\mathrm{R}}^\dagger(s,t)(iL(s) - iL_0(s))U_{\mathrm{R0}}^\dagger(0,s) \quad (8.40)$$

$$U_{\mathrm{R}}^\dagger(0,t) = U_{\mathrm{R0}}^\dagger(0,t) - \int_0^t \mathrm{d}s\, U_{\mathrm{R0}}^\dagger(s,t)(iL(s) - iL_0(s))U_{\mathrm{R}}^\dagger(0,s) \quad (8.41)$$

We will give a proof for one of these equations, (8.39). Proofs for the other equations are very similar. If we let LHS denote $U_{\mathrm{R}}(0,t)$, the left-hand side of (8.39), we know that,

$$\frac{\partial}{\partial t}\mathrm{LHS} = U_{\mathrm{R}}(0,t)\mathrm{iL}(t) = \mathrm{LHS}\,\mathrm{iL}(t) \quad (8.42)$$

However, we see that

$$\begin{aligned}\frac{\partial}{\partial t}\mathrm{RHS} &= U_{\mathrm{R0}}(0,t)\mathrm{iL}_0(t) + U_{\mathrm{R0}}(0,t)(\mathrm{iL}(t) - \mathrm{iL}_0(t)) \\ &\quad + \int_0^t \mathrm{d}s\, U_{\mathrm{R0}}(0,s)(\mathrm{iL}(s) - \mathrm{iL}_0(s))U_{\mathrm{R}}(s,t)\mathrm{iL}(t) \\ &= \mathrm{RHS}\,\mathrm{iL}(t)\end{aligned} \quad (8.43)$$

Thus since both sides of (8.38) satisfy the same differential equation with the same initial condition, both sides must be the same for all time.

8.9 RELATION BETWEEN p- AND f-PROPAGATORS

In order to be able to manipulate propagators for thermostatted systems, it is useful to be able to relate p-propagators and f-propagators. The relation we shall derive is a time-dependent generalization of (7.24). It is a relatively straightforward application of the Dyson equation. Let $U_R(0,t) = \exp_R \int_0^t iL(s)\,ds$, be the test propagator and $U_{R0}(0,t) = \exp_R \int_0^t \mathrm{iL}(s)\,ds$, be the reference propagator, then $iL(s)A(\mathbf{\Gamma}) = \partial(A(\mathbf{\Gamma})\,d\mathbf{\Gamma}/dt)/\partial\mathbf{\Gamma}$ and $\mathrm{iL}(s)A(\mathbf{\Gamma}) = d\mathbf{\Gamma}/dt \cdot \partial(A(\mathbf{\Gamma}))/\partial\mathbf{\Gamma}$.

Substitution into the Dyson equation gives

$$U_R(0,t) = U_{R0}(0,t) + \int_0^t ds\, U_R(0,s)(iL(s) - \mathrm{iL}(s))U_{R0}(s,t) \quad (8.44)$$

We define,

$$iL(s) - \mathrm{iL}(s) \equiv \Lambda(\mathbf{\Gamma}, s) = \left[\frac{\partial}{\partial\mathbf{\Gamma}} \cdot \dot{\mathbf{\Gamma}}(s)\right] \quad (8.45)$$

It is important to realize that Λ is a phase variable and *not* an operator. Λ is the phase-space compression factor since $d \ln f(t)/dt = -\Lambda = 3N\alpha(t) + O(1)$ (see (7.17)).

One can recursively substitute for U_R in (8.44) to eliminate U_R from the right-hand side. This gives

$$U_R(0,t) = U_{R0}(0,t) + \int_0^t ds_1\, U_{R0}(0,s_1)\Lambda(s_1)U_{R0}(s_1,t)$$
$$+ \int_0^t ds_1 \int_0^{s_1} ds_2\, U_{R0}(0,s_2)\Lambda(s_2)U_{R0}(s_2,s_1)\Lambda(s_1)U_{R0}(s_1,t) + \ldots \quad (8.46)$$

Using the fact that Λ is a phase variable rather than an operator, we see that

$$U_R(0,t) = U_{R0}(0,t) + \int_0^t ds_1\, \Lambda(\mathbf{\Gamma}(s_1), s_1)U_{R0}(0,t)$$
$$+ \int_0^t ds_1 \int_0^{s_1} ds_2\, \Lambda(\mathbf{\Gamma}(s_2), s_2)\Lambda(\mathbf{\Gamma}(s_1), s_1)U_{R0}(0,t) + \ldots \quad (8.47)$$

so that

$$U_{\mathrm{R}}(0,t) = \exp\left[\int_0^t \mathrm{d}s\, \Lambda(\mathbf{\Gamma}(s),s)\right] U_{\mathrm{R}0}(0,t) \tag{8.48}$$

or

$$\exp_{\mathrm{R}}\left[\int_0^t \mathrm{d}s\, iL(s)\right] = \exp\left[\int_0^t \mathrm{d}s\, \Lambda(\mathbf{\Gamma}(s),s)\right] \exp_{\mathrm{R}}\left[\int_0^t \mathrm{d}s\, \mathrm{i}\mathrm{L}(s)\right] \tag{8.49}$$

This result is fundamental to our understanding of the dynamic behaviour of thermostatted systems. Its correctness can easily be checked by verifying that the left- and right-hand sides satisfy the same differential equation with the same initial condition. At zero time both sides are equal to unity. The derivative of the left-hand side is

$$\frac{\partial}{\partial t}[\mathrm{LHS}] = \exp_{\mathrm{R}}\left[\int_0^t \mathrm{d}s\, iL(s)\right] iL(t) = \mathrm{LHS}\; iL(t) \tag{8.50}$$

While the derivative of the right-hand side is

$$\begin{aligned}\frac{\partial}{\partial t}[\mathrm{RHS}] &= \exp\left[\int_0^t \mathrm{d}s\, \Lambda(\mathbf{\Gamma}(s),s)\right] \Lambda(\mathbf{\Gamma}(t),t) \exp_{\mathrm{R}}\left[\int_0^t \mathrm{d}s\, \mathrm{iL}(s)\right] + \mathrm{RHS}\, \mathrm{iL}(\mathbf{\Gamma},t) \\ &= [\mathrm{RHS}]\Lambda(\mathbf{\Gamma},t) + [\mathrm{RHS}]\, \mathrm{iL}(\mathbf{\Gamma},t) \\ &= \mathrm{RHS}\; iL(\mathbf{\Gamma},t)\end{aligned} \tag{8.51}$$

Thus the right-hand side and the left-hand side are identical.

8.10 TIME-DEPENDENT RESPONSE THEORY

Consider an equilibrium ensemble of systems, characterized by a distribution function, f_0, subject at $t = 0$, to an external time-dependent field $F_e(t)$. We assume that the equilibrium system ($t < 0$) has evolved under the influence of the Gaussian isokinetic Liouvillean iL_0. This Liouvillean has no explicit time dependence. The equilibrium distribution could be the canonical or the isokinetic distribution. These assumptions are summarized by the equation

$$\frac{\partial f_0}{\partial t} = -iL_0 f_0 = 0 \tag{8.52}$$

The equations of motion for the system can be written as,

$$\dot{\mathbf{q}}_i = \frac{\mathbf{p}_i}{m} + \mathbf{C}_i(\boldsymbol{\Gamma})F_e(t)$$

$$\dot{\mathbf{p}}_i = \mathbf{F}_i + \mathbf{D}_i(\boldsymbol{\Gamma})F_e(t) - \alpha(\boldsymbol{\Gamma}, t)\mathbf{p}_i \tag{8.53}$$

Provided that the temperature can be obtained from the expression, $3Nk_B T/2 = \Sigma \mathbf{p}_i^2/2m$, the term $\alpha\mathbf{p}_i$ represents the Gaussian thermostat. α is chosen so that $\Sigma \mathbf{p}_i^2/2m$ is a constant of the motion.

$$\alpha = \frac{\sum \mathbf{p}_i \cdot \mathbf{F}_i + \sum \mathbf{p}_i \cdot \mathbf{D}_i F_e}{\sum \mathbf{p}_i^2} \tag{8.54}$$

The terms **C** and **D** couple the external field $F_e(t)$ to the system. The adiabatic, unthermostatted equations of motion need not be derivable from a Hamiltonian (i.e. **C** and **D** do not have to be perfect differentials). We assume that $AI\boldsymbol{\Gamma}$ holds,

$$\frac{\partial}{\partial \boldsymbol{\Gamma}} \cdot \mathrm{iL}^{\mathrm{ad}}(s)\boldsymbol{\Gamma} = 0 \tag{8.55}$$

The dissipative flux is defined in the usual way,

$$\mathrm{iL}^{\mathrm{ad}}(s)H_0 \equiv -J(\boldsymbol{\Gamma})F_e(s) \tag{8.56}$$

where

$$H_0 = \sum_i \frac{\mathbf{p}_i^2}{2m} + \tfrac{1}{2}\sum_{i,j} \phi_{ij} \tag{8.57}$$

The response of an arbitrary phase variable $B(\boldsymbol{\Gamma})$ can obviously be written as

$$\langle B(t)\rangle = \int \mathrm{d}\boldsymbol{\Gamma}\, f_0\, \mathrm{e}_{\mathrm{R}}^{\int_0^t \mathrm{d}s\, \mathrm{iL}(s)} B(\boldsymbol{\Gamma}) = \int \mathrm{d}\boldsymbol{\Gamma}\, f_0\, U_{\mathrm{R}}(0,t)B \tag{8.58}$$

In this equation $\mathrm{iL}(t)$ is the p-Liouvillean for the field-dependent Gaussian thermostatted dynamics, $t > 0$. If we use the Dyson decomposition of the field-dependent p-propagator in terms of the equilibrium thermostatted propagator, we find that

$$\langle B(t)\rangle = \langle B(0)\rangle + \int_0^t \mathrm{d}s \int \mathrm{d}\boldsymbol{\Gamma}\, f_0\, U_{\mathrm{R}0}(0,s)(\mathrm{iL}(s) - \mathrm{iL}_0)U_{\mathrm{R}}(s,t)B \tag{8.59}$$

By successive integrations we unroll the U_{R0} propagator onto the distribution function.

$$\langle B(t)\rangle = \langle B(0)\rangle + \int_0^t ds \int d\Gamma\, [U_{R0}^{\dagger}(0,s) f_0](iL(s) - iL_0) U_R(s,t) B \tag{8.60}$$

However, $U_{R0}^{\dagger}$ is the equilibrium f-propagator and, by equation (8.52) it has no effect on the equilibrium distribution f_0.

$$\langle B(t)\rangle = \langle B(0)\rangle + \int_0^t ds \int d\Gamma\, f_0(iL(s) - iL_0) U_R(s,t) B \tag{8.61}$$

We can now unroll the Liouvilleans to attack the distribution function rather than the phase variables. The result is

$$\langle B(t)\rangle = \langle B(0)\rangle - \int_0^t ds \int d\Gamma\, [(iL(s) - iL_0) f_0] U_R(s,t) B \tag{8.62}$$

From equation (8.52) it is obvious that it is only the operation of the field-dependent Liouvillean which needs to be considered. Provided $AI\Gamma$ is satisfied, we know from (7.29, *et seq.*) that,

$$(iL(s) - iL_0) f_0 iL(s) f_0 = \beta f_0 J(\Gamma) F_e(s) \tag{8.63}$$

For either the canonical or Gaussian isokinetic ensembles, therefore,

$$\langle B(t)\rangle = \langle B(0)\rangle - \beta \int_0^t ds \int d\Gamma\, f_0 J F_e(s) U_R(s,t) B \tag{8.64}$$

Thus far the derivation has followed the same procedures used for the time-dependent linear response and time-independent nonlinear response. The operation of $U_R(s,t)$ on B, however, presents certain difficulties. No simple meaning can be attached to $U_R(s,t)B$. We can now use the composition and the inverse theorems to break up the incremental p-propagator $U_R(s,t)$. Using equation (8.18),

$$U_R(s,t) = U_R^{-1}(0,s) U_R(0,t) \tag{8.65}$$

Substituting this result into (8.64) we find

$$\langle B(t)\rangle = \langle B(0)\rangle - \beta \int_0^t ds \int d\Gamma\, F_e(s) f_0 J U_R^{-1}(0,s) U_R(0,t) B \tag{8.66}$$

Using the inverse theorem (8.7), and integrating by parts we find

$$\langle B(t)\rangle = \langle B(0)\rangle - \beta \int_0^t ds \int d\Gamma\, F_e(s) B(t) \exp_R\left[\int_0^s ds_1\, iL(s_1)\right] J f_0 \tag{8.67}$$

where after unrolling $U_R^{-1}(0, s)$ we attack B with $U_R(0, t)$ giving $B(t)$. As it stands the exponential in this equation has the right time ordering of a p-propagator, but the argument of the exponential contains an f-Liouvillean. We obviously have some choices here. We choose to use (8.48) to rewrite the exponential in terms of a p-propagator. This equation gives

$$\exp_R\left[\int_0^s ds_1\, iL(s_1)\right] = \exp\left[\int_0^s ds_1\, \Lambda(s_1)\right] U_R(0, s) \tag{8.68}$$

where

$$\Lambda(s_1) = -3N\alpha(\mathbf{\Gamma}(s_1), s_1) + O(1) \tag{8.69}$$

$\alpha(\mathbf{\Gamma}, s)$ is the Gaussian isokinetic multiplier required to maintain a fixed kinetic energy. Substituting these results into (8.67), using the fact that

$$\mathrm{i}L(s)H_0(\mathbf{\Gamma}) = -J(\mathbf{\Gamma})F_e(s) - 3Nk_BT\alpha(\mathbf{\Gamma}, s) \tag{8.70}$$

gives

$$\langle B(t)\rangle = \langle B(0)\rangle - \beta\int_0^t ds_1 \int d\mathbf{\Gamma}\; f_0 B(t)J(s_1)\exp\left[\int_0^{s_1} ds_2\, \beta J(s_2)F_e(s_2)\right]F_e(s_1) \tag{8.71}$$

or

$$\langle B(t)\rangle = \langle B(0)\rangle - \beta\int_0^t ds_1 \left\langle B(t)J(s_1)\exp\left[\beta\int_0^{s_1} ds_2\, J(s_2)F_e(s_2)\right]\right\rangle F_e(s_1) \tag{8.72}$$

This equation is the fundamental result of this chapter. It must be remembered that all time evolution is governed by the field-dependent thermostatted equations of motion implicit in the Liouvillean, $\mathrm{iL}(t)$.

8.11 RENORMALIZATION

We can apply our fundamental result, equation (8.72), to a number of known special cases. In the linear regime our equation obviously becomes

$$\langle B(t)\rangle = \langle B(0)\rangle - \beta\int_0^t ds\, \langle B(t)J(s)\rangle_0 F_e(s) \tag{8.73}$$

The notation '$\langle \ldots \rangle_0$' denotes an equilibrium average over the field-free thermostatted dynamics implicit in the Liouvillean, iL_0. This equation is the well-known result of time-dependent linear response theory (see Section 5.3).

Another special case that can be examined is the time-independent nonlinear response. In this circumstance the Liouvillean $\mathrm{i}L(t)$ is independent of time and the propagator, $U_R(0, t)$ becomes much simpler,

$$U_R(0, t) = \mathrm{e}^{\mathrm{i}Lt} \tag{8.74}$$

One does not need to use time-ordered exponentials. In this case the response is

$$\langle B(t) \rangle = \langle B(0) \rangle - \beta F_e \int_0^t \mathrm{d}s \left\langle B(t) J(s) \exp\left[\beta F_e \int_0^s \mathrm{d}s_1 \, J(s_1) \right] \right\rangle \tag{8.75}$$

Again all time propagation is generated by the field-dependent thermostatted Liouvillean, iL. This equation is new. As was the case for the Kawasaki form of the nonequilibrium distribution function, explicit normalization can be easily achieved.

Comparing (8.75) with the identity that can be obtained using the equivalence of the Schrödinger and Heisenberg representations (Section 8.7) namely,

$$\langle B(t) \rangle = \langle B(0) \rangle + \int \mathrm{d}\Gamma \, B(t)(f(0) - f(-t)) \tag{8.76}$$

implies that

$$f(0) - f(-t) = -\beta F_e \int_0^t \mathrm{d}s \, f(0) J(s) \exp\left[\beta \int_0^s \mathrm{d}s_1 \, J(s_1) F_e \right] \tag{8.77}$$

The integral $(0, t)$, on the right-hand side of the equation, can be performed and gives

$$f(-t) = f(0) \exp\left[\beta F_e \int_0^t \mathrm{d}s \, J(s) \right] \tag{8.78}$$

The correctness of this equation can easily be checked by differentiation. Furthermore, it is clear that this expression is just the unnormalized form of the Kawasaki distribution function (7.25).

This equation can be used to renormalize our expression for the time-independent nonlinear response. Clearly,

$$f(-t) = \frac{f(0) \exp\left[\beta F_e \int_0^t \mathrm{d}s \, J(s) \right]}{\int \mathrm{d}\Gamma \, f(0) \exp\left[\beta F_e \int_0^t \mathrm{d}s \, J(s) \right]} \tag{8.79}$$

is an explicitly normalized distribution function. By differentiating this distribution in time and then reintegrating, we find that

$$f(0) - f(-t) = -\beta F_e \int_0^t ds_1 \left\{ \frac{f(0)J(s_1)\exp\left[\beta F_e \int_0^{s_1} ds_2\, J(s_2)\right]}{\left\langle \exp\left[\beta F_e \int_0^{s_1} ds_2\, J(s_2)\right]\right\rangle} - \frac{f(0)\exp\left[\beta F_e \int_0^{s_1} ds_2\, J(s_2)\right]\left\langle J(s_1)\exp\left[\beta F_e \int_0^{s_1} ds_2\, J(s_2)\right]\right\rangle}{\left\langle \exp\left[\beta F_e \int_0^{s_1} ds_2\, J(s_2)\right]\right\rangle^2} \right\} \tag{8.80}$$

To simplify the notation we define the brace $\{\quad\}_s$ as

$$\{B(t)\}_s \equiv \frac{\int d\mathbf{\Gamma}\, B(t) f(0) \exp\left[\beta F_e \int_0^s ds_1\, J(s_1)\right]}{\int d\mathbf{\Gamma}\, f(0) \exp\left[\beta F_e \int_0^s ds_1\, J(s_1)\right]} \tag{8.81}$$

Using this definition our renormalized expression for the response is (Evans and Morriss, 1988).

$$\langle B(t)\rangle = \langle B(0)\rangle - \beta F_e \int_0^t ds\, \{[B(t) - \{B(t)\}_s][J(s) - \{J(s)\}_s]\}_s \tag{8.82}$$

8.12 DISCUSSION

We have described a consistent formalism for the nonlinear response of many-body systems to time-dependent external perturbations. This theory reduces to the standard results of linear response theory in the linear regime and can be used to derive the Kawasaki form of the time-independent nonlinear response. It also is easy to show that these results lead to the transient time-correlation function expressions for the time-independent nonlinear case.

If we consider equation (8.64) in the time-independent case and remember that

$$U_{\mathrm{R}}(s,t) = \exp_{\mathrm{R}} \int_s^t \mathrm{d}s_1 \, \mathrm{iL}(s_1) = \exp[(t-s)\mathrm{iL}] \tag{8.83}$$

then we can see immediately that

$$\begin{aligned} \langle B(t) \rangle &= \langle B(0) \rangle - \beta F_{\mathrm{e}} \int_0^t \mathrm{d}s \int \mathrm{d}\boldsymbol{\Gamma} \, f_0 J B(t-s) \\ &= \langle B(0) \rangle - \beta F_{\mathrm{e}} \int_0^t \mathrm{d}s \, \langle J(0) B(s) \rangle \end{aligned} \tag{8.84}$$

This is the standard transient time-correlation function expression for the nonlinear response (7.33).

It may be thought that we have complete freedom to move between the various forms of the nonlinear response: the Kawasaki form equation (8.78), the transient correlation function expression equation (8.84) and the new formulation described in this chapter, (8.82). These various formulations can be characterized by noting the times at which the test variable B and the dissipative flux J are evaluated. In the Kawasaki form B is evaluated at time zero, in the transient correlation approach J is evaluated at time zero, and in the new form developed in this chapter B is evaluated at time t. These manipulations are essentially trivial for the linear response.

As we have shown, these forms are all equivalent for the nonlinear response to time-independent external fields. However, for the time-dependent nonlinear case only our new form, (8.82), seems to be valid. One can develop a Kawasaki version of the nonlinear response to time-dependent fields, but it is found that the resulting expression is not very useful. Like the corresponding transient correlation form, such a version involves convolutions of incremental propagators, Liouvilleans and phase variables which have no directly interpretable meaning. None of the operators in the convolution chains commute with one another and the resulting expressions are intractable and formal.

REFERENCES

Evans, D.J. and Morriss, G.P. (1988). *Mol. Phys.*, **64**, 521.
Holian, B.L. and Evans, D.J. (1985). *J. Chem. Phys.*, **83**, 3560.
Parry, W.E. (1973). *The Many-body Problem.* Oxford: Clarendon Press.
Raimes, S. (1972). *Many-electron Theory.* Amsterdam: North-Holland.

9 Steady-State Fluctuations

9.1 INTRODUCTION

Nonequilibrium steady states are fascinating systems to study. Although there are many parallels between these states and equilibrium states, a convincing theoretical description of steady states, particularly far from equilibrium, has yet to be found. Close to equilibrium, linear response theory and linear irreversible thermodynamics provide a relatively complete treatment (Sections 2.1 to 2.3). However, in systems where local thermodynamic equilibrium has broken down, and thermodynamic properties are not the same local functions of thermodynamic state variables such that they are at equilibrium, our understanding is very primitive indeed.

In Section 7.3 we gave a statistical mechanical description of thermostatted, nonequilibrium steady states far from equilibrium—the *transient time-correlation function* (TTCF) and *Kawasaki* formalisms. The transient time-correlation function is the nonlinear analogue of the Green–Kubo correlation functions. For linear transport processes the Green–Kubo relations play a role which is analogous to that of the partition function at equilibrium. Like the partition function, the Green–Kubo relations are highly nontrivial to evaluate. They do, however, provide an exact starting point from which one can derive exact interrelations between thermodynamic quantities. The Green–Kubo relations also provide a basis for approximate theoretical treatments as well as being used directly in equilibrium molecular dynamics simulations.

The TTCF and Kawasaki expressions may be used as nonlinear, nonequilibrium partition functions. For example, if a particular derivative commutes with the thermostatted, field-dependent propagator then one can formally differentiate the TTCF and Kawasaki expressions for steady-state phase averages, yielding fluctuation expressions for the so-called *derived properties*. The key point in such derivations is that the particular derivative should commute with the relevant propagators. If this is not so one cannot derive tractable or useful results.

In order to constrain thermodynamic variables, two basic feedback mechanisms can be employed: the integral feedback mechanism, employed for example in the Nose–Hoover thermostat (Section 5.2), and the differential mechanism employed in the Gaussian thermostat. A third mechanism, the proportional mechanism, has not found much use either in simulations or in theory because it necessarily employs *irreversible* equations of motion.

In this chapter we derive fluctuation expressions for the derivatives of steady-state phase averages. We derive expressions for derivatives with respect to temperature, pressure and the mean value of the dissipative flux. Applying these derivatives in turn to averages of the internal energy, the volume and the thermodynamic driving force yields expressions for the specific heats, the compressibility and the inverse Burnett coefficients, respectively. In order to ensure the commutivity of the respective derivatives and propagators, we employ the Gaussian feedback mechanism exclusively. Corresponding derivations using Nose–Hoover feedback are presently unknown.

Rather than giving a general but necessarily formal derivation of the fluctuation formulae, we concentrate instead on two specific systems: planar Couette flow and colour conductivity. By concentrating on specific systems we hope to make the discussion more concrete and simultaneously to illustrate particular applications of the theory of nonequilibrium steady states discussed in Chapter 7.

9.2 THE SPECIFIC HEAT

In this section we illustrate the use of the Kawasaki distribution function and the transient time correlation function (TTCF) formalism by deriving formally exact expressions for the temperature derivative of nonequilibrium averages. Applying these expressions to the internal energy, we obtain two formulae (Evans and Morriss, 1987) for the isochoric specific heat. One of these formulae shows that the specific heat can be calculated by analysing fluctuations in the steady state. The second formulae relates the steady-state specific heat to the transient response observed when an ensemble of equilibrium systems is perturbed by the field.

Transient time-correlation function approach

For a system undergoing planar Couette flow the transient correlation function expression for the canonical ensemble average of a phase variable B is

$$\langle B(t)\rangle = \langle B(0)\rangle - \beta\gamma V \int_0^t ds \langle B(s) P_{xy}(0)\rangle \tag{9.1}$$

This expression relates the nonequilibrium value of a phase variable B at time t to the integral of a TTCF (the correlation between P_{xy} in the equilibrium starting state, $P_{xy}(0)$, and B at time s after the field is turned on). The temperature implied by the β term is the temperature of the initial ensemble. The steady state is *tied* to the initial ensemble by the constraint of constant peculiar kinetic energy. Therefore, for systems that exhibit mixing (9.1) can be rewritten as

$$\langle B(t)\rangle = \langle B(0)\rangle - \beta\gamma V \int_0^t \mathrm{d}s \, \langle \Delta B(s) P_{xy}(0)\rangle \tag{9.2}$$

where the difference variable $\Delta B(s)$ is defined as the difference between the phase variable at s and its *average value at s*

$$\Delta B(s) \equiv B(s) - \langle B(s)\rangle \tag{9.3}$$

Systems which are not expected to exhibit mixing are turbulent systems or systems which execute quasi-periodic oscillations.

An important property of the Gaussian thermostat is that, although it fixes the kinetic energy of a system, the Gaussian isokinetic Liouville operator is *independent* of the temperature of the initial distribution. For each member of the ensemble, the Gaussian thermostat simply constrains the peculiar kinetic energy to be constant. As the Liouvillean and the propagator in (9.2) are independent of the *value* of the temperature, we can calculate the temperature derivative very easily:

$$\frac{\partial}{\partial T}\langle B(t)\rangle = k_B\beta^2\langle \Delta(B(0))\Delta(H_0(0))\rangle - k_B\beta(\langle B(t)\rangle - \langle B(0)\rangle)$$
$$- k_B\beta^3 F_e \int_0^t \mathrm{d}s \, \langle \Delta(B(s)J(0))\Delta(H_0(0))\rangle \tag{9.4}$$

The first term on the right-hand side of (9.4) is the equilibrium contribution. This is easily seen by setting $t = 0$. The second and third terms are nonequilibrium terms. In deriving the second term on the right-hand side of (9.4) we use equation (9.3) to simplify a number of terms. It is worth noting that (9.4) is not only valid in the steady-state limit $t \to \infty$, but is also correct for all intermediate times t, which correspond to the transients which take the system from the initial equilibrium state to the final nonequilibrium steady state.

If we choose to evaluate the temperature derivative of the internal energy H_0, we can calculate the specific heat at constant volume and external field, C_{V,F_e}. The result is (Evans and Morriss, 1987)

$$C_{V,F_e}(t) = k_B\beta^2\langle \Delta(H_0(0)^2)\rangle - k_B\beta(\langle H_0(t)\rangle - \langle H_0(0)\rangle)$$
$$- k_B\beta^3 F_e \int_0^t \mathrm{d}s \, \langle \Delta(H_0(s)J(0))\Delta(H_0(0))\rangle \tag{9.5}$$

Again the first term on the right-hand side is easily recognized as the equilibrium specific heat. The second and third terms are nonlinear nonequilibrium terms; they signal the breakdown of local thermodynamic equilibrium. In the linear regime for which linear response theory is valid, these terms are of course both zero. The third term takes the form of a TTCF; it measures the correlations of equilibrium energy fluctuations, $\Delta H_0(0)$, with the transient fluctuations in the composite-time variable, $\Delta(H_0(s)J(0))$. The second term can of course be rewritten as the integral of a TTCF using (9.1).

Kawasaki representation

Consider the Schrödinger form

$$\langle B(t)\rangle = \int \mathrm{d}\boldsymbol{\Gamma}\, B(\boldsymbol{\Gamma}) f(t) \tag{9.6}$$

The thermostatted Kawasaki form for the N-particle distribution function is

$$f(t) = \exp\left[-\beta F_{\mathrm{e}} \int_0^t \mathrm{d}s\, J(-s) \right] f(0) \tag{9.7}$$

Since $f(t)$ is a distribution function it must be normalized. We guarantee this by dividing the right-hand side of (9.7) by its phase integral. If we take the initial ensemble to be *canonical*, we find

$$f(t) = \frac{\exp\left[-\beta\left(H_0 + F_{\mathrm{e}} \int_0^t \mathrm{d}s\, J(-s) \right) \right]}{\int \mathrm{d}\boldsymbol{\Gamma} \exp\left[-\beta\left(H_0 + F_{\mathrm{e}} \int_0^t \mathrm{d}s\, J(-s) \right) \right]} \tag{9.8}$$

The exponents contains a divergence due to the fact that the time average of $J(-s)$ is nonzero. This secular divergence can be removed by multiplying the numerator and the denominator of the explicitly normalized form by $\exp[+\beta F_{\mathrm{e}} \int_0^t \mathrm{d}s \langle J(-s)\rangle]$. This has the effect of changing the dissipative flux that normally appears in the Kawasaki exponent from $J(-s)$ to $\Delta J(-s)$, in both the numerator and denominator.

$$f(t) = \frac{\exp\left[-\beta\left(H_0 + F_{\mathrm{e}} \int_0^t \mathrm{d}s\, \Delta J(-s) \right) \right]}{\int \mathrm{d}\boldsymbol{\Gamma} \exp\left[-\beta\left(H_0 + F_{\mathrm{e}} \int_0^t \mathrm{d}s\, \Delta J(-s) \right) \right]} \tag{9.9}$$

The average of an arbitrary phase variable $B(\boldsymbol{\Gamma})$ in the renormalized Kawasaki representation is

$$\langle B(t)\rangle = \langle B(0)\rangle + \frac{\int \mathrm{d}\boldsymbol{\Gamma}\, \Delta(B) \exp\left[-\beta\left(H_0 + \int_0^t \mathrm{d}s\, \Delta J(-s)F_{\mathrm{e}}\right)\right]}{\int \mathrm{d}\boldsymbol{\Gamma} \exp\left[-\beta\left(H_0 + \int_0^t \mathrm{d}s\, \Delta J(-s)F_{\mathrm{e}}\right)\right]} \tag{9.10}$$

To obtain the temperature derivative of (9.10) we differentiate with respect to β. This gives

$$\frac{\partial\langle B(t)\rangle}{\partial\beta} = -\int \mathrm{d}\boldsymbol{\Gamma}\, B\left(H_0 + F_{\mathrm{e}}\int_0^t \mathrm{d}s\, \Delta J(-s)\right) f(t) + \left(\int \mathrm{d}\boldsymbol{\Gamma}\, Bf(t)\right)\left(\int \mathrm{d}\boldsymbol{\Gamma}\left(H_0 + F_{\mathrm{e}}\int_0^t \mathrm{d}s\, \Delta J(-s)\right) f(t)\right) \tag{9.11}$$

Using the Schrödinger–Heisenberg equivalence we transfer the time dependence function to the phase variable in each of the terms in (9.11). This gives

$$\frac{\partial\langle B(t)\rangle}{\partial\beta} = -\left\langle \Delta B(t)\Delta\left(H_0(t) + F_{\mathrm{e}}\int_0^t \mathrm{d}s\, \Delta J(t-s)\right)\right\rangle \tag{9.12}$$

Substituting the internal energy for B in (9.12) and making a trivial change of variable in the differentiation ($\beta \to T$) and integration ($t - s \to s$), we find that the specific heat can be written as

$$C_{V,F_{\mathrm{e}}}(t) = k_{\mathrm{B}}\beta^2\langle \Delta(H_0(t)^2)\rangle + k_{\mathrm{B}}\beta^2 F_{\mathrm{e}}\int_0^t \mathrm{d}s\,\langle \Delta H_0(t)\Delta J(s)\rangle \tag{9.13}$$

The first term gives the steady-state energy fluctuations and the second term is a steady-state time correlation function. As $t \to \infty$, the only times s, which contribute to the integral are times within a relaxation time of t, so that in this limit the time correlation function has no memory of the time at which the field was turned on.

These theoretical results for the specific heat of nonequilibrium steady states have been tested in nonequilibrium molecular dynamics (NEMD) simulations of isothermal planar Couette flow (Evans, 1986; Evans and Morriss, 1987). The system studied was the Lennard–Jones fluid at its triple point, ($k_{\mathrm{B}}T/\varepsilon = 0.722$, $\rho\sigma^3 = 0.8442$). One hundred and eight particles were employed with a cutoff of 2.5σ.

The steady-state specific heat was calculated in three ways: from the transient correlation function expression (9.5), from the Kawasaki expression equation (9.13) and by direct numerical differentiation of the internal energy with respect to the initial temperature. The results are shown in Table 9.1. Although we have been unable to prove the result theoretically, the numerical results suggest that the integral appearing on the right-hand side of (9.5) is zero. All our simulation results, within error bars, are consistent with this. As can be seen in Table 9.1, the transient correlation expression for the specific heat predicts that it decreases as we go away from equilibrium. The predicted specific heat at a reduced strain rate $(\gamma\sigma(m/\varepsilon)^{1/2}) = 1$ is some 11% smaller than the equilibrium value. This behaviour of the specific heat was first observed in 1983 by Evans (Evans, 1983).

The results obtained from the Kawasaki formula show that, although the internal energy fluctuations are greater than at equilibrium, the specific heat decreases as the strain rate is increased. The integral of the steady-state energy–stress fluctuations more than compensates for the increase in internal energy fluctuations. The Kawasaki prediction for the specific heat is in statistical agreement with the transient correlation results. Both sets of results also agree with the specific heat obtained by direct numerical differentiation of the internal energy. Table 9.2 shows a similar set of comparisons based on published data (Evans, 1983). Once again there is good agreement between

Table 9.1 Lennard–Jones specific heat data.‡
Potential: $\Phi(r) = 4\varepsilon[(r/\sigma)^{-12} - (r/\sigma)^{-6}]$.
State point: $T^* = 0.722$, $\rho^* = 0.8442$, $\gamma^* = 1.0$, $N = 108$, $r_c^* = 2.5$.

Transient correlation results: 200 K time-steps	
$C^*_{v,\gamma=0}/N$	**2.662 ± 0.004**
$(\langle E^*_\gamma\rangle_{ss} - \langle E^*_\gamma\rangle_{\gamma=0})/NT^*$	0.287 ± 0.0014
$\gamma^* T^{*-3}\rho^{*-1}\int_0^\infty ds\, \langle \Delta(H_0^*(s)\, P_{xy}^*(0))\Delta(H_0^*(0))\rangle$	−0.02 ± 0.05
$C^*_{v,\gamma=1}/N$	**2.395 ± 0.06**
Kawasaki correlation results: 300 K time-steps	
$\langle \Delta(E)^{*2}\rangle_{ss}/NT^{*2}$	3.307 ± 0.02
$\gamma^* T^{*-2}\rho^{*-1}\int_0^\infty ds\, \langle \Delta E^*(s)\Delta P_{xy}^*(0)\rangle_{ss}$	−1.050 ± 0.07
$C^*_{v,\gamma=1}/N$	**2.257 ± 0.09**
Direct NEMD calculation: 100 K time-steps	
$C^*_{v,\gamma=1}/N$	2.35 ± 0.05

‡ Reduced units are denoted by *. Units are reduced to dimensionless form in terms of the Lennard–Jones parameters, m, σ and ε; $\gamma^* = \partial u_x/\partial y\ \sigma(m/\varepsilon)^{1/2}$; and $\Delta t^* = 0.004$. $\langle\ \rangle_{ss}$ denotes nonequilibrium steady-state average.

Table 9.2 Comparison of soft sphere specific heats as a function of strain rate.‡ Potential: $\Phi(r) = \varepsilon(r/\sigma)^{-12}$. State point: $T^* = 1.0877$, $\rho^* = 0.7$, $N = 108$, $r_c^* = 1.5$.

γ^*	E^*/NT^*	$C^*_{V,\gamma}/N$ Direct	$C^*_{V,\gamma}/N$ Transient correlation
0.0	4.400	2.61	2.61
0.4	4.441	2.56	2.57
0.6	4.471	2.53	2.53
0.8	4.510	2.48	2.49
1.0 ± 0.01	4.550 ± 0.01	2.43	2.46 ± 0.002

‡ Note: In these calculations the TTCF integral, (9.5), was assumed to be zero. Data from Evans (1983, 1986) and Evans and Morriss (1987).

results predicted on the basis of the transient correlation formalism and the direct NEMD method.

As a final comment on this section we should stress that the specific heat as we have defined it, refers only to the derivative of the internal energy with respect to the temperature of the *initial* ensemble (or, equivalently, with respect to the nonequilibrium *kinetic* temperature). Thus far, our derivations say nothing about the thermodynamic temperature ($\equiv \partial E/\partial S$) of the steady state. We will return to this subject in Chapter 10.

9.3 THE COMPRESSIBILITY AND ISOBARIC SPECIFIC HEAT

In this section we calculate formally exact fluctuation expressions for other derived properties including the specific heat at constant pressure and external field, C_{p,F_e}, and the compressibility, $\chi_{T,F_e} \equiv (-\partial \ln V/\partial p)_{T,F_e}$. The expressions are derived using the isothermal Kawasaki representation for the distribution function of an isothermal isobaric steady state.

The results indicate that the compressibility is related to nonequilibrium volume fluctuations in exactly the same way that it is at equilibrium. The isobaric specific heat, C_{p,F_e}, however, is not simply related to the mean square of the enthalpy fluctuations as it is at equilibrium. In a nonequilibrium steady state these enthalpy fluctuations must be supplemented by the integral of the steady-state time cross-correlation function of the dissipative flux and the enthalpy.

We begin by considering the isothermal–isobaric equations of motion considered in Section 6.7. The obvious nonequilibrium generalization of these equations is

$$\frac{\mathrm{d}\mathbf{q}_i}{\mathrm{d}t} = \frac{\mathbf{p}_i}{m} + \varepsilon\mathbf{q}_i + \mathbf{C}(\mathbf{\Gamma})F_e(t)$$

$$\frac{\mathrm{d}\mathbf{p}_i}{\mathrm{d}t} = \mathbf{F}_i - \varepsilon\mathbf{p}_i + \mathbf{D}(\mathbf{\Gamma})F_e(t) - \alpha(\mathbf{\Gamma}, t)\mathbf{p}_i \tag{9.14}$$

$$\frac{\mathrm{d}V}{\mathrm{d}t} = 3V\dot{\varepsilon}$$

In the above equations $\mathrm{d}\varepsilon/\mathrm{d}t$ is the dilation rate required to fix precisely the value of the hydrostatic pressure, $p = \Sigma(p^2/m + \mathbf{q}\cdot\mathbf{F})/3V$; α is the usual Gaussian thermostat multiplier used to fix the peculiar kinetic energy, K. Simultaneous equations must be solved to yield explicit expressions for both multipliers. We do not give these expressions here since they are straightforward generalizations of the field-free ($F_e = 0$) equations given in Section 6.7.

The external-field terms are assumed to be such that they satisfy the usual adiabatic incompressibility of phase space ($AI\Gamma$) condition. We define the dissipative flux, J, as the obvious generalization of the usual isochoric case.

$$\left(\frac{\mathrm{d}I_0}{\mathrm{d}t}\right)^{\mathrm{ad}} \equiv -J(\mathbf{\Gamma})F_e \tag{9.15}$$

This definition is consistent with the fact that in the field-free adiabatic case the enthalpy $I_0 \equiv H_0 + pV$, is a constant of the equations of motion given in (9.14). It is easy to see that the isothermal isobaric distribution, f_0, is preserved by the field-free thermostatted equations of motion.

$$f_0 = \frac{\mathrm{e}^{-\beta I_0}}{\int_0^\infty \mathrm{d}V \int \mathrm{d}\mathbf{\Gamma}\, \mathrm{e}^{-\beta I_0}} \tag{9.16}$$

It is a straightforward matter to derive the Kawasaki form of the N-particle distribution for the isothermal–isobaric steady state. The normalized version of the distribution function is

$$f(t) = \frac{\exp -\beta\left[I_0 + \int_0^t \mathrm{d}s\, J(-s)F_e\right]}{\int_0^\infty \mathrm{d}V \int \mathrm{d}\mathbf{\Gamma} \exp -\beta\left[I_0 + \int_0^t \mathrm{d}s\, J(-s)F_e\right]} \tag{9.17}$$

The calculation of derived quantities is a simple matter of differentiation with respect to the variables of interest. As was the case for the isochoric specific heat, the crucial point is that the field-dependent isothermal–isobaric propagator implicit in the notation $f(t)$ is independent of the pressure and the temperature of the ensemble. This means that the differential operators $\partial/\partial T$ and $\partial/\partial p_0$ commute with the propagator.

The pressure derivative is easily calculated as

$$\left(\frac{\partial\langle B(t)\rangle}{\partial p_0}\right)_{T,F_e} = -\beta\langle B(t)V(t)\rangle + \beta\langle B(t)\rangle\langle V(t)\rangle \tag{9.18}$$

If we choose B to be the phase variable corresponding to the volume then the expression for the isothermal, fixed field compressibility takes on a form which is formally identical to its equilibrium counterpart.

$$\chi_{T,F_e} = \lim_{t\to\infty} -\frac{\beta}{V}\langle \Delta V(t)^2\rangle \tag{9.19}$$

The limit appearing in (9.19) implies that a steady-state average should be taken. This follows from the fact that the external field was 'turned on' at $t = 0$.

The isobaric temperature derivative of the average of a phase variable can again be calculated from (9.17)

$$\frac{\partial}{\partial\beta}\langle B(t)\rangle = \int_0^\infty dV \int d\Gamma\, f(t)B(0)\left(I_0 - \int_0^t ds\, J(-s)F_e\right)$$
$$-\langle I_0(t)\rangle \int_0^\infty dV \int d\Gamma\, f(t)\left(I_0 + \int_0^t ds\, J(-s)F_e\right) \tag{9.20}$$

In deriving (9.20) we have used the fact that $\int dV \int d\Gamma\, f(t)B(0) = \langle B(t)\rangle$. Equation (9.20) can clearly be used to derive expressions for the thermal expansion coefficient. However, setting the test variable B to be the enthalpy and remembering that

$$C_{p,F_e} = \left(\frac{\partial I_0}{\partial T}\right)_{p,F_e} \tag{9.21}$$

leads to the isobaric specific heat

$$C_{p,F_e} = \lim_{t\to\infty} \frac{1}{k_B T^2}\left\{\langle \Delta(I_0(t))^2\rangle + F_e\int_0^t ds\, \langle \Delta I_0(t)\Delta J(s)\rangle\right\} \tag{9.22}$$

This expression is of course very similar to the expression derived for the isochoric specific heat in Section 9.2.

In contrast to the situation for the compressibility, the expressions for the specific heats are not simple generalizations of the corresponding equilibrium fluctuation formulae. Both specific heats also involve integrals of steady-state time-correlation functions involving cross-correlations of the appropriate energy with the dissipative flux. Although the time integrals in (9.13) and (9.22) extend back to $t = 0$ when the system was at equilibrium, for systems which exhibit mixing, only the steady-state portion of the integral contributes. This is because in such systems, $\lim(t \to \infty)\langle \Delta B(t)\Delta J(0)\rangle = \langle \Delta B(t)\rangle \langle \Delta J(0)\rangle = 0$. These correlation functions are therefore comparatively easy to calculate in computer simulations.

9.4 DIFFERENTIAL SUSCEPTIBILITY

In Section 2.3 we introduced the linear transport coefficients as the first term in an expansion, about equilibrium, of the thermodynamic flux in terms of the driving thermodynamic forces. The nonlinear Burnett coefficients are the coefficients of this Taylor expansion. Before we address the question of the nonlinear Burnett coefficients we will consider the *differential susceptibility* of a nonequilibrium steady state. Suppose we expand the irreversible fluxes in powers of the forces, about a nonequilibrium steady state. The leading term in such an expansion is called the differential susceptibility. As we will see, difficulties with commutation relations force us to work in the Norton rather than the Thévenin ensemble. This means that we will always be considering the variation of the thermodynamic forces which result from possible changes in the thermodynamic fluxes.

Consider an ensemble of N-particle systems satisfying the following equations of motion. For simplicity we assume that each member of the ensemble is electrostatically neutral and consists only of univalent ions of charge, $\pm e = \pm 1$. This system is formally identical to the colour conductivity system which we considered in Section 6.2.

$$\dot{\mathbf{q}}_i = \frac{\mathbf{p}_i}{m} \equiv \mathbf{v}_i \tag{9.23}$$

$$m\dot{\mathbf{v}}_i = \mathbf{F}_i + \mathbf{i}\lambda e_i - \alpha(\mathbf{v}_i - \mathbf{i}e_i J) \tag{9.24}$$

In these equations, λ and α are Gaussian multipliers chosen so that the x component of the current per particle, $J = \Sigma\, e_i v_{xi}/N$ and the temperature $T = \Sigma\, m(\mathbf{v}_i - \mathbf{i}e_i J)^2/3Nk_B$ are constants of the motion. This will be the case provided that

$$\lambda = -\frac{\sum F_{xi} e_i}{N} \tag{9.25}$$

and

$$\alpha = \frac{\sum \mathbf{F}_i \cdot (\mathbf{v}_i - \mathbf{i}e_i J)}{\sum \mathbf{v}_i \cdot (\mathbf{v}_i - \mathbf{i}e_i J)} \tag{9.26}$$

In more physical terms λ can be thought of as an external electric/colour field which takes on precisely those values required to ensure that the current J is constant. Because it precisely fixes the current, it is a phase variable. It is clear from (9.25) that the form of the phase variable λ is independent of the average value of the current. Of course the ensemble average of λ will depend on the average value of the current. It is also clear that the expression for α is similarly independent of the *average* value of the current for an ensemble of such systems.

These points can be clarified by considering an initial ensemble characterized by the canonical distribution function, $f(0)$,

$$f(0) = \frac{\exp\left\{-\beta\left[\sum \frac{m}{2}(\mathbf{v}_i - e_i\mathbf{J}_0)^2 + \Phi\right]\right\}}{\int \mathrm{d}\mathbf{\Gamma} \exp\left\{-\beta\left[\sum \frac{m}{2}(\mathbf{v}_i - e_i\mathbf{J}_0)^2 + \Phi\right]\right\}} \tag{9.27}$$

In this equation $\mathbf{J}_0$ is a constant which is equal to the canonical average of the current

$$\langle \mathbf{J}(0) \rangle = \mathbf{J}_0 = \mathbf{i}J_0 \tag{9.28}$$

If we now subject this ensemble of systems which we will refer to as the J-ensemble, to the equations of motion ((9.23) and (9.24)), the electrical current and the temperature will remain fixed at their initial values and the mean value of the field multiplier, λ, will be determined by the electrical conductivity of the system.

It is relatively straightforward to apply the theory of nonequilibrium steady states to this system. It is easily seen from the equations of motion that the condition known as the adiabatic incompressibility of phase space ($AI\Gamma$) holds. Using equations (9.23) to (9.27), the adiabatic time derivative of the energy functional is easily seen to be

$$(\dot{H})^{\mathrm{ad}} \equiv \left(\frac{\mathrm{d}}{\mathrm{d}t}\right)^{\mathrm{ad}} \sum_i \frac{m}{2}(\mathbf{v}_i - e_i\mathbf{J}_0)^2 + \Phi = N\lambda(\mathbf{\Gamma})J(\mathbf{\Gamma}) \tag{9.29}$$

This equation is unusual in that the adiabatic derivative does not factorize into the product of a dissipative flux and the magnitude of a perturbing external field. This is because in the J-ensemble the obvious external field, λ, is in fact a phase variable and the current, J, is a constant of the motion. As we shall see this causes no particular problems. The last equation that we need for the application of nonlinear response theory is the derivative

$$\frac{\partial}{\partial \mathbf{J}_0} H = -mN(\mathbf{J} - \mathbf{J}_0) \tag{9.30}$$

Kawasaki representation

If we use the isothermal generalization of the Kawasaki expression for the average of an arbitrary phase variable, B, we find that

$$\langle B(t)\rangle = \frac{\left\langle B(0)\exp\left[\beta N\int_0^{-t} \mathrm{d}s\, J(s)\lambda(s)\right]\right\rangle}{\left\langle \exp\left[\beta N\int_0^{-t} \mathrm{d}s\, J(s)\lambda(s)\right]\right\rangle} \tag{9.31}$$

In distinction to the usual case which we considered in Section 7.2, the Kawasaki exponent involves a product of two phase variables, J and λ, rather than the usual product of a dissipative flux (i.e. a phase variable) and a time-dependent external field. The propagator used in (9.31) is the field-dependent thermostatted propagator implicit in the equations of motion ((9.23) to (9.26)). The only place that the ensemble averaged current appears in (9.31) is in the initial ensemble averages. We can, therefore, easily differentiate (9.31) with respect to $\mathbf{J}_0$ to find that (Evans and Lynden-Bell, 1988)

$$\frac{\partial\langle B(t)\rangle}{\partial \mathbf{J}_0} = \beta m N\langle \Delta B(t)\Delta\mathbf{J}(0)\rangle \tag{9.32}$$

where $\Delta(B(t)) \equiv B(t) - \langle B(t)\rangle$ and $\Delta(\mathbf{J}(t)) \equiv \mathbf{J}(t) - \langle \mathbf{J}(t)\rangle = \mathbf{J}(0) - \mathbf{J}_0$. This is an exact canonical ensemble expression for the J-derivative of the average of an arbitrary phase variable. If we let t tend toward infinity we obtain a steady-state-fluctuation formula which complements the ones we derived earlier for the temperature and pressure derivatives. Equation (9.32) gives a steady-state-fluctuation relation for the differential susceptibility of B.

One can check that this expression is correct by rewriting the right-hand side of (9.32) as an integral of responses over a set of Norton ensembles in which the current takes on specific values. Using equation (9.27) we can write the average of $B(t)$ as

$$\langle B(t)\rangle = \frac{\int \mathrm{d}J \exp[-\beta m N\Delta J^2/2]\langle B(t); J\rangle}{\int \mathrm{d}J \exp[-\beta m N\Delta J^2/2]} \tag{9.33}$$

We use the notation $\langle B(t); J\rangle$ to denote that subset of the canonical ensemble, (9.27), in which the current takes on the exact value of J. The probability of the J-ensemble taking on an initial x current of J is easily calculated from (9.27) to be proportional to $\exp[-\beta m N\Delta J^2/2]$. Since the current is a constant of the motion we do not need to specify a time at which the current takes on the specified value.

Differentiating (9.33) we can write the derivative with respect to the average current as a superposition of ΔJ-ensemble contributions

$$\frac{\partial\langle B(t)\rangle}{\partial J_0} = \beta mN \frac{\int \mathrm{d}J \exp[-\beta mN\Delta J^2]\Delta\mathbf{J}\langle B(t);J\rangle}{\int \mathrm{d}J \exp[-\beta mN\Delta J^2/2]}$$

$$= \beta mN\langle \Delta B(t)\Delta\mathbf{J}(0)\rangle \tag{9.34}$$

This expression is of course identical to equation (9.32) which was derived using the Kawasaki distribution. (9.34) was derived, however, without the use of perturbative mechanical considerations such as those implicit in the use of the Kawasaki distribution. This second derivation is based on two points: the initial distribution is a normal distribution of currents about J_0; and the dynamics preserves the value of the current for each member of the ensemble. Of course the result is still valid even when J is not exactly conserved provided that the time-scale over which it changes is much longer than the time-scale for the decay of steady-state fluctuations. This derivation provides independent support for the validity of the renormalized Kawasaki distribution function.

We will now derive relations between the J-derivatives in the J-ensemble and in the constrained ensemble in which J takes on a precisely fixed value (the ΔJ-ensemble). In the thermodynamic limit, the spread of possible values of ΔJ will become infinitely narrow, suggesting that we can write a Taylor expansion of $\langle B(t);J\rangle$ in powers of ΔJ about J_0.

$$\langle B(t);J\rangle = \langle B(t);J_0\rangle + \Delta J\frac{\partial\langle B(t);J_0\rangle}{\partial J} + \frac{\Delta J^2}{2!}\frac{\partial^2\langle B(t);J_0\rangle}{\partial J^2} + \ldots \tag{9.35}$$

Substituting (9.35) into (9.34) and performing the Gaussian integrals over J, we find that

$$\frac{\partial\langle B(t)\rangle}{\partial J_0} = \frac{\partial\langle B(t);J_0\rangle}{\partial J} + \frac{1}{2\beta mN}\frac{\partial\langle B(t);J_0\rangle}{\partial J^3} + \ldots \tag{9.36}$$

This is a very interesting equation. It shows the relationship between the derivative computed in a canonical ensemble and a ΔJ-ensemble. It shows that differences between the two ensembles arise from nonlinearities in the local variation of the phase variable with respect to the current. It is clear that these ensemble corrections are of order $1/N$ compared with the leading terms.

9.5 THE INVERSE BURNETT COEFFICIENTS

We will now use the transition time-correlation function (TTCF) formalism in the Norton ensemble to derive expressions for the inverse Burnett coefficients. The Burnett coefficients, L_i, give a Taylor series representation of a nonlinear transport coefficient $L(X)$, defined by a constitutive relation between a thermodynamic force X, and a thermodynamic flux $J(\mathbf{\Gamma})$,

$$\begin{aligned}\langle J \rangle &= L(X)X \\ &= L_1 X + \frac{1}{2!} L_2 X^2 + \frac{1}{3!} L_3 X^3 + \dots\end{aligned} \tag{9.37}$$

It is clear from this equation the Burnett coefficients are given by the appropriate partial derivatives of $\langle J \rangle$, evaluated at $X = 0$. As mentioned in Section 9.4 we will actually be working in the Norton ensemble in which the thermodynamic force X, is the dependent rather than the independent variable. So we will in fact derive expressions for the inverse Burnett coefficients, $\mathscr{L}_i$.

$$\begin{aligned}\langle X \rangle &= \mathscr{L}(J)J \\ &= \mathscr{L}_1 J + \frac{1}{2!} \mathscr{L}_2 J^2 + \frac{1}{3!} \mathscr{L}_3 J^3 + \dots\end{aligned} \tag{9.38}$$

The TTCF representation for a steady-state phase average for our electrical/colour diffusion problem is easily seen to be

$$\langle B(t) \rangle = \beta N \int_0^t \mathrm{d}s \, \langle \Delta B(s) \lambda(0) J(0) \rangle \tag{9.39}$$

We assume that $\langle B(0) \rangle = 0$. We expect that the initial values of the current will be clustered about J_0. If we write

$$\langle \Delta B(s) \lambda(0) J(0) \rangle = \langle \Delta B(s) \lambda(0) \rangle J_0 + \langle \Delta B(s) \lambda(0) \Delta J(0) \rangle \tag{9.40}$$

it is easy to see that if B is extensive then the two terms on the right-hand side of (9.40) are $O(1)$ and $O(1/N)$, respectively. For large systems we can therefore write

$$\langle B(t) \rangle = \beta N \int_0^t \mathrm{d}s \, \langle \Delta B(s) \lambda(0) \rangle J_0 \tag{9.41}$$

It is now a simple matter to calculate the appropriate J-derivatives

$$\frac{\partial\langle B(t)\rangle}{\partial J_0} = \beta N \int_0^t \mathrm{d}s\, \langle B(s)\lambda(0)\rangle + \beta J_0 N \int_0^t \mathrm{d}s \int \mathrm{d}\mathbf{\Gamma}\, B(s)\lambda(0) \frac{\partial f(0)}{\partial J_0}$$

$$= \beta N \int_0^t \mathrm{d}s\, \langle B(s)\lambda(0)\rangle + \beta^2 J_0 m N^2 \int_0^t \mathrm{d}s\, \langle B(s)\lambda(0)\Delta J(0)\rangle \tag{9.42}$$

This equation relates the J-derivative of phase variables to TTCFs. If we apply these formulae to the calculation of the leading Burnett coefficient we of course evaluate the derivatives at $J_0 = 0$. In this case the TTCFs become equilibrium time-correlation functions. The results for the leading Burnett coefficients are (Evans and Lynden-Bell, 1988):

$$\left(\frac{\partial\langle B(t)\rangle}{\partial J_0}\right)_{J_0=0} = \beta N \int_0^t \mathrm{d}s\, \langle B(s)\lambda(0)\rangle_{\mathrm{eq}} \tag{9.43}$$

$$\left(\frac{\partial^2\langle B(t)\rangle}{\partial J_0^2}\right)_{J_0=0} = 2\beta^2 m N^2 \int_0^t \mathrm{d}s\, \langle B(s)\lambda(0)\Delta J(0)\rangle_{\mathrm{eq}} \tag{9.44}$$

$$\left(\frac{\partial^3\langle B(t)\rangle}{\partial J_0^3}\right)_{J_0=0} = 3\beta^3 m^2 N^3 \int_0^t \mathrm{d}s\, \langle B(s)\lambda(0)\{\Delta^2[J(0)] - \langle\Delta^2[J(0)]\rangle\}\rangle_{\mathrm{eq}} \tag{9.45}$$

Surprisingly, the expressions for the Burnett coefficients only involve equilibrium, two-time correlation functions. At long times assuming that the system exhibits mixing they each factor into a triple product $\langle B(s \to \infty)\rangle\langle\lambda(0)\rangle\langle \mathrm{cum}(J(0))\rangle$. The terms involving $\lambda(0)$ and the cumulants of $J(0)$ factor because at time zero the distribution function (9.27), factors into kinetic and configurational parts. Of course these results for the Burnett coefficients could have been derived using the ΔJ-ensemble methods discussed in Section 9.4.

It is apparent that our discussion of the differential susceptibility and the inverse Burnett coefficients has relied heavily on features unique to the colour conductivity problem. It is not obvious how one should carry out the analogous derivations for other transport coefficients. General fluctuation expressions for the inverse Burnett coefficients have recently been derived by Standish and Evans (1989). The general results are of the same form as the corresponding colour conductivity expressions. We refer the reader to the above reference for details.

REFERENCES

Evans, D.J. (1983). *J. Chem. Phys.*, **78**, 3297.

Evans, D.J. (1986). *Proceedings of the Fourth Australian National Congress on Rheology*, June 1986 (R.R. Huilgol, ed.), p. 23. Australian Society of Rheology, Bedford Park, South Australia.

Evans, D.J. and Lynden-Bell, R.M. (1988). *Phys. Rev., Ser. A*, **38**, 5249.

Evans, D.J. and Morriss, G.P. (1987). *Mol. Phys.*, **61**, 1151.

Standish, R.K. and Evans, D.J. (1989). *Phys. Rev., Ser. A*, to appear.

10 Towards a Thermodynamics of Steady States

10.1 INTRODUCTION

In the previous three chapters we have described a theory which can be applied to calculate the nonlinear response of an arbitrary phase variable to an applied external field. We have described several different representations for the N-particle, nonequilibrium distribution function, $f(\mathbf{\Gamma}, t)$: the Kubo representation (Section 7.1) which is only useful from a formal point of view; and two related representations, the transient time correlation function (TTCF) formalism (Section 7.3) and the Kawasaki representation (Section 7.2), both of which can be applied to obtain useful results. We now turn our interest towards thermodynamic properties which are not simple phase averages but rather are functionals of the distribution function itself. We will consider the entropy and free energy of nonequilibrium steady states. At this point it is useful to recall the connections between equilibrium statistical mechanics, the thermodynamic entropy (Gibbs, 1902), and Boltzmann's famous H theorem (Boltzmann, 1872). Gibbs pointed out that *at equilibrium* the entropy of a classical N-particle system can be calculated from the relation

$$S(t) = -k_B \int d\mathbf{\Gamma}\, f(\mathbf{\Gamma}) \log f(\mathbf{\Gamma}) \tag{10.1}$$

where $f(\mathbf{\Gamma})$ is a time-independent equilibrium distribution function. Using the same equation, Boltzmann calculated the *nonequilibrium* entropy of gases in the low-density limit. He showed that if one uses the single particle distribution of velocities obtained from the irreversible Boltzmann equation, the entropy of a gas at equilibrium is greater than that of any nonequilibrium gas with the same number of particles, volume and energy. Furthermore, he showed that the Boltzmann equation predicts a *monotonic* increase in the entropy of an isolated gas as it relaxes towards equilibrium. These results are the content of his famous H theorem (Huang, 1963). The results are in accord with our intuition that the increase in entropy is the driving force behind the relaxation to equilibrium.

One can use the reversible Liouville equation to calculate the change in the entropy of a dense many-body system. Suppose we consider a Gaussian isoenergetic system subject to a time-independent external field, F_e (8.53). We expect that the entropy of a nonequilibrium steady state will be finite and less than that of the corresponding equilibrium system with the same energy. From (10.1) we see that

$$\dot{S} = -k_B \int d\mathbf{\Gamma}\, [1 + \ln f] \frac{\partial f}{\partial t} \tag{10.2}$$

Using successive integrations by parts one finds that for an N-particle system in three dimensions

$$\begin{aligned} \dot{S} &= -k_B \int d\mathbf{\Gamma}\, f \dot{\mathbf{\Gamma}} \cdot \frac{\partial}{\partial \mathbf{\Gamma}} [1 + \ln f] \\ &= -k_B \int d\mathbf{\Gamma}\, \dot{\mathbf{\Gamma}} \cdot \frac{\partial f}{\partial \mathbf{\Gamma}} \\ &= k_B \int d\mathbf{\Gamma}\, f(t) \frac{\partial}{\partial \mathbf{\Gamma}} \cdot \dot{\mathbf{\Gamma}} = -3Nk_B \langle \alpha(t) \rangle \end{aligned} \tag{10.3}$$

Now for any nonequilibrium steady state, the average of the Gaussian multiplier, α, is positive. The external field does work on the system which must be removed by the thermostat. This means that the Liouville equation predicts that the Gibbs entropy (10.1) diverges to negative infinity! After the decay of initial transients, (10.3) shows that the rate of decrease of the entropy is constant. This paradoxical result was first derived by Evans (1985). If there is no thermostat, the Liouville equation predicts that the Gibbs entropy of an arbitrary system, satisfying $AI\Gamma$ and subject to an external dissipative field, is constant! This result was known to Gibbs (1902).

Gibbs went on to show that if one computes a coarse-grained entropy, by limiting the resolution with which we compute the distribution function, then the coarse-grained entropy based on (10.1) obeys a generalized H theorem. He showed that the coarse-grained entropy cannot decrease (Gibbs, 1902). We shall return to the question of coarse graining in Section 10.5.

The reason for the divergence in (10.3) is not difficult to find. Consider a small region of phase space, $d\mathbf{\Gamma}$, at $t = 0$, when the field is turned on. If we follow the phase trajectory of a point originally within $d\mathbf{\Gamma}$, the local relative density of ensemble points in phase space about $\mathbf{\Gamma}(t)$ can be calculated from the Liouville equation

$$\frac{1}{f(t)} \frac{df(t)}{dt} = -3N\alpha(t) \tag{10.4}$$

If the external field is sufficiently large we know that there will be some trajectories along which the multiplier, $\alpha(t)$, is positive for *all* time. For such trajectories (10.4) predicts that the local density of the phase-space distribution function must diverge in time, towards positive infinity. The distribution function of a steady state will be singular at long times. One way in which this could happen would be for the distribution function to evolve into a space of lower dimension that the *ostensible* $6N$ dimensions of phase space. If the dimension of the phase space which is *accessible* to nonequilibrium steady states is lower than the ostensible dimension, the volume of accessible phase space (as computed from within the ostensible phase space) will be zero. If this were so, the Gibbs entropy of the system (which occupies zero volume in ostensible phase space) would be $-\infty$.

At this stage these arguments are not at all rigorous. We have yet to define what we mean by a continuous change in the dimension. In the following sections we show that a reduction in the dimension of accessible phase space is a universal feature of nonequilibrium steady states. The phase-space trajectories are chaotic and separate exponentially with time, and for nonequilibrium systems the accessible steady-state phase space is a strange attractor whose dimension is less than that of the initial equilibrium phase space. These ideas are new and the relations between them and nonlinear response theory are yet to develop. We feel, however, that the ideas and insights already gleaned are sufficiently important to present here.

Before we start a detailed analysis it is instructive to consider two classic problems from the new science of dynamical systems: the quadratic map and the Lorenz model. This will introduce many of the concepts needed later to characterize quantitatively nonequilibrium steady states.

10.2 CHAOTIC DYNAMICAL SYSTEMS

The study of low-dimensional dynamical systems which exhibit chaos is a very active area of current research. A very useful introductory account can be found in Schuster (1988). It was long thought that the complex behaviour of systems of many degrees of freedom was inherently different to that of simple mechanical systems. It is now known that simple one-dimensional nonlinear systems can indeed show very complex behaviour. For example, the family of quadratic maps $F_\mu(x) = \mu x(1-x)$ demonstrates many of these features. This is very well described in a recent book by Devaney (1986). The connection between a discrete mapping, and the solution of a system of ordinary differential equations in a molecular dynamics simulation is clear when we realize that the numerical solution of the equations of motion for a system involves an iterative mapping of points in phase space. Although we are solving a problem which is continuous in time, the differential equation solver transforms this into a discrete time problem. The result is that if the mapping f takes $\mathbf{\Gamma}(0)$ to $\mathbf{\Gamma}(\Delta)$ where Δ is the time step, then $\mathbf{\Gamma}(n\Delta) = f^n[\mathbf{\Gamma}(0)]$. Here f^n means the composite mapping consisting of n repeated operations of f, $f[f[\ldots f[\mathbf{\Gamma}(0)]]\ldots]$.

An important difference exists between difference equations and similar differential equations. For example, if we consider the differential equation

$$\frac{\mathrm{d}x}{\mathrm{d}t} = \mu x(1 - x) \tag{10.5}$$

the solution can easily be obtained

$$x(t) = \frac{x_0 \mathrm{e}^{\mu t}}{1 - x_0 + x_0 \mathrm{e}^{\mu t}} \tag{10.6}$$

where $x_0 = x(t = 0)$. The trajectory for this system is now quite straightforward to understand. The solution of the quadratic map difference equation is a much more difficult problem which is still not completely understood.

The quadratic map

The quadratic map is defined by the equation

$$x_{n+1} = F_\mu(x_n) = \mu x_n(1 - x_n) \tag{10.7}$$

If we iterate this mapping for $\mu = 4$, starting with a random number in the interval between 0 and 1, then we obtain dramatically different behaviour depending upon the initial value of x. Sometimes the values repeat; other times they do not; and usually they wander about in the range 0 to 1. Initial values of x which are quite close together can have dramatically different iterates. This unpredictability or sensitive dependence on initial conditions is a property familiar in statistical mechanical simulations of higher dimensional systems. If we change the map to $x_{n+1} = 3.839x_n(1 - x_n)$, then a random initial value of x leads to a repeating cycle of three numbers (0.149888…, 0.489172…, 0.959299…). This mapping includes a set of initial values which behave just as unpredictably as those in the $\mu = 4$ example, but due to round-off error this randomness is not seen. The map is depicted in Fig. 10.1.

Before we look at the more complicated behaviour, we consider some of the simpler properties of the family of quadratic maps. First we require some definitions; x_1 is called a fixed point of the map f if $f(x_1) = x_1$. The x_1 term is a periodic point, of period n, if $f^n(x_1) = x_1$, where f^n represents n applications of the mapping f. Clearly, a fixed point is a periodic point of period one. The fixed point at x_1 is stable if $|f'(x_1)| < 1$. We will consider

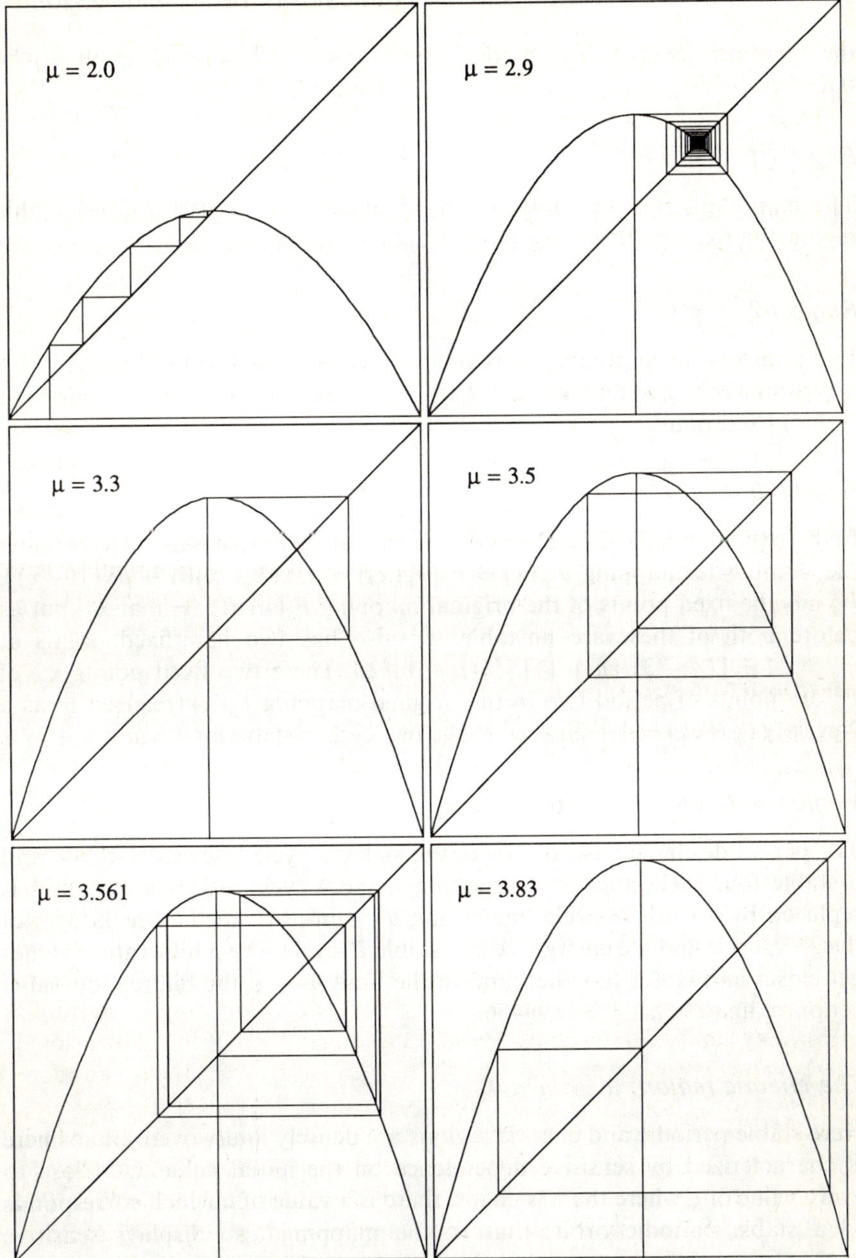

Fig. 10.1 The iterates of the quadratic map for some particular values of the parameter μ. The horizontal axis is x_n and the vertical axis is x_{n+1}. For $\mu = 2$ and 2.9 there is a single stable fixed point. For $\mu = 3.3$ there is a stable 2-cycle; for $\mu = 3.5$ a stable 4-cycle and for $\mu = 3.561$ a stable 8-cycle. The value $\mu = 3.83$ is in the period three window.

the quadratic map $F_\mu(x)$ on the interval $0 < x < 1$, as a function of the parameter μ.

Region 1: $0 < \mu < 1$

The mapping $F_\mu(x)$ has only one fixed point $x = 0$. $F'_\mu(0) = \mu$, so in this region the fixed point at $x = 0$ is attracting (or stable).

Region 2: $1 < \mu < 3$

$F_\mu(x)$ has two fixed points $x = 0$ and $x_p = (\mu - 1)/\mu$. The fixed point $x = 0$ is repelling (or unstable), while $F'_\mu(x_p) = 2 - \mu$ so that x_p is an attracting (or stable) fixed point.

Region 3: $3 < \mu < 1 + \sqrt{6}$

In this region both the fixed points of $F_\mu(x)$ are unstable so it is useful to consider the composite mapping $F_\mu^2(x) = F_\mu(F_\mu(x)) = \mu^2 x(1 - x)[1 - \mu x(1 - x)]$. F_μ^2 has the fixed points of the original mapping $F_\mu(x)$ at $x = 0$ at x_p, but as before both of these are unstable, F_μ^2 also has two new fixed points at $x_\pm = \{1 \pm [(\mu - 3)/(\mu + 1)]^{1/2}\}(\mu + 1)/2\mu$. These two fixed points $x_\pm$ of F_μ^2 are points of period two in the original mapping $F_\mu(x)$ (referred to as a 2-cycle). $(F_\mu^2)'(x_\pm) = 4 + 2\mu - \mu^2$ so the two-cycle is stable for $3 < \mu < 1 + \sqrt{6}$.

Region 4, 5, etc.: $1 + \sqrt{6} < \mu < \mu_\infty$

The period doubling cascade where the stable 2-cycle loses its stability, and a stable four-cycle appears; increasing μ the 4-cycle loses stability and is replaced by a stable 8-cycle; increasing μ again leads to the breakdown of the 2^{n-1}-cycle and the emergence of a stable 2^n-cycle. The μ bifurcation values get closer and closer together, and in the limit $n \to \infty$ the bifurcation value is approximately $\mu_\infty = 3.5699456$.

The chaotic region: $\mu_\infty < \mu < 4$

Here stable periodic and chaotic regions are densely interwoven. Chaos here is characterized by sensitive dependence on the initial value x_0. Close to every value of μ where there is chaos, there is a value of μ which corresponds to a stable periodic orbit; that is, the mapping also displays sensitive dependence on the parameter μ. The windows of period three, five and six are examples. From the mathematical perspective the sequence of cycles in a unimodal map is completely determined by the Sarkovskii theorem (Sarkovskii, 1964). If $f(x)$ has a point x which leads to a cycle of period p

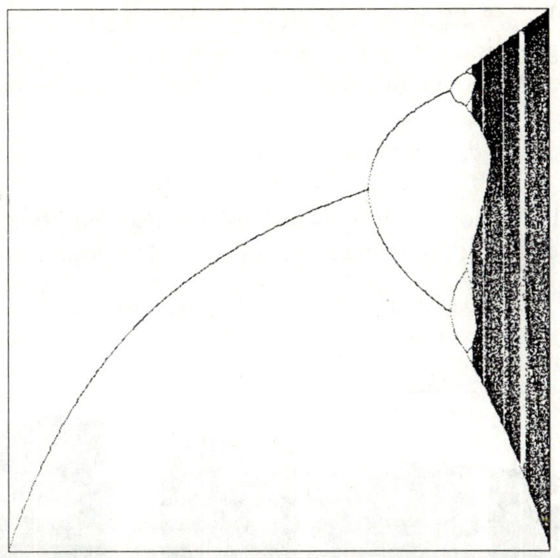

Fig. 10.2 The iterates of the quadratic map as a function of the parameter μ. The horizontal axis is the parameter $1 \le \mu \le 4$, and the vertical axis is the iterate $0 \le x_n \le 1$.

then it must have a point x' which leads to a q-cycle for every $q \leftarrow p$ where p and q are elements of the following sequence (here we read $\leftarrow$ as *precedes*)

$$
\begin{aligned}
&1 \leftarrow 2 \leftarrow 4 \leftarrow 8 \leftarrow 16 \leftarrow 32 \leftarrow \ldots \leftarrow 2^m \leftarrow \ldots \\
&\ldots \leftarrow 2^m.9 \leftarrow 2^m.7 \leftarrow 2^m.5 \leftarrow 2^m.3 \leftarrow \ldots \\
&\ldots \leftarrow 2^2.9 \leftarrow 2^2.7 \leftarrow 2^2.5 \leftarrow 2^2.3 \leftarrow \ldots \\
&\ldots \leftarrow 2^1.9 \leftarrow 2^1.7 \leftarrow 2^1.5 \leftarrow 2^1.3 \leftarrow \ldots \\
&\ldots \leftarrow \quad 9 \leftarrow \quad 7 \leftarrow \quad 5 \leftarrow \quad 3
\end{aligned}
$$

This theorem applies to values of x at a fixed parameter μ, but says nothing about the stability of the cycle or the range of parameter values for which it is observed.

Region ∞^-: $\mu = 4$

Surprisingly for this special value of μ it is possible to solve the mapping exactly (Kadanoff, 1983). Making the substitution $x_n = (1 - \cos 2\pi\theta_n)/2$

$$
\begin{aligned}
x_{n+1} = \tfrac{1}{2}(1 - \cos 2\pi\theta_{n+1}) &= 4\frac{(1 - \cos 2\pi\theta_n)}{2}\left[1 - \frac{(1 - \cos 2\pi\theta_n)}{2}\right] \\
&= \tfrac{1}{2}(1 - \cos 4\pi\theta_n) \qquad (10.8)
\end{aligned}
$$

A solution is $\theta_{n+1} = 2\theta_n \bmod 1$, or $\theta_n = 2^n\theta_0 \bmod 1$. Since x_n is related to $\cos 2\pi\theta_n$ adding an integer to θ_n leads to the same value of x_n. Only the fractional part of θ_n has significance. If θ_n is written in binary (base 2) notation

$$\theta_n = 0.a_1 a_2 a_3 a_4 a_5 \ldots = \sum_{i=1}^{\infty} a_i 2^{-i} \tag{10.9}$$

then the mapping is done simply by shifting the decimal point one place to the right and removing the integer part of θ_{n+1}. The equivalent mapping is

$$f(0.a_1 a_2 a_3 a_4 a_5 \ldots) = 0.a_2 a_3 a_4 a_5 \ldots. \tag{10.10}$$

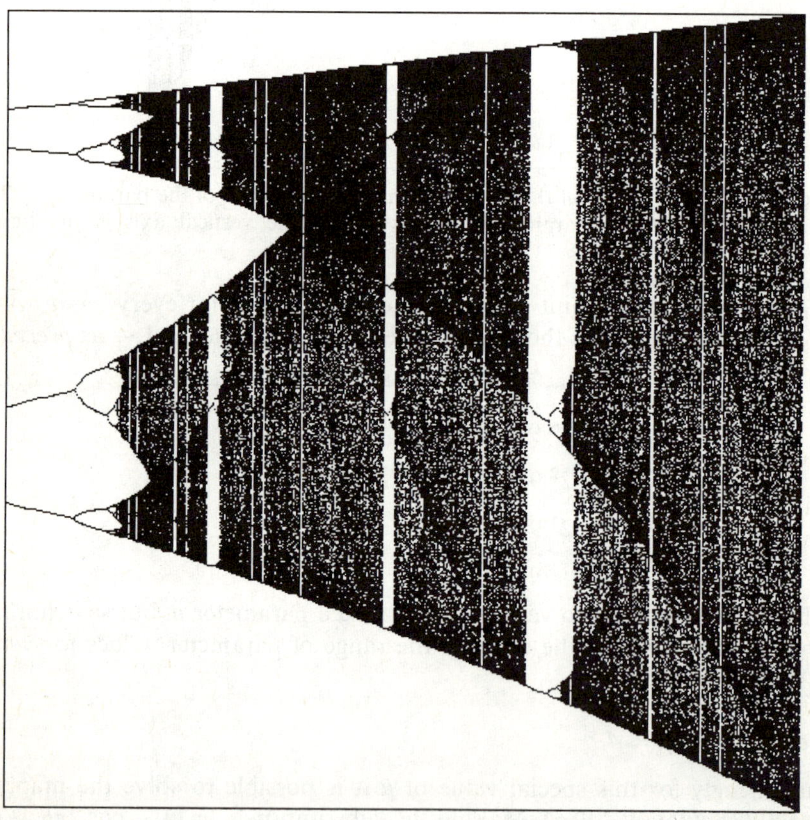

Fig. 10.3 The iterates of the quadratic map as a function of the parameter μ. This is an expanded version of Fig. 10.2 to include more detail in the chaotic region. The horizontal axis is the parameter $3.5 \leq \mu \leq 4$, and the vertical axis is the iterate $0 \leq x_n \leq 1$. The windows of period three (at about $\mu = 3.83$), period five (at about $\mu = 3.74$), and period six (at about 3.63) are clearly visible.

It is easy to see that any finite precision approximation to the initial starting value θ_0 consisting of N digits will lose all its significant digits in N iterations.

If x_0 evolves to $f(x_0)$ after one iteration, then the distribution $\delta(x - x_0)$ evolves to $\delta(x - f(x_0))$ after one iteration. This can be written as

$$\delta(x - f(x_0)) = \int_0^1 \mathrm{d}y\, \delta(x - f(y))\delta(y - x_0) \qquad (10.11)$$

An arbitrary density $\rho_n(x)$ constructed from a normalized sum of (perhaps infinitely many) delta functions, satisfies an equation of the form

$$\rho_{n+1}(x) = \int_0^1 \mathrm{d}y\, \delta(x - f(y))\rho_n(y) \qquad (10.12)$$

The invariant measure $\rho(x)$, or steady-state distribution, is independent of time (or iteration number n), so

$$\rho(x) = \int_0^1 \mathrm{d}y\, \delta(x - f(y))\rho(y) \qquad (10.13)$$

There is no unique solution to this equation, as $\rho(x) = \delta(x - x^*)$, where x^* is an unstable fixed point of the map, is always a solution. However, in general there is a physically relevant solution and it corresponds to the one that is obtained numerically. This is because the set of unstable fixed points is measure zero in the interval $[0, 1]$, so the probability of choosing to start a numerical calculation from an unstable fixed point x^* and remaining on x^* is zero due to round off and truncation errors.

Figures 10.2 and 10.3 show the iterates of the quadratic map. In Fig. 10.4 we present the invariant measure of the quadratic map in the chaotic region. The parameter value is $\mu = 3.65$. The distribution contains a number of dominant peaks which are in fact fractional power law singularities.

For the transformed mapping $\theta_{n+1} = 2\theta_n \bmod 1$, it is easy to see that the continuous loss of information about the initial starting point with each iteration of the map means that the invariant measure as a function of θ is uniform on $[0, 1]$ (that is $g(\theta) = 1$). From the change of variable $x = (1 - \cos 2\pi\theta)/2$ it is easy to see that x is a function of θ, $x = q(\theta)$ (but not the reverse). If $x_1 = q(\theta_1)$, then the number of counts in the distribution function histogram bin centered at x_1 with width $\mathrm{d}x_1$ is equal to the number of counts in the bins centered at θ_1 and $1 - \theta_1$ with widths $\mathrm{d}\theta_1$. That is

$$f(x_1) = \frac{g(\theta_1) + g(1 - \theta_1)}{|\mathrm{d}x/\mathrm{d}\theta|} \qquad (10.14)$$

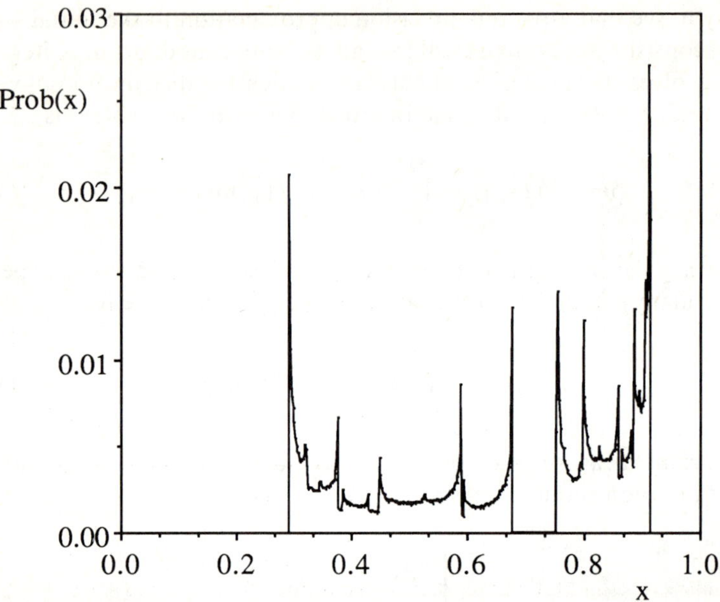

Fig. 10.4 The distribution function for the iterates of the quadratic map in the chaotic region, at $\mu = 3.65$. The horizontal axis is the value of the iterate, and the vertical axis is the probability. Note the distribution of narrow peaks which dominate the probability distribution.

It is then straightforward to show that the invariant measure as a function of x is given by

$$f(x) = \frac{1}{\pi} \frac{1}{\sqrt{x(1-x)}} \tag{10.15}$$

In Fig. 10.5 we present the invariant measure for the quadratic map at $\mu = 4$. The two singularities of type $(x - x_0)^{-1/2}$ at $x_0 = 0$ and $x_0 = 1$ are clearly shown.

Region ∞: $\mu > 4$

Here the maximum of $F_\mu(x)$ is greater than one. Once the iterate leaves the interval $0 < x < 1$ it does not return. The mapping $F_\mu^2(x)$ has two maxima, both of which are greater than one. If $\mathbf{I}$ is the interval 0 to 1, and $\mathbf{A}_1$ is the region of $\mathbf{I}$ mapped out of $\mathbf{I}$ by the mapping $F_\mu(x)$, $\mathbf{A}_2$ the region of $\mathbf{I}$ mapped out of $\mathbf{I}$ by $F_\mu^2(x)$, etc., then the trajectory wanders the interval defined by $\mathbf{I} - (\mathbf{A}_0 \cup \mathbf{A}_1 \cup \mathbf{A}_2 \cup \ldots)$. It can be shown that this set is a Cantor set.

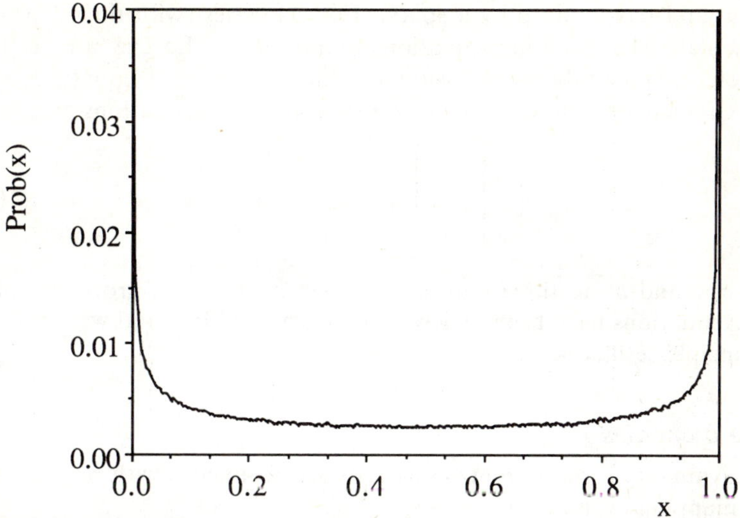

Fig. 10.5 The distribution of iterates for the quadratic map at $\mu = 4$. The horizontal axis is the iterate, and the vertical axis is the probability. When correctly normalized, this agrees well with (10.15).

This example of a seemingly very simple iterative equation exhibits very complex behaviour as a function of the parameter μ. As μ is increased the stable fixed point becomes unstable and is replaced by stable 2^n-cycles (for $n = 1, 2, 3$, etc.) until chaotic behaviour develops at μ_∞ (about 3.5699456). For $\mu > 3.5699456$ the behaviour of the quadratic map shows sensitive dependence upon the parameter μ, with an infinite number of islands of periodic behaviour immersed in a sea of chaos. This system is not atypical, and a wide variety of nonlinear problems show this same behaviour. We will now consider a simple model from hydrodynamics which has a potential impact on the practical limitations of weather forecasting.

The Lorenz model

Consider two flat plates, separated by a liquid layer. The lower plate is heated and the fluid is assumed to be two dimensional and incompressible. A coupled set of nonlinear field equations must be solved in order to determine the motion of the fluid between the plates (the continuity equation, the Navier–Stokes equation and the energy equation). These equations are simplified by introducing the stream function in place of the two velocity components. Saltzman (1961) and Lorenz (1963) proceeded by making the field equations dimensionless and then representing the dimensionless stream function and temperature by a spatial Fourier series (with time-dependent coefficients). The resulting equations obtained by Lorenz are a three-parameter family of three-dimensional ordinary-differential equations which have extremely complicated numerical solutions. The equations are

$$\begin{bmatrix} \dot{x} \\ \dot{y} \\ \dot{z} \end{bmatrix} = \begin{bmatrix} -\sigma(x-y) \\ (r-z)x-y \\ xy-bz \end{bmatrix} \tag{10.16}$$

where σ, r and b are three real positive parameters. The properties of the Lorenz equations have been reviewed by Sparrow (1982) and we summarize the principle results below.

Simple properties

(1) Symmetry—the Lorentz equations are symmetric with respect to the mapping $(x, y, z) \to (-x, -y, z)$.

(2) The z axis is invariant. All trajectories which start on the z axis remain there and move toward the origin. All trajectories which rotate around the z axis do so in a clockwise direction (when viewed from above the $z = 0$ plane). This can be seen from the fact that if $x = 0$, then $\mathrm{d}x/\mathrm{d}t > 0$ when $y > 0$, and $\mathrm{d}x/\mathrm{d}t < 0$ when $y < 0$.

(3) Existence of a bounded attracting set of zero volume, that is the existence of an attractor. The divergence of the flow, is given by

$$\frac{\partial \dot{x}}{\partial x} + \frac{\partial \dot{y}}{\partial y} + \frac{\partial \dot{z}}{\partial z} = -(1 + b + \sigma) \tag{10.17}$$

The volume element V is contracted by the flow into a volume element $V \exp[-(1 + b + \sigma)t]$ in time t. We can show that there is a bounded region E, such that every trajectory eventually enters E and remains there forever. There are many possible choices of Lyapunov function which describe the surface of the region E. One simple choice is $V = rx^2 + \sigma y^2 + \sigma(z - 2r)^2$. Differentiating with respect to time and substituting the equations of motion gives

$$\frac{\mathrm{d}V}{\mathrm{d}t} = -2\sigma(rx^2 + y^2 + bz^2 - 2brz) \tag{10.18}$$

Another choice of Lyapunov function is $E = r^2x^2 + \sigma y^2 + \sigma(z - r(r-1))^2$ for $b \le r + 1$. This shows that there exists a bounded ellipsoid, and together with the negative divergence shows that there is a bounded set of zero volume within E towards which all trajectories tend.

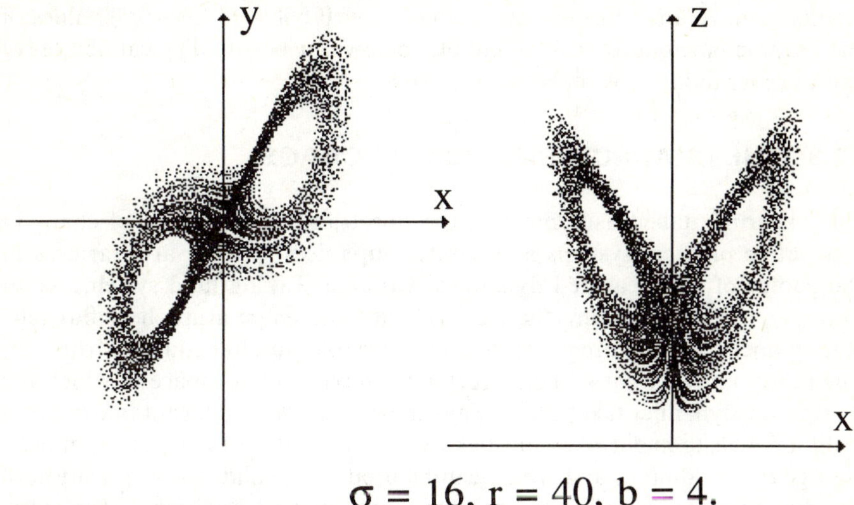

Fig. 10.6 The iterates of the Lorentz model for a typical set of parameters which leads to chaotic behaviour. The iterates are the values obtained at the end of each fourth-order Runge–Kutta step.

(4) Fixed points. The Lorentz equations have three fixed points; one at the origin, the other two area at $C_1 = (-\sqrt{b(r-1)}, -\sqrt{b(r-1)}, r-1)$ and $C_2 = (\sqrt{b(r-1)}, \sqrt{b(r-1)}, r-1)$.

(5) Eigenvalues for linearized flow about the origin are

$$\lambda_1 = -b$$

$$\lambda_2 = \frac{-(\sigma+1) - \sqrt{(\sigma+1)^2 - 4\sigma(1-r)}}{2}$$

$$\lambda_3 = \frac{-(\sigma+1) + \sqrt{(\sigma+1)^2 - 4\sigma(1-r)}}{2}$$

(6) Stability

$0 < r < 1$	The origin is stable.
$r > 1$	The origin is non-stable. Linearized flow about the origin has two negative and one positive, real eigenvalues.
$1 < r < \frac{470}{19}$	C_1 and C_2 are stable. All three eigenvalues of the linearized flow about C_1 and C_2 have negative real parts. For $r > 1.346$ ($\sigma = 10$, $b = \frac{8}{3}$) there is a complex-conjugate pair of eigenvalues.
$r > \frac{470}{19}$	C_1 and C_2 are non-stable. Linearized flow about C_1 and C_2 has one negative real eigenvalue and a complex-conjugate pair of eigenvalues with positive real parts.

Again we have a nonlinear system which is well behaved for small values of the parameter r, but for $r > \frac{470}{19}$ chaotic behaviour begins. Typical iterates of the Lorentz model are shown in Fig. 10.6.

10.3 THE CHARACTERIZATION OF CHAOS

The experimental measurement of the onset and development of chaos in dissipative physical systems is often accompanied by some arbitrariness in the choice of the measured dynamical variable. Taking fluid systems as an example, one can measure the fluid velocity, its temperature, heat flux, etc. Rarely does one measure more than one variable simultaneously. Moreover, one rarely knows what is the correct, or complete, phase space in which the dissipative dynamics takes place. Thus the extraction of relevant information calls for measurement of quantities that remain invariant under a smooth change of coordinates and which can be used for a valid characterization of the dynamical system. There are two classes of these invariants. The static ones, dependent primarily on the invariant measure (the underlying distribution function for the attractor) and appear as the dimension of the attractor (either fractal, information, correlation) and as other mass exponents which have to do with various static correlation functions. The dynamic ones depend on properties of trajectories and include various entropies (topological, metric, etc.), the Lyapunov exponents, and moments of the fluctuations in the Lyapunov exponents. Here we present a short review of the theory of these invariants and the interrelations between them.

Studies of simple dissipative systems have shown that if we begin with a Euclidian space of initial phase positions then, as time passes, transients relax, some modes may damp out, and the point in phase space that describes the state of the system approaches an *attractor*. In this process it is common for the number of degrees of freedom to be reduced, and hence the dimension of the system is lowered. This change in dimension is a continuous process and to describe such systems we have to generalize the concept of dimension (Farmer, 1982; Farmer *et al.*, 1983). We distinguish three intuitive notions of dimension: direction, capacity and measurement. These lead to the definition of: topological dimension (Hurewicz and Wallman, 1948), fractal dimension (Mandelbrot, 1983), and information dimension (Balatoni and Renyi, 1956). As we will see the fractal and information dimensions allow the dimension to be a continuous positive variable.

The fractal and information dimensions

The fractal dimension of an attractor can be defined by the following construction. Let $b(\varepsilon)$ be the minimum number of balls of diameter ε needed to cover the attractor. The fractal dimension is defined by the limit

$$D_F = \lim_{\varepsilon \to 0} \frac{\ln b(\varepsilon)}{|\ln \varepsilon|} \tag{10.19}$$

As the length scale, ε, is reduced, the number of balls required to cover the attractor increases. As $b(\varepsilon)$ is a positive integer, its logarithm is positive. The term $\ln \varepsilon$ is negative as soon as the length scale, ε, is less than one (in the appropriate units), so the dimension is a positive real quantity.

To obtain the information dimension we suppose an observer makes an isolated measurement of the coarse-grained probability distribution function p_i. Coarse graining implies a length scale ε for the observation, and an associated number of cells, $N(\varepsilon)$. The discrete, or coarse-grained, entropy, $S(\varepsilon)$, as a function of the length scale is given by

$$S(\varepsilon) = -\sum_{i=1}^{N(\varepsilon)} p_i \ln p_i \tag{10.20}$$

Note that $S(\varepsilon)$ is positive since for each i, $-p_i \ln p_i \geqslant 0$. The information dimension D_I is then defined as

$$D_I = \lim_{\varepsilon \to 0} \frac{S(\varepsilon)}{|\ln \varepsilon|} \tag{10.21}$$

This dimension is a property of any distribution function since nothing in the definition is specific to attractors, or to some underlying dynamics.

If all the $N(\varepsilon)$ elements have the same probability, then $S(\varepsilon) = \ln N(\varepsilon) = \ln \varepsilon^d$. Further, if $b(\varepsilon)$ is a minimal covering, then a smaller covering can be formed by removing the overlapping parts of circles so that $\ln b(\varepsilon) \geq \ln N(\varepsilon) = S(\varepsilon)$. It is then straightforward to see that the fractal dimension is an upper bound on the information dimension. (We will generalize this result later.) From a computational point of view it is easier to tabulate the steady-state distribution function and calculate D_I, rather than to attempt to identify the attractor and construct a covering to calculate D_F.

Correlation dimension

The correlation dimension, D_C, introduced by Grassberger and Procaccia (1983a,b) is a scaling relation on the correlation function, $C(\varepsilon)$, where

$$C(\varepsilon) = \frac{1}{N^2} \sum_{i \neq j} \theta(\varepsilon - |\mathbf{\Gamma}_i - \mathbf{\Gamma}_j|) \tag{10.22}$$

Here $\theta(x)$ is the heavyside step function, and $C(\varepsilon)$ is the correlation integral which counts the number of pairs of points whose distance of separation

$|\boldsymbol{\Gamma}_i - \boldsymbol{\Gamma}_j|$ is less than ε. The correlation dimension is

$$D_C = \lim_{\varepsilon\to 0}\lim_{N\to\infty}\frac{\ln C(\varepsilon)}{|\ln \varepsilon|} \tag{10.23}$$

It has been argued that the correlation dimension can be calculated numerically, more easily and more reliably than either the information dimension or the fractal dimension.

Generalized dimensions

In a series of papers by Grassberger (1983), Hentschel and Procaccia (1983), Halsey *et al.* (1986), and Procaccia (1985) it has been shown that the concept of dimension can be generalized further. The above authors introduced a generating function D_q which provides an infinite spectrum of dimensions depending upon the value of q. We will show that all previous dimensions are related to special values of q. Again we divide the attractor into boxes of linear dimension ε, and let p_i be the probability that the trajectory on the strange attractor visits box i. By averaging powers of the p_is over all boxes, the generalized dimension D_q is obtained.

$$D_q = -\lim_{\varepsilon\to 0}\frac{1}{q-1}\frac{1}{|\ln \varepsilon|}\ln\left(\sum_i p_i^q\right) \tag{10.24}$$

There are formal similarities between the D_q and the free energy per particle, F_β, in the thermodynamic limit,

$$F_\beta = -\lim_{N\to\infty}\frac{1}{\beta}\frac{1}{N}\ln\left(\sum_i (\mathrm{e}^{-E_i})^\beta\right) \tag{10.25}$$

where E_i are the energy levels in the system, N is the number of particles and $\beta = (k_B T)^{-1}$ is the inverse temperature. The analogy is not a strict one as the probability of state, i, is $\exp(-\beta E_i)$ rather than simply $\exp(-E_i)$ as implied above. Also the probabilities p_i are normalized, while neither $\exp(-\beta E_i)$ nor $\exp(-E_i)$ are. This is crucial in statistical mechanics since if normalized probabilities are inserted into (10.25) in place of $\exp(-\beta E_i)$, the free energy, F_β, is trivially zero!

It is straightforward to see that D_q gives each of the previously defined dimensions. For $q = 0$, $p_i^q = 1$ for all values of i, such that $p_i \neq 0$,

$$D_0 = \lim_{\varepsilon\to 0}\frac{\ln \sum_{i,p_i\neq 0} 1}{|\ln \varepsilon|} = \lim_{\varepsilon\to 0}\frac{\ln b(\varepsilon)}{|\ln \varepsilon|} \tag{10.26}$$

This is the fractal or Hausdorff dimension equation (10.19).

For $q = 1$ write p_i^q as $\exp(q \ln p_i)$ and consider the limit

$$\lim_{q\to 1} \frac{\ln \sum_i p_i^q}{q-1} = \lim_{q\to 1} \frac{\dfrac{\mathrm{d}}{\mathrm{d}q} \ln \sum_i p_i^q}{\dfrac{\mathrm{d}}{\mathrm{d}q}(q-1)} = \sum_i p_i \ln p_i = -S(\varepsilon) \tag{10.27}$$

Substituting this limit into the expression for D_q gives

$$\lim_{q\to 1} D_q = \lim_{\varepsilon\to 0} \frac{S(\varepsilon)}{|\ln \varepsilon|} = D_1 \tag{10.28}$$

This is simply the information dimension. For $q = 2$ it is easy to show that the generalized dimension is the correlation dimension.

The generalized dimension, D_q, is a non-increasing function of q. To show this we consider the generalized mean, $M(t)$, of the set of positive quantities $\{a_1, \ldots, a_n\}$, where p_k is the probability of observing a_k. The generalized mean is defined as

$$M(t) = \left(\sum_{k=1}^{n} p_k a_k^t \right)^{1/t} \tag{10.29}$$

This reduces to the familiar special cases: $M(1)$ is the arithmetic mean and the limit as $t \to 0$ is the geometric mean. It is not difficult to show that if $a_k = p_k(\varepsilon)$, where the $p_k(\varepsilon)$ are a set of discrete probabilities calculated using a length scale of ε, then the generalized dimension in equation (10.24) is related to the generalized mean by

$$D_q = -\lim_{\varepsilon\to 0} \frac{\ln M(q-1)}{|\ln \varepsilon|} \tag{10.30}$$

Using a theorem concerning generalized means, namely if $t < s$ then $M(t) \leq M(s)$ (Hardy *et al.*, 1934: 26) it follows that if $s > t$ then $D_s \leq D_t$.

The probability distribution on the attractor

If we consider the quadratic map for $\mu = 4$, the distribution of the iterates shown in Fig. 10.5 is characterized by the two singularities at $x = 0$ and $x = 1$. For $\mu = 3.65$, the distribution of iterates, shown in Fig. 10.4, has approximately ten peaks which also appear to be singularities. It is common to find a probability distribution on the attractor which consist of sets of singularities with differing fractional power law strengths. This distribution of singularities can be calculated from the generalized dimension D_q. To illustrate the connection between the generalized dimensions D_q and the singularities of the distribution function, we consider a one-dimensional system whose underlying distribution function is

$$\rho(x) = \tfrac{1}{2}x^{-1/2} \qquad \text{for } 0 \leq x \leq 1 \tag{10.31}$$

First note that, despite the fact that $\rho(x)$ is singular, $\rho(x)$ is integrable on the interval $0 \leq x \leq 1$ and it is correctly normalized. The generalized dimension D_q is defined in terms of discrete probabilities, so we divide the interval into bins of length ε – $[0, \varepsilon)$ is bin 0, $[\varepsilon, 2\varepsilon)$ is bin 1, etc. The probability of bin 0 is given by

$$p_0 = \int_0^{\varepsilon} \mathrm{d}x\, \tfrac{1}{2}x^{-1/2} = \varepsilon^{1/2} \tag{10.32}$$

and in general the probability of bin i is given by

$$p_i + \int_{x_i}^{x_i+\varepsilon} \mathrm{d}x\, \tfrac{1}{2}x^{-1/2} = (x_i + \varepsilon)^{1/2} - x_i^{1/2} \tag{10.33}$$

where $x_i = i\varepsilon$. As $(x_i + \varepsilon)^{1/2}$ is analytic for $i \neq 0$, we can expand this term to obtain

$$p_i = \tfrac{1}{2}x_i^{-1/2}\varepsilon + O(\varepsilon^2) = \rho(x_i)\varepsilon + O(\varepsilon^2) \tag{10.34}$$

So for $i = 0$, $p_i \simeq \varepsilon^{1/2}$ but for all nonzero values of i, $p_i \simeq \varepsilon$. To construct D_q we need to calculate

$$\sum_{i=0} p_i^q = p_0^q + \sum_{i\neq 0} p_i^q = \varepsilon^{q/2} + \varepsilon^{q-1} \sum_{i \neq 0} \rho(x_i)^q \varepsilon \tag{10.35}$$

We can replace the last sum in this equation by an integral,

$$\sum_{i\neq 0} \rho(x_i)^q \varepsilon \cong \int_{\varepsilon}^{1} \mathrm{d}x\, \rho(x)^q = a(1 - \varepsilon^{1-q/2}) \tag{10.36}$$

where $a = (1/2)^q(1 - q/2)^{-1}$. Combining this result with that for $i = 0$ we obtain

$$\sum_i p_i^q = (1 - a)\varepsilon^{q/2} + a\varepsilon^{q-1} \tag{10.37}$$

The distribution function $\rho(x)$ in (10.31) gives rise to singularities in the discrete probabilities p_i.

If the discrete probabilities scale with exponent α, so that $p_i \simeq \varepsilon_{\alpha_i}$ and

$$p_i^q \simeq \varepsilon^{\alpha_i q} \tag{10.38}$$

then α can take on a range of values corresponding to different regions of the underlying probability distribution. In particular, if the system is divided into pieces of size ε, then the number of times α takes on a value between α' and $\alpha' + d\alpha'$ will be of the form

$$d\alpha' \, \rho(\alpha')\varepsilon^{-f(\alpha')} \tag{10.39}$$

where $f(\alpha')$ is a continuous function. The exponent $f(\alpha')$ reflects the differing dimensions of the sets whose singularity strength is α'. Thus fractal probability distributions can be modelled by an interwoven set of singularities of strength α, each characterized by its own dimension $f(\alpha)$.

In order to determine the function $f(\alpha)$ for a given distribution function, we must relate it to observable properties, in particular we relate $f(\alpha)$ to the generalized dimensions D_q. As q is varied, different subsets associated with different scaling indices become dominant. Using (10.39) we obtain

$$\sum_i p_i^q = \sum_i \varepsilon_i^{\alpha q} = \int d\alpha' \, \rho(\alpha')\varepsilon^{-f(\alpha')}\varepsilon^{\alpha' q} \tag{10.40}$$

Since ε is very small, the integral will be dominated by the value of α' which makes the exponent $q\alpha' - f(\alpha')$ smallest, provided that $\rho(\alpha')$ is nonzero. The condition for an extremum is

$$\frac{d}{d\alpha'}[q\alpha' - f(\alpha')] = 0$$

and

$$\frac{d^2}{d\alpha'^2}[q\alpha' - f(\alpha')] > 0 \tag{10.41}$$

If $\alpha(q)$ is the value of α' which minimizes $q\alpha' - f(\alpha')$ then $f'(\alpha(q)) = q$ and $f''(\alpha(q)) < 0$. If we approximate the integral in (10.40) by its maximum value, and substitute this into (10.24), then

$$D_q = \frac{1}{q-1}[q\alpha(q) - f(\alpha(q))] \tag{10.42}$$

so that

$$f(\alpha) = q\alpha(q) - (q-1)D_q \tag{10.43}$$

Thus if we know $f(\alpha)$ and the spectrum of α values we can find D_q. Alternatively, given D_q we can find $\alpha(q)$, since $f'(\alpha) = q$ implies that

$$\alpha(q) = \frac{d}{dq}[(q-1)D_q] \tag{10.44}$$

and knowing $\alpha(q)$, $f(\alpha(q))$ can be obtained.

Dynamic invariants

Grassberger and Procaccia (1983) and Eckmann and Procaccia (1986) have shown that it is possible to define a range of scaling indices for the dynamical properties of chaotic systems. Suppose that phase space is partitioned into boxes of size ε, and that a measured trajectory $\mathbf{X}(t)$ is in the basin of attraction. The state of the system is measured at intervals of time τ. Let $p(i_1, \ldots, i_n)$ be the joint probability that $X(t = \tau)$ is in box i_1, $X(t = 2\tau)$ is in box $i_2, \ldots$, and $X(t = n\tau)$ is in box i_n. The generalized entropies K_q are defined as

$$K_q = -\lim_{\tau \to 0} \lim_{\varepsilon \to 0} \lim_{n \to \infty} \frac{1}{n\tau} \frac{1}{q-1} \ln \sum_{i_1, i_2, \ldots, i_n} p^q(i_1, i_2, \ldots, i_n) \tag{10.45}$$

where the sum is over all possible sequences $i_1, \ldots, i_n$. As before the most interesting K_q for experimental applications are the low-order ones. The limit $q \to 0$ $K_q = K$ is the Kolmogorov or metric entropy, whereas K_2 has been suggested as a useful lower bound on the metric entropy. For a regular dynamical system $K = 0$, and for a random signal $K = \infty$. In general for a chaotic system K is finite and is related to the inverse predictability time and to the sum of the positive Lyapunov exponents. The Legendre transform of $(q-1)K_q$, that is $g(\Lambda)$, is the analogue of singularity structure quantities $f(\alpha)$ introduced in the last section (for more details see Jensen *et al.*, 1987).

Lyapunov exponents

In Section 3.4 we introduced the concept of Lyapunov exponents as a quantitative measure of the mixing properties of a system. Here we will develop these ideas further, but first we review the methods which can be used to calculate the Lyapunov exponents. The standard method of calculating Lyapunov exponents for dynamical systems is that due to Benettin *et al.* (1976) and Shimada and Nagashima (1979). These authors linearized the equations of motion and studied the time evolution of a set of orthogonal vectors. To avoid problems with rapidly-growing vector lengths they periodically renormalized the vectors using a Gram–Schmidt procedure. This allows one vector to follow the fastest-growing direction in phase space, and the second to follow the next-fastest direction, while remaining orthogonal to the first vector, etc. The Lyapunov exponents are given by the average rates of growth of each of the vectors.

A new method of calculating Lyapunov exponents has been developed by Hoover and Posch (1985) and extended to multiple exponents by Morriss (1988) and Posch and Hoover (1988). The method uses Gauss' principle of least constraint to fix the length of each tangent vector, and to maintain the orthogonality of the set of tangent vectors. The two extensions of the method differ in the vector character of the constraint forces—the Posch–Hoover method uses orthogonal forces, while the Morriss method uses non-orthogonal constraint forces. In earlier chapters we have used Gauss' principle to change from one ensemble to another. This application of Gauss' principle to the calculation of Lyapunov exponents exactly parallels this situation. In the Benettin method one monitors the divergence of a pair of trajectories, with periodic rescaling. In the Gaussian scheme we monitor the *force* required to keep two trajectories a fixed distance apart in phase space.

Lyapunov dimension

The rate of exponential growth of a vector $\delta x(t)$ is given by the largest Lyapunov exponent. The rate of growth of a surface element $\delta\sigma(t) = \delta x_1(t) \times \delta x_2(t)$ is given by the sum of the two largest Lyapunov exponents. In general, the exponential rate of growth of a k-volume element is determined by the sum of the largest k Lyapunov exponents $\lambda_1 + \ldots + \lambda_k$. This sum may be positive implying growth of the k-volume element, or negative implying shrinkage of the k-volume element.

A calculation of the Lyapunov spectrum gives as many Lyapunov exponents as phase-space dimensions. All the previous characterizations of chaos that we have considered have led to a single scalar measure of the dimension of the attractor. From a knowledge of the complete spectrum of Lyapunov exponents Kaplan and Yorke (1979) have conjectured that the effective dimension of an attractor is given by that value of k for which the k-volume element neither grows nor decays. This requires some generalization of the idea of a k-dimensional volume element as the result is almost always non-integer. The Kaplan and Yorke conjecture is that the Lyapunov dimension can be calculated from

$$D_{\mathrm{L}}^{\mathrm{KY}} = n + \frac{\sum_{i=1}^{n} \lambda_i}{|\lambda_{n+1}|} \tag{10.46}$$

where n is the largest integer for which $\Sigma_{i=1}^{n} \lambda_i > 0$.

Essentially, the Kaplan–Yorke conjecture corresponds to plotting the sum of Lyapunov exponents $\Sigma_i^n \lambda_i$ versus n, and the dimension is estimated by finding where the curve intercepts the n-axis by linear interpolation (Fig. 10.7).

There is a second postulated relation between Lyapunov exponents and dimension due to Mori (1980).

$$D_{\mathrm{L}}^{\mathrm{M}} = m_0 + m^+\left(1 + \frac{|\lambda^+|}{|\lambda^-|}\right) \tag{10.47}$$

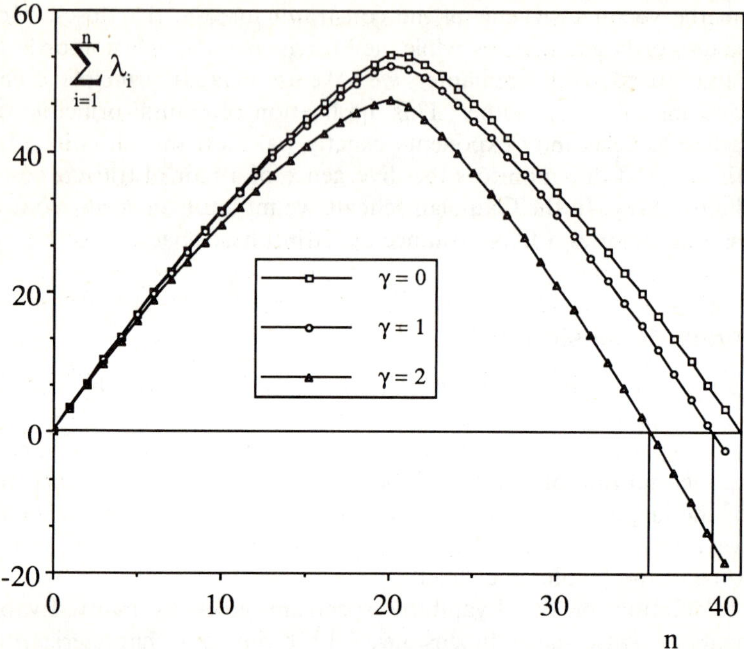

Fig. 10.7 We show the sum of the largest n exponents, plotted as a function of n, for three-dimensional eight-particle Couette flow at three difference shear rates $\gamma = 0$, 1 and 2. The Kaplan–Yorke dimension is the n-axis intercept.

where m_0 and m^+ are the number of zero and positive exponents, respectively, and $\lambda^{\pm}$ is the mean value of the positive or negative exponents (depending upon the superscript). Farmer (1982) gives a modified form of the Mori dimension which is found to give integer dimensions for systems of an infinite number of degrees of freedom.

10.4 CHAOS IN PLANAR COUETTE FLOW

We have seen in Section 10.2 that in apparently simple dynamical systems, such as the quadratic map and the Lorenz model, a single trajectory or sequence of iterates can have quite unusual behaviour. In Section 10.3 we introduced a number of techniques for characterizing the dynamical behaviour of a system with a strange attractor. Here we will apply those techniques to the SLLOD planar Couette flow algorithm that was introduced in Chapter 6. The first difficulty is that to apply the various techniques that determine the dimension of an attractor, the dimension of the initial phase space must be small enough to make the numerical calculations feasible. To calculate the static dimensions D_q we need to calculate the discrete probability distribution function. To do this we divide phase space up into boxes of size ε. The number of boxes needed varies as $(1/\varepsilon)^{6N}$, for a $6N$-dimensional phase space. Such a calculation quickly becomes impractical as the phase-space dimension increases. A typical statistical mechanical system has a phase space of $2dN$ dimensions (where d is the dimension of the translational coordinate space of a single particle), so clearly N must be small, but also N must be large enough to give nontrivial behaviour. Surprisingly enough both of these considerations can be satisfied with $d = 2$ and $N \geq 2$ (Ladd and Hoover, 1985; Morriss *et al.*, 1985, 1986).

The SLLOD equations of motion for Gaussian thermostatted planar Couette flow are:

$$\dot{\mathbf{q}}_i = \frac{\mathbf{p}_i}{m} + \mathbf{i}\gamma y_i$$
$$\dot{\mathbf{p}}_i = \mathbf{F}_i - \mathbf{i}\gamma p_{yi} - \alpha \mathbf{p}_i \tag{10.48}$$

$$\alpha = \frac{\sum_{i=1}^{N} (\mathbf{F}_i \cdot \mathbf{p}_i - \gamma p_{xi} p_{yi})}{\sum_{i=1}^{N} \mathbf{p}_i^2} \tag{10.49}$$

$\mathbf{i}$ is the unit vector in the x direction, and γ is the strain rate. The dissipative flux $J(\mathbf{\Gamma})$ due to the applied field is found from the adiabatic time derivative of the internal energy H_0. Here $J(\mathbf{\Gamma})$ is the shear stress $P_{xy}(\mathbf{\Gamma})$ times the volume V;

$$P_{xy}(\mathbf{\Gamma})V = \sum_{i=1}^{N} \left(\frac{p_{xi} p_{yi}}{m} + y_i F_{xi} \right) \tag{10.50}$$

and the shear-rate-dependent viscosity $\eta(\gamma)$ is related to the shear stress in the usual way $\eta(\gamma)\gamma = -\langle P_{xy} \rangle$.

If we consider a two-dimensional, two-body, planar Couette flow system we find that the total phase space has eight degrees of freedom: $\{x_1, y_1, x_2, y_2, p_{x1}, p_{y1}, p_{x2}, p_{y2}\}$. We then construct an infinite system made up of periodic replications of the central two-particle square, using the usual *sliding brick* periodic boundary conditions (see Section 6.3). We choose an origin for the coordinate axis where $\Sigma_i \mathbf{p}_i = 0$ and $\Sigma_i y_i = 0$. In this case both the centre of mass and the total momentum are constants of the motion. If the total kinetic energy (kinetic temperature) is also fixed, the accessible phase space has three dimensions. A convenient choice for these three variables is: the relative separation of the two particles $(x_{12}, y_{12}) = (x_2 - x_1, y_2 - y_1)$, and the direction of the momentum vector of particle one (p_{x1}, p_{y1}) with respect to the x axis, which we call θ. The magnitude of the momentum is fixed by the total kinetic energy constraint and the fact that $\mathbf{p}_1 + \mathbf{p}_2 = 0$. For $N > 2$ we find the total phase space reduces from $4N$ degrees of freedom to $4N - 5$, when the fixed centre of mass, fixed linear momentum and the constant value of kinetic energy are taken into account. The sliding brick periodic boundary conditions in the Couette flow algorithm induce an explicit time dependence into the equations of motion for Couette flow. This is most easily seen by removing the potential cutoff. The force on particle i due to particle j is then given by a lattice sum where the positions of the lattice points are explicit functions of time. The equations of motion are then nonautonomous and hence do not have a zero Lyapunov exponent. These $4N - 5$ equations can be transformed into $4N - 4$ autonomous equations by the introduction of a trivial extra variable whose time derivative is the relative velocity of the lattice points one layer above the central cell. In this form there is a zero Lyapunov exponent associated with this extra variable (see Haken, 1983). Here we work with the $4N - 5$ nonautonomous equations of motion and we ignore this extra zero Lyapunov exponent.

Information dimension

The first evidence for the existence of a strange attractor in the phase space of the two-dimensional, two-body planar Couette flow system was obtained by Morriss (1987). He showed numerically that the information dimension of two-body planar Couette flow is a decreasing function of the strain rate, dropping steadily from three towards a value near two, before dropping dramatically at a critical value of the strain rate to become asymptotic to one. These results are for the WCA potential (equation (6.5)) at a reduced temperature of 1 and a reduced density of 0.4. The sudden change in dimension, from a little greater than two to near one, is associated with the onset of the *string phase* for this system (see Section 6.4). A change in effective dimensionality for shearing systems of 896 particles, under large shear rates, has been observed. In this case the vector separation between two atoms $\mathbf{r}_{ij} = (x_{ij}, y_{ij})$ has components whose sign is independent of time. This arises because within *strings* the atoms are ordered, and the strings themselves once formed remain forever intact (and in the same order). It has been shown that the string phase is an artifact of the definition of the temperature with respect to an *assumed* streaming velocity profile (Section 6.4), so it is likely that this decrease in dimensionality is pathological, and is not associated with the attractor which is found at intermediate strain rates.

Generalized dimensions

Morriss (1989a,b) calculated the generalized dimension D_q and the spectrum of singularities $f(\alpha)$ for the steady-state phase-space distribution function of two-dimensional two-body planar Couette flow using the WCA potential at a reduced temperature of 1 and a reduced density of 0.4. This system is identical to that considered in the information dimension calculations referred to above. The maximum resolution of the distribution function was 3×2^6 bins in each of the three degrees of freedom, leading to more accurate results than the previous information dimension calculations. He found that at equilibrium the discrete probabilities $p_i(\varepsilon)$ scale with the dimension of the initial phase space. Away from equilibrium the $p_i(\varepsilon)$ scale with a range of indices, extending from the fully-accessible phase-space dimension to a lower limit which is controlled by the value of the shear rate γ.

In Fig. 10.8 we present the singularity distribution $f(\alpha)$ for a number of values of the strain rate γ. The results linear $\gamma = 0$ depend significantly on the values of grid size used, and could be improved by considering finer meshes (the minimum grid size is limited by computer memory size). At higher values of γ (say $\gamma = 1$) the values of $f(\alpha)$ above the shoulder in Fig. 10.8 are insensitive to grid size. However, the position of the shoulder does change with grid size. In the limit $q \to \infty$, the value of D_q, and hence the value of $\alpha = \alpha_{\min}$ for which $f(\alpha) \to 0$, is controlled by the scaling of the most probable p_i in the histogram $p_{\max}$. It is easy to identify $p_{\max}$ and determine its scaling as an independent check on the value of $\alpha_{\min}$. Just as large positive values of q weight the most probable p_i most strongly, large negative values of q weight the least probable p_i most strongly. The accuracy with which the least probable p_i can be determined limits the minimum value of q for which the calculation of D_q is accurate. This is reflected in poor values of D_q for $q \leq 0.5$, and we believe that this is a contributing factor in obtaining inconsistent values of the fractal dimension D_0.

We interpret the results shown in Fig. 10.8 as follows. The value of $f(\alpha)$ is the dimension of the set of points on the attractor which scale as ε^α in the discrete phase-space distribution function $\{p_i\}$. For this system it implies singularities of the form $|\mathbf{\Gamma} - \mathbf{\Gamma}_0|^{\alpha-3}$ in the underlying (continuous) phase-space distribution function $f(\mathbf{\Gamma}, \gamma)$. At equilibrium most p_is scale as ε^3, with a very narrow spread of lower α values. Indeed, with finer grid sizes this distribution may narrow still further. Away from equilibrium two effects are clearly discernible. First the dimension of the set of p_is which scale as ε^3 drops with increasing γ. Second the distribution of values of α increases downwards with the lower limit $\alpha_{\min}$ controlled by the value of γ. This distribution is monotonic with the appearance of a shoulder at an intermediate value of α.

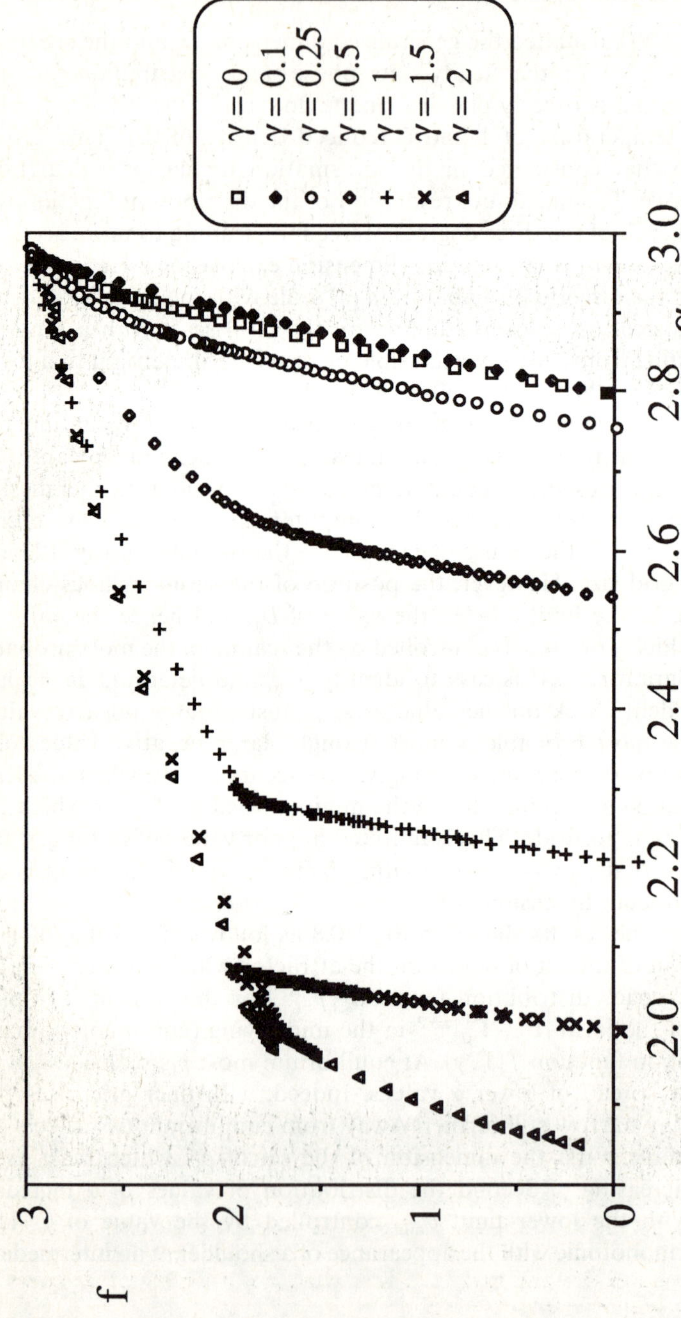

Fig. 10.8 The spectrum of phase-space singularities for two-dimensional two-particle planar Couette flow at $T^* = 1$ and $\rho^* = 0.4$ as a function of γ. The function $f(\alpha)$ is the dimension of the set of points on the attractor that scale with exponent α. The range of singularities extends from 3 to α_{min} where the value of α_{min} decreases with increasing strain rate.

Having calculated the full phase-space distribution function on a resolution ε, we can investigate the behaviour of the various reduced distributions; for example, we may consider the coordinate space distribution function $f_2(r, \phi)$, or the distribution of the momentum angle θ. Each of these reduced distributions is obtained by integrating (or summing) over the redundant coordinates or momenta. Perhaps the most interesting of these reduced distribution functions is the coordinate space distribution $f_2(x_{12}, y_{12})$, shown in Fig. 10.9. If the underlying distribution function has a singularity of the

Fig. 10.9 The coordinate space distribution function for the relative position coordinate (x_{12}, y_{12}) at $\gamma = 1.25$. The centre of the plot is the position of particle 1 (x_1, y_1), that is $x_{12} = y_{12} = 0$. Note that there is a preference for collisions to occur in the top right-hand side and lower left-hand side, and a significant depletion of counts near $x_{12} = 0$, $\rho^* = 0.4$, $e^* = 0.25$.

form $|\mathbf{\Gamma} - \mathbf{\Gamma}_0|^{\alpha-3}$, then f_2 can have singularities of the form $|\mathbf{\Gamma} - \mathbf{\Gamma}_0|^{\alpha-2}$. However, if $2 \le \alpha \le 3$ then these points are no longer singularities, and the reduced distribution f_2 has a different character to the full phase-space distribution. If the exponent $\alpha - 2$ is positive, then f_2 is zero at $\mathbf{\Gamma}_0$ and the discrete probability $p_i(\varepsilon)$ which includes $\mathbf{\Gamma}_0$ will scale as ε^2, whereas if $\alpha - 2$ is negative then f_2 is singular.

In this study *all* the two variable distribution functions, although being highly structured in many cases, did not show any evidence of singularity sets of nonzero measure. This observation has important ramifications for the Green entropy which we will meet in Section 10.5.

Lyapunov exponents

The complete set of Lyapunov exponents for two- and three-dimensional planar Couette flow have been calculated for two-, four- and eight-particle systems by Morriss (1988, 1989a,b). For the two-particle system the Lyapunov dimension, D_L, has been calculated using both the Mori and Kaplan–Yorke conjectures (equations (10.46) and (10.47)). This requires the complete set of Lyapunov exponents (that is three exponents for $N = 2$) and has the advantage over static dimensions that no subsequent extrapolation procedure is needed. Table 10.1 contains the results for the two-body, two-dimensional Couette flow system at the same state point as that used in the information and generalized dimension calculations.

For both the Kaplan–Yorke and Mori forms, the Lyapunov dimension is found to be a decreasing function of the shear rate. This is consistent with the contraction of phase-space dimension that we have already seen from the numerical evaluated static dimensions D_q. It confirms that the nonequilibrium distribution function is a fractal attractor whose dimension is less than that of the equilibrium phase space. When the shear rate γ is zero, both methods of calculating the Lyapunov dimension agree. However, as soon as the shear rate changes from zero, differences appear. In the Kaplan–Yorke formula (equation (10.47)), the value of n is 2 from $\gamma = 0$, until the magnitude of λ_2 exceeds that of λ_1 (somewhere between $\gamma = 2$ and 2.5). This means that $2 < D_L^{KY} < 3$ in this range. For $\gamma > 2$, $1 < D_L^{KY} < 2$ as long as λ_1 remains positive. The value of λ_3 is irrelevant as soon as $|\lambda_2| > \lambda_1$. Then as λ_1 becomes negative the dimension is equal to zero. The Kaplan–Yorke formula can never give fractional values between zero and one. In the Mori formula the value of λ_3 always contributes to the dimension, and its large negative value tends to dominate the denominator, reducing D_L^M. The transition from $D_L^M > 2$ to $D_L^M < 2$ is somewhere between $\gamma = 1$ and 1.5. Indeed the Mori dimension is systematically less than the Kaplan–Yorke dimension.

Table 10.1 Lyapunov exponents for the two-body, two-dimensional Couette flow system.

	Lyapunov exponents for $N = 2$			Dimension	
γ	λ_1	λ_2	λ_3	D_L^{KY}	D_L^M
0	2.047(2)	0.002(2)	−2.043(2)	3.003	3.00
0.25	2.063(3)	−0.046(2)	−2.1192(3)	2.952	2.90
0.5	1.995(3)	−0.187(4)	−2.242(3)	2.81	2.64
0.75	1.922(4)	−0.388(3)	−2.442(3)	2.62	2.36
1.0	1.849(5)	−0.63(1)	−2.74(1)	2.445	2.10
1.25	1.807(4)	−0.873(5)	−3.17(1)	2.295	1.89
1.5	1.800(5)	−1.121(2)	−4.12(5)	2.14	1.68
1.75	1.733(4)	−1.424(3)	−5.63(6)	2.058	1.49
2.0	1.649(9)	−1.54(1)	−7.36(8)	2.015	1.37
2.25	1.575(3)	−1.60(1)	−9.25(9)	1.981	1.29
2.5	1.61(2)	−2.14(1)	−11.5(1)	1.75	1.24
2.75	0.2616(8)	−2.12(1)	−19.84(3)	1.123	1.02
3.0	0.678(5)	−2.69(1)	−19.85(2)	1.252	1.06
3.5	−0.111(4)	−2.62(1)	−17.49(4)	0	0
4.0	0.427(4)	−4.25(1)	−14.43(5)	1.10	1.05
4.5	−0.674(5)	−2.96(1)	−10.78(3)	0	0
5.0	−0.132(2)	−1.97(1)	−8.152(3)	0	0

In Table 10.2 we compare the values of D_q for two-particle two-dimensional planar Couette flow for several values of q, with the Kaplan–Yorke Lyapunov dimension for this system obtained from the full spectrum of Lyapunov exponents. Of the two routes to the Lyapunov dimension, the Kaplan–Yorke method agrees best with the information dimension results listed in Table 10.2, whereas the Mori method does not. In particular, the Kaplan–Yorke method and the information dimension both give a change from values of greater than two, to values of less than two at about $\gamma = 2.5$. There are a number of points to note about the results in this table. First, it can be shown that D_1 is a lower bound for D_0; however, the numerical results for D_0 and D_1 are inconsistent with this requirement as $D_0 < D_1$. As we remarked previously, the results for D_q when $q < 0.5$ are poor. It has been argued that the fractal (Hausdorff) dimension and Kaplan–Yorke Lyapunov dimension should yield the same result, at least for homogeneous attractors. In this work we find that D_L^{KY} is significantly lower than D_1 (which is itself a lower bound on D_0) for all values of the strain rate. Indeed D_L^{KY} is approximately equal to D_q, where q is somewhat greater than 3.

It is possible to calculate the Lyapunov exponents of systems with more than two particles, whereas extending the distribution-function histogramming

Table 10.2 Generalized dimensions for the two-body, two-dimensional Couette flow systems.

γ	Generalized dimensions				
	D_0	D_1	D_2	D_3	D_L^{KY}
0.0	2.90(1)	2.98(2)	2.98(2)	2.98(2)	3.003
0.1	2.91	2.98	2.98	2.98	—
0.25	2.91	2.98	2.98	2.97	2.95
0.5	2.91	2.97	2.95	2.91	2.81
1.0	2.89(1)	2.90(3)	2.67(3)	2.49(3)	2.445
1.5	2.87	2.75	2.290	2.15	2.14
2.0	2.80(3)	2.65(3)	2.20(2)	2.10(3)	2.015

algorithms for the information dimension or generalized dimension is much more difficult. The full Lyapunov spectrum has been calculated for four- and eight-particle planar Couette flow systems in both two and three dimensions.

In Fig. 10.10 we show the Lyapunov spectra for the four-particle system at $\rho = 0.4$ for a range of values of the shear rate. For the equilibrium spectrum

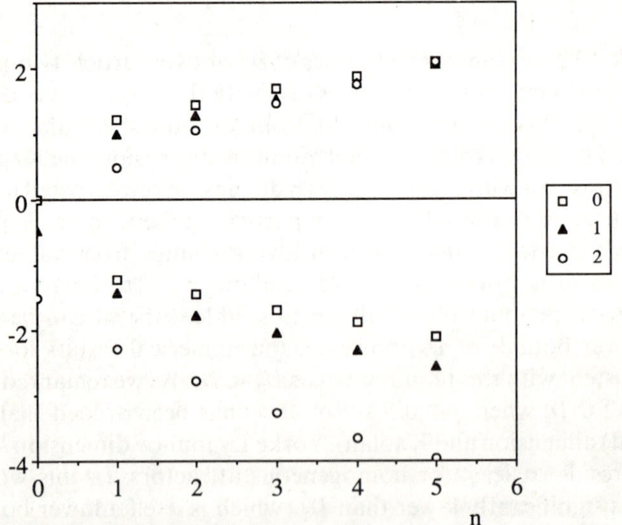

Fig. 10.10 The Lyapunov spectra for two-dimensional four-particle planar Couette flow at $T^* = 1$ and $\rho^* = 0.4$. The Lyapunov spectra shifts downwards with increasing strain rate with the largest exponent shifting least. The sum of the exponents is zero at equilibrium and become more negative with increasing strain rate.

($\gamma = 0$) one exponent is zero, while the others occur in Smale pairs $\{\lambda_{-i}, \lambda_{+i}\}$, where $\lambda_{-i} = -\lambda_i$. This symmetry is a consequence of the time reversibility of the equations of motion and the conservation of phase-space volume from the Liouville theorem. For the two-dimensional system the exponents appear to be essentially linear in exponent number, but a linear fit to the positive exponents is not consistent with an exponent of zero for exponent number zero. As the external field is increased systematic changes in the Lyapunov spectrum occur.

The positive branch decreases, with the smallest positive exponent decreasing most. The largest positive exponent seems almost independent of the external field. We expect that the most vigorous *mixing* in phase space, which is controlled by the positive exponents, is first a function of the curvature of the particles trajectories (the higher the curvature, the more defocusing is each collision), and second depends on the collision frequency (and hence the density). It could be argued that the insensitivity of the largest exponent is associated with only a small change in collision frequency with strain rate, at this density. The zero exponent becomes more negative with increasing field, as does the negative branch of the Lyapunov spectrum. The change in the negative branch is larger than the change in the positive branch. The change in the sum of each exponent pair is the same, that is $\lambda_{-i} + \lambda_{+i} = c$, where c is constant independent of i and related directly to the dissipation. The change in the exponent which is zero at equilibrium is $\frac{1}{2}c$.

The idea of being able to characterize the Lyapunov spectrum without having to calculate all of the exponents is very attractive, as the computation time for the Gaussian constraint method depends on the fourth power of the number of particles, N. We decided to compare the Lyapunov spectra as a function of system size, at the same state point. It is well known that the bulk properties will have some system-size dependence, but the trends as a function of density and temperature should be reliable. In Fig. 10.11 we present the Lyapunov spectra for an equilibrium system at $\rho = 0.4$, for a range of system sizes $N = 2$, 4 and 8. Each spectrum is scaled so that the largest positive and negative exponents have the same exponent number regardless of system size. These results look very encouraging as the spectra of the three systems are very similar. The linear fit to the positive branch for $N = 4$ and $N = 8$ have slightly different slopes but the qualitative features are the same.

In Fig. 10.12 we present the Lyapunov spectra for a strain rate of $\gamma = 1.0$ at $\rho = 0.4$ for system sizes of $N = 2$, 4 and 8. This shows that there is also a close correspondence between the results at different system sizes away from equilibrium.

In Fig. 10.13 we show the Lyapunov dimension of the planar Couette flow system at $\rho = 0.4$ as a function of strain rate, for a range of system sizes. For

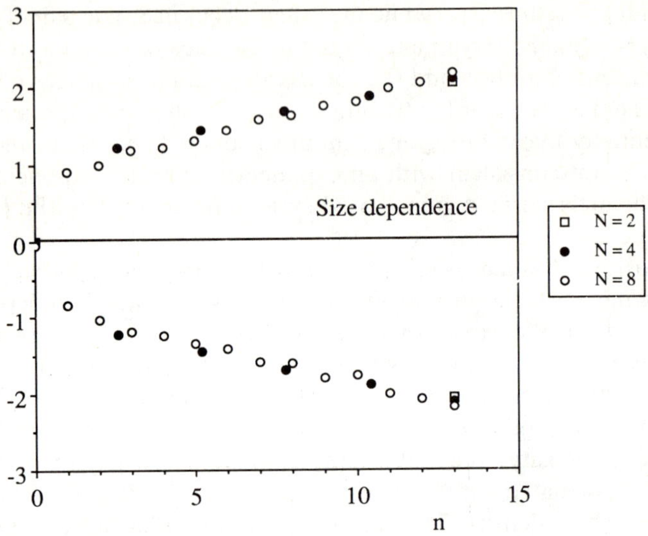

Fig. 10.11 The Lyapunov spectra for two-dimensional two-, four- and eight-particle equilibrium simulations at $T^* = 1$ and $\rho^* = 0.4$. The spectra are scaled so that the largest positive exponent occurs at the same exponent number regardless of system size.

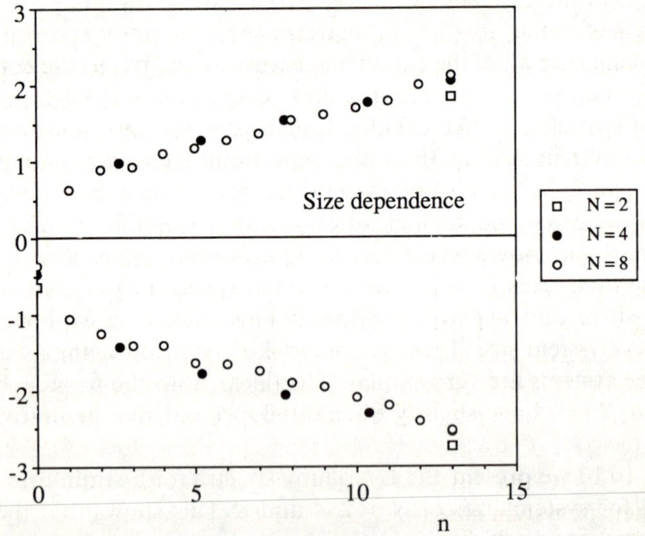

Fig. 10.12 The Lyapunov spectra for two-dimensional two-, four- and eight-particle planar Couette flow at $T^* = 1$, $\rho^* = 0.4$ and $\gamma = 1.0$. The spectra are scaled so that the largest positive exponent occurs at the same exponent number regardless of system size.

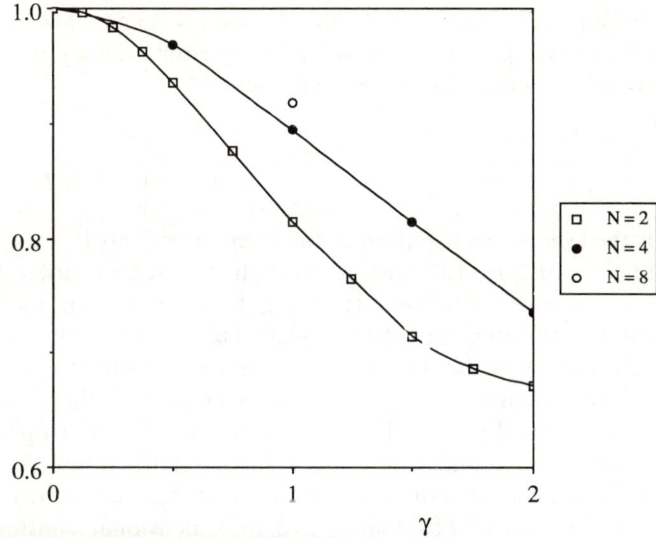

Fig. 10.13 The Lyapunov dimension for two-dimensional two-, four- and eight-particle Couette flow at $T^* = 1$ and $\rho^* = 0.4$, as a function of strain rate. The values of dimension are scaled with respect to the equilibrium dimension so that the y axis represents the proportional change in dimension.

each system size the Lyapunov dimension is scaled by the equilibrium value, so that the plotted results represent the proportional reduction in dimension. The qualitative trends are the same. There is a decrease in dimension with increasing strain rate. The proportional change in dimension is greatest for the two-particle system and smallest for the eight-particle system, whereas the absolute changes are in the opposite order.

In summary, the results confirm the dimensional contraction observed previously in two-body, two-dimensional planar Couette flow simulations. The initial phase-space dimension of $D = 2dN - 2d - 1$ contracts with increasing external field, and the distribution function is only nonzero on a fractal attractor of dimension less than $2dN - 2d - 1$. Although the results for these systems differ in detail from the generalized dimension results, the observation of significant dimensional contraction is universal. An approach which may help reduce the magnitude of the numerical calculations is the observation that the qualitative features of the spectra are essentially independent of system size.

If we consider the volume element V_{2dN} where $2dN$ is the phase-space dimension of the initial system (d is the spatial dimension and N is the number of particles), then the phase-space compression factor gives the rate of change of phase-space volume (see equation (3.78)), so that the average of the divergence is equal to the sum of the Lyapunov exponents. A careful calculation of the divergence for the SLLOD algorithm, taking into account the precise number of degrees of freedom gives

$$\sum_{i=1}^{2dN-2d-1} \lambda_i = -(dN-d-1)\alpha + \frac{\gamma P_{xy}^{K} V}{(dN-d-1)/kT} \tag{10.51}$$

where P_{xy}^{K} is the kinetic contribution to the shear stress and V is the volume. The term involving P_{xy}^{K} is order one, whereas the first term is order N, so for many-particle systems the second term can be ignored. For the systems considered here both terms must be included. This is a valuable consistency check on the accuracy of the numerical calculation of Lyapunov exponents.

We have now identified two effects associated with the phase-space distribution functions of nonequilibrium systems: the first was dimensional contraction, and the second is a range of sets of fractional power law singularities. The results are consistent in the sense that as each distribution is normalized, the loss of probability due to dimensional contraction, is compensated for by the appearance of singularities in the distribution function.

Studies of two- and three-dimensional colour diffusion systems by Posch and Hoover (1988) have produced an impressive calculation—the full Lyapunov spectrum for a three-dimensional system of 32 repulsive Lennard–Jones atoms (185 Lyapunov exponents) as well as results for the same system with eight atoms. Lyapunov spectra are sensitive to ensemble, both at and away from equilibrium. All indications are that nonequilibrium systems are also insensitive to the details of the ensemble or thermostatting mechanism. However, boundary effects do have a significant influence on the shape of spectra of small systems. In particular, the homogeneous algorithms for shear flow (such as SLLOD) give different Lyapunov exponents to boundary reservoir methods (Posch and Hoover, 1989).

As small nonequilibrium molecular dynamics (NEMD) simulations of planar Couette flow and colour diffusion are dominated by a fractal attractor whose dimension is determined by the strength of the applied field, this behaviour can be expected for all nonequilibrium steady-state simulations. The existence of a fractal attractor is a vital clue to understanding the nonequilibrium entropy, but as yet we only have information concerning the rate of approach of a trajectory to the attractor, and measures of its effective dimension. We know a good deal about the structure of the attractor, and the singularities of the nonequilibrium distribution function. Some recent work in the study of dynamical systems (Takahashi and Oono, 1984) shows that modelling chaotic behaviour with statistical mechanical analogues is a useful approach, however to date the approach parallels irreversible thermodynamics with a continuous production of entropy. For a theory of nonequilibrium steady states we need to be able to calculate an entropy *shift* from equilibrium to the steady state which is finite. The appearance of an attractor and the relative stability of *entropy-producing trajectories* provides a plausible explanation for the observation of irreversibility and a mechanism for the resolution of Löschmidt's paradox (Holian *et al.*, 1987).

It is interesting to make a connection between the results given here and the numerical calculations based on the Kawasaki distribution function. In Section 7.7 we described some very recent numerical studies of the Kawasaki form for the full nonlinear response of an equilibrium system subject to the sudden application of a fixed shear rate. From a theoretical point of view there are two points of interest in this stress growth experiment. First, is the renormalized Kawasaki shear stress equal to that observed directly? Second, how does the Kawasaki normalization behave as a function of time? The results show that the renormalized Kawasaki shear stress is in quite good agreement with the direct result, and that the Kawasaki normalization, which is onc initially, decreases with time. The results obtained here for the two-body system suggest that the Kawasaki distribution function may have singularities which compensate for the observed decrease in both the individual probabilities and the normalization, and that these singularities are not adequately represented in the phase-space sampling used.

Equation (10.4) implies that if we consider a co-moving phase-space volume element containing a fixed number of trajectories, then the local density of phase space increases indefinitely because the associated Lagrangian volume is constantly decreasing (because the sum of the Lyapunov exponents is negative). Since the contraction of the accessible phase space is *continuous* there is, in a sense, no possibility of generating a steady-state distribution function. Computed from the ostensible phase space the volume of accessible phase space shrinks at a constant rate becoming zero at infinite time. A steady state is in fact characterized by a constant rate of decrease in the relative volume occupied by accessible phase space. This is in spite of the fact that in a steady state averages of phase variables are constant. This apparent contradiction can be understood by considering the following example.

Suppose we consider a system which at $t = 0$ occupies a two-dimensional phase space $0 < x, y < L$. Suppose that by some means this thermostatted system is subject to a dissipative external field which, after initial transients, causes the distribution function, $f(x, y)$, to collapse towards a one-dimensional attractor, $x^2 + y^2 = r^2$. At some time t, the distribution function is given by the equation

$$\begin{aligned} f(x, y, t) &\simeq \frac{1}{2\pi r \Delta(t)}; \qquad r^2 < x^2 + y^2 < (r + \Delta(t))^2 \\ &= 0; \qquad \text{otherwise} \end{aligned} \tag{10.52}$$

Further, we suppose that the width of the annulus which forms the distribution function satisfies an equation of motion,

$$\frac{d\Delta(t)}{dt} = -\alpha\Delta(t) \tag{10.53}$$

for some positive constant value of α. It is easy to see that in the steady state, $df/dt = \alpha f$, which is the analogue of (10.4). The phase-space distribution function diverges at a constant rate, α. In spite of this, if we compute the phase average of a nonsingular phase variable $B(x, y)$, time averages will *converge* exponentially fast towards their steady-state values, $\langle B(t)\rangle - \langle B(\infty)\rangle \simeq e^{-\alpha t}$. This example points out that, although the distribution function, as computed from the ostensible phase space, may be diverging at a constant rate, steady-state phase averages may still be well defined and convergent. The distribution function computed from within the accessible phase space has no singularities, $f_{acc}(x, y, t) \equiv f(x, y, t)/(2\pi r\Delta(t)) = 1$, $\forall t$, provided, $r^2 < x^2 + y^2 < (r + \Delta(t))^2$. In our example it is always uniform and constant in time. Phase averages are fundamentally functions of phase-space distances not of volumes. Indeed the notion of a phase-space volume is somewhat arbitrary.

10.5 GREEN'S EXPANSION FOR THE ENTROPY

Since the dimension of the accessible phase space decreases to less than the ostensible $2dN$ dimensions, the volume of the accessible phase space, as measured from the ostensible space, is zero. The entropy of a system is proportional to the logarithm of the accessible phase volume. Since that volume as determined from the ostensible phase space is zero, the entropy will diverge to negative infinity. These simple observations explain the divergence of entropy as computed in the ostensible space. Presumably the thermodynamic entropy should be arrived at by integrating over the accessible phase space only. This would remove the apparent divergence. However, the determination of the topology of the phase space which is accessible to nonequilibrium steady states is exceedingly complex. Even the dimension of the accessible space is only known approximately. Such a program for the calculation of the nonequilibrium entropy would, therefore, appear quite hopeless.

The fine-grained entropy as computed from the ostensible phase-space dimension has a number of further difficulties. From a quantum mechanical point of view, if a system such as the one depicted in Fig. 10.9 is meant to represent argon, it is in violation of the Heisenberg uncertainty principle. The uncertainty principle puts an absolute limit on the degree to which a distribution function can be fractal. There is a lower limit imposed by Planck's constant to the scale of features that can be found in phase space. The extreme fineness of the filaments depicted in Fig. 10.9 implies extreme sensitivity to external perturbations. The finer the length scale of the phase-space structures the more sensitive those structures will be to external perturbations. If the distribution function is fractal there is no limit to the smallness of the phase-space structures and, therefore, no limit to the sensitivity of the full distribution function to uncontrolled external perturbations. In an experiment, averaging over an ensemble of possible external fluctuations would of course *wash out* the fine structure below a critical length scale. The precise cutoff value would be determined by the amplitude and spectrum of the external fluctuations. This *washing out* of fine structure provides an ansatz for the computation of the entropy of nonequilibrium steady states.

Evans (1989) has described a systematic method for computing the coarse-grained entropy of nonequilibrium steady states. The coarse graining is introduced by decomposing the Gibbs (1902) entropy into terms arising from the partial distribution functions involving correlations of successive numbers of particles. If the expansion is carried out to order N, the total number of particles in the system, the results will of course be identical to the fine-grained Gibbs entropy. The expansion has been tested at equilibrium and it has been found that, for densities less than ca. 75% of the freezing density, the singlet and pair contributions to the entropy appear to be accurate to more than ca. 90%. At equilibrium, therefore, the expansion appears to converge rapidly. Away from equilibrium the expansion will consist of a series of finite terms until the dimension of the partial distribution function exceeds the dimension of the accessible phase space. Once this occurs all succeeding terms will be infinite. The method yields finite terms below this dimension because all the lower-dimensional integrals are carried out in the accessible phase space.

Green (1952) used Kirkwood's factorization of the N-particle distribution function to write an expansion for the entropy. If we define z-functions in an infinite hierarchy as

$$\begin{aligned} \ln f_1^{(i)} &\equiv z_1^{(i)} \\ \ln f_2^{(ij)} &\equiv z_2^{(ij)} + z_1^{(i)} + z_1^{(j)} \\ \ln f_3^{(ijk)} &\equiv z_3^{(ijk)} + z_2^{(jk)} + z_2^{(ki)} + z_2^{(ij)} + z_1^{(i)} + z_1^{(j)} + z_1^{(k)} \end{aligned} \tag{10.54}$$

where the various f-functions are the partial one-, two-, three-body, etc., distribution functions, then Green showed that Gibbs' fine grained entropy (10.1) can be written as an infinite series,

$$S = -k_B\left\{\frac{1}{1!}\int d\mathbf{\Gamma}_1\, f_1 z_1 + \frac{1}{2!}\int\int d\mathbf{\Gamma}_1\, d\mathbf{\Gamma}_2\, f_2 z_2 + \ldots\right\} \tag{10.55}$$

Using equation (10.54) one can easily show that the entropy per particle is given by

$$\frac{S}{N} = -\frac{k_B}{\rho} \int d\mathbf{p}_1 \, f_1(\mathbf{p}_1) \ln f_1(\mathbf{p}_1) - \frac{k_B}{2N} \iint d\mathbf{\Gamma}_1 \, d\mathbf{\Gamma}_2 \, f_2(\mathbf{\Gamma}_1, \mathbf{\Gamma}_2) \ln\left[\frac{f_2^{(12)}}{f_1^{(1)} f_1^{(2)}}\right] + \ldots \tag{10.56}$$

In deriving this equation we have assumed that the fluid is homogeneous. This enables a spatial integration to be performed in the first term. This equation is valid away from equilibrium. Using the fact that at equilibrium the two-body distribution function factors into a product of kinetic and configurational parts, equation (10.56) for two-dimensional fluids reduces to

$$\frac{S}{N} = 1 - k_B \ln\left(\frac{\rho}{2\pi m k_B T}\right) - \frac{k_B \rho}{2} \int d\mathbf{r}_{12} \, g(r_{12}) \ln g(r_{12}) + \ldots \tag{10.57}$$

where $g(r_{12})$ is the equilibrium radial distribution function. Equation (10.57) has been tested using experimental radial distribution function data by Mountain and Raveché (1971) and by Wallace (1987). They found that the Green expansion for the entropy, terminated at the pair level, gives a surprisingly accurate estimate of the entropy from the dilute gas to the freezing density. As far as we know, prior to Evans' work in 1989, the Green expansion had never been used in computer simulations. This was because, in the canonical ensemble, Green's entropy expansion is non-local. In Evans' calculations the entropy was calculated by integrating the relevant distribution functions over the *entire* simulation volume. A recent reformulation of (10.57) by Baranyai and Evans (1989) have succeeded in developing a local expression for the entropy of a canonical ensemble of systems. Furthermore, the Baranyai–Evans expression for the entropy is ensemble independent.

Evans (1989) used a simulation of 32 soft discs ($\phi(r) = \varepsilon(\sigma/r)^{12}$ truncated at $r/\sigma = 1.5$) to test (10.57) truncated at the pair level. All units were expressed in dimensionless form by expressing all quantities in terms of the potential parameters σ, ε and the particle mass m. Table 10.3 shows some of the equilibrium data gathered for the soft disc fluid. All units are expressed in reduced form. Each state point was generated from a 10 million time-step simulation run using a reduced time-step of 0.002. The energy per particle is denoted e, and the total one- and two-body entropy per particle are denoted by s. The entropy was calculated by forming histograms for both $g(\mathbf{r})$ and $f(\mathbf{p})$. These numerical approximations to the distribution functions were then integrated numerically. The radial distribution function was calculated over the minimum image cell to include exactly the long-range, nonlocal, contributions arising from the fact that at long range, $g(r) = (N-1)/N$. The equipartition, or kinetic, temperature corrected for $O(1/N)$ factors, is denoted

Table 10.3 Equilibrium moderate density data.‡

ρ	e	s	T_k	T_{th}
0.6	1.921	3.200		
0.6	2.134	3.341	1.552	1.614
0.6	2.347	3.464		
0.625	1.921	3.034		
0.625	2.134	3.176	1.499	1.500
0.625	2.347	3.318		
0.65	1.921	2.889		
0.65	2.134	3.044	1.445	1.454
0.65	2.347	3.182		
0.675	1.921	2.754		
0.675	2.134	2.919	1.306	1.374
0.675	2.347	3.064		
0.7	1.921	2.889		
0.7	2.134	3.044	1.326	1.291
0.7	2.347	3.182		

‡ The uncertainties in the entropies are ± 0.005.

by T_k. The thermodynamic temperature, T_{th} was calculated from equation (10.57) using the thermodynamic relation, $T_{th} = (\partial e/\partial s)_V$. For each density the three state points were used to form a simple finite-difference approximation for the derivative.

The analytical expression for the kinetic contribution to the entropy was not used, but rather this contribution was calculated from simulation data by histogramming the observed particle velocities and numerically integrating the single particle contribution. The numerical estimate for the kinetic contribution to the entropy was then compared to the theoretical expression (basically the Boltzmann H function) and agreement was observed within the estimated statistical uncertainties.

By using the entropies calculated at $\rho = 0.6$ and 0.7 to form a finite-difference approximation to the derivative $\partial s/\partial \rho^{-1}$ one can compare the pressure calculated from the relation $p = (T\partial S/\partial V)_E$, with the virial expression calculated directly from the simulation. The virial pressure at $e = 2.134$ and $\rho = 0.65$ is 3.85, whereas the pressure calculated exclusively by numerical differentiation of the entropy is 3.72 ± 0.15. The largest source of error in these calculations is likely to be in the finite-difference approximation for the various partial derivatives.

Away from equilibrium the main difficulty in using even the first two terms in (10.56) is the dimensionality of the required histograms. The nonequilibrium pair distribution function does *not* factorize into a product of kinetic and configurational parts. One has to deal with the full function of six variables for translationally invariant two-dimensional fluid. In his work, Evans reduced the density to $\rho \simeq 0.1$ where the configurational contributions to the entropy should be small. He evaluated the entropy of the same system of 32 soft discs, but now the system was subject to isoenergetic planar Couette flow, using the SLLOD equations of motion. In this simulation a constant thermodynamic internal energy $H_0 \equiv \Sigma\, p^2/2m + \Phi$ was maintained. The thermostatting multiplier, α, takes the form (see (5.21)),

$$\alpha = -\frac{P_{xy}\gamma V}{\sum \frac{p_i^2}{m}} \qquad (10.58)$$

where P_{xy} is the xy element of the pressure tensor.

To check the validity of our assumption that at these low densities the configurational parts of the entropy may be ignored, he performed some checks on the equilibrium thermodynamic properties of the system. Table 10.4 shows the thermodynamic temperature computed using a finite-difference approximation to the derivative, $\partial e/\partial s, (e = \langle H_0 \rangle/N, s = S/N)$. It also shows the kinetic temperature computed using the equipartition expression. At equilibrium, the data at a reduced density of 0.1 predicts a thermodynamic temperature which is in statistical agreement with the kinetic temperature, 2.12 ± 0.04 as against 2.17, respectively. The equilibrium data at $e = 2.134$ and $\rho = 0.1$ gives a thermodynamic pressure of 0.22, which is in reasonably good agreement with the virial pressure (including both kinetic and configurational components) of 0.24. The disagreement between the thermodynamic and the kinetic expressions for both the temperature and the pressure arise from two causes: the absence of the configurational contributions, and the finite-difference approximations for the partial derivatives.

Figure 10.14 shows the analogue of Fig. 10.9 for a 32-particle system under shear. The nonequilibrium pair distribution function is free of the singularities apparent in the two-particle system. The reason why the function is smooth is that for one- and two-particle distributions in systems of many particles one averages over all possible positions and momenta for the other $N - 2$ particles. This averaging *washes out* the fine structure. These distributions, even at very high strain rates, are *not* fractal. If the Green expansion converges rapidly we will clearly arrive at a finite value for the entropy.

Table 10.4 gives the computed kinetic contribution to the entropy as a function of energy, density and strain rate. At low densities the increased mean free paths of particles relative to the corresponding situation in dense fluids means that considerably longer simulation runs are required to achieve

Table 10.4 Low-density data.‡

ρ	γ	e	s	T_k	T_{th}
0.075	0.0	2.134	6.213		
0.1	0.0	1.921	5.812		
0.1	0.0	2.134	5.917(27)	2.175	2.12(6)
0.1	0.0	2.346	6.013		
0.125	0.0	2.134	5.686		
0.075	0.5	1.921	5.744		
0.075	0.5	2.134	5.852	2.190	2.088
0.075	0.5	2.347	5.948		
0.1	0.5	1.921	5.539		
0.1	0.5	2.134	5.653	2.171	2.048
0.1	0.5	2.346	5.747		
0.125	0.5	1.921	5.369		
0.125	0.5	2.134	5.478	2.153	2.088
0.125	0.5	2.347	5.573		
0.075	1.0	1.921	5.380		
0.075	1.0	2.134	5.499	2.188	1.902
0.075	1.0	2.347	5.604		
0.1	1.0	1.921	5.275		
0.1	1.0	2.134	5.392	2.169	1.963
0.1	1.0	2.346	5.492		
0.125	1.0	1.921	5.157		
0.125	1.0	2.134	5.267	2.149	2.019
0.125	1.0	2.347	5.368		

‡ Away from equilibrium the uncertainties in the entropy rate are ± 0.005.

an accuracy comparable with that for dense fluids. The data given in Table 10.4 are taken from 15-million time-step simulation runs. Away from equilibrium the strain rate tends to increase the *mixing* of trajectories in phase space so that the errors actually decrease as the strain rate is increased.

For a given energy and density, the entropy is observed to be a monotonically *decreasing* function of the strain rate. As expected from thermodynamics, the equilibrium state has the maximum entropy. Although there is no generally agreed upon framework for thermodynamics far from equilibrium, it is clear that the entropy can be written as a function, $S = S(N, V, E, \gamma)$. *Defining* T_{th} as $(\partial E/\partial S)_{V,\gamma}$, p_{th} as $(T\partial S/\partial V)_{E,\gamma}$ and ζ_{th} as $(-T\partial S/\partial\gamma)_{E,V}$, we can write

$$dE = T_{th}\, dS - p_{th}\, dV + \zeta_{th}\, d\gamma \tag{10.59}$$

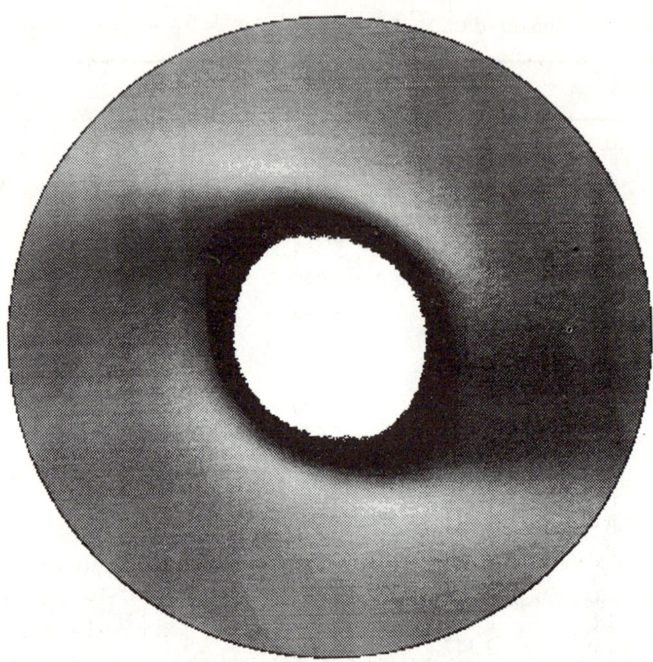

Fig. 10.14 Shows the pair distribution function for the 32-particle soft disc fluid at a relatively high reduced strain rate of 2.0. The reduced density and total energy per particle are 0.1 and 1.921, respectively. The run length was 24 million time-steps. The distribution is, as far as can be told from the simulation data, completely smooth. In spite of the high anisotropy of this distribution, the configurational contribution to the system entropy is only about 0.4%.

Some years ago Evans and Hanley (1980) proposed (10.59) as a generalized Gibbs relation; however, at that time there was no way of directly computing the entropy or any of the free energies. This forced Evans and Hanley to postulate that the thermodynamic temperature was equal to the equipartition or kinetic temperature, $T_k \equiv 2K/(dNk_B)$, for systems in d dimensions. Evans and Hanley observed that, although the pressure tensor is anisotropic, away from equilibrium, the thermodynamic pressure must be independent of the manner in which a virtual volume change is performed. The thermodynamic pressure must, therefore, be a scalar. They assumed that the thermodynamic pressure would be equal to the *simplest scalar invariant of the pressure tensor* that was also consistent with equilibrium thermodynamics. In two-dimensional systems they assumed that $p = (P_{xx} + P_{yy})/2$.

Since we can now calculate the coarse-grained Gibbs entropy directly, we can check the correctness of these postulates. We assume that the internal energy is given by the sum of the peculiar kinetic energy and the potential energy, that we know the system volume and strain rate and that the thermodynamic entropy is equal to the coarse-grained Gibbs entropy which at low densities can be approximated by the first term of (10.56). Table 10.4 gives a comparison of kinetic and thermodynamic temperatures for the 32-particle soft disc system.

As has been known for some time (Evans, 1983), $(\partial T_k/\partial \gamma)_{V,E}$ is negative leading to a decrease in the kinetic temperature with increasing strain rate. For this low-density system the effect is far smaller than has been seen for moderately-dense systems. At a density of 0.1 the kinetic temperature drops by 0.3% as the shear rate is increased to unity. The precision of the kinetic temperature for these runs is about 0.01%. The thermodynamic temperature also decreases as the strain rate is increased but in a far more dramatic fashion: it decreases by 10% over the same range of strain rates. The results clearly show that away from equilibrium the thermodynamic temperature is smaller than the kinetic or equipartition temperature. As the strain rate increases the discrepancy becomes larger.

Using the simulation data at $e = 2.134$, one can estimate the thermodynamic pressure as a function of strain rate. Table 10.5 shows the finite-difference approximation for the thermodynamic pressure, p_{th}, the hydrostatic pressure, $p_{tr} = (P_{xx} + P_{xy})/2$ and the largest and smallest eigenvalues of the pressure tensor p_1 and p_2, respectively. As expected the hydrostatic pressure increases with shear rate. This effect, known as shear dilatancy, is very slight at these low densities. The thermodynamic pressure shows a much larger effect but it *decreases* as the strain rate is increased. In an effort to give a mechanical interpretation to the thermodynamic pressure we calculated the two eigenvalues of the pressure tensor. Away from equilibrium, the diagonal elements of the pressure tensor differ from one another and from their equilibrium values, these are termed normal stress effects. The eigenvalues are influenced by all the elements of the pressure tensor, including the shear stress. One of the eigenvalues increases with strain rate while the other decreases and within statistical uncertainties the latter is equal to the thermodynamic pressure.

Evans (1989) conjectured that the thermodynamic pressure is equal to the minimum eigenvalue of the pressure tensor, that is $p_{th} = p_2$. This relation is

Table 10.5 Nonequilibrium pressure: $e = 2.134$, $\rho = 0.1$.

γ	p_{th}	p_{tr}	p_1	p_2
0.0	0.215(7)	0.244	0.244	0.244
0.5	0.145	0.245	0.361	0.130
1.0	0.085	0.247	0.397	0.096

exact at equilibrium and is in accord with our numerical results. It is also clear that if the entropy is related to the minimum reversible work required to accomplish a virtual volume change in a nonequilibrium steady-state system, then $p_2\,dV$ is the minimum pV work that is possible. If one imagines carrying out a virtual volume change by moving walls inclined at arbitrary angles with respect to the shear plane, then the minimum virtual pV work (minimized over all possible inclinations of the walls) will be $p_2\,dV$.

Figure 10.15 shows the kinetic contribution to the entropy as a function of strain rate for the 32-particle system at an energy $e = 2.134$ and a density $\rho = 0.1$. The entropy seems to be a linear function of strain rate for the range of strain rates covered by the simulations. Combining these results with those from Table 10.4 allows us to compute ζ_{th} as a function of strain rate. For $\gamma = 0.0$, 0.5 and 1.0 we find that $\zeta_{th}/N = 1.22$, 1.08, and 0.91, respectively. Most of the decrease in ζ is due to the decrease in the thermodynamic temperature with increasing strain rate. We have assumed that asymptotically s is linear in strain rate as the strain rate tends to zero. It is always possible that at strain rates which are too small for us to simulate, that this linear dependence gives way to a quadratic variation.

Although these calculations are restricted to the low-density gas regime, the results suggest that a sensible definition for the nonequilibrium entropy

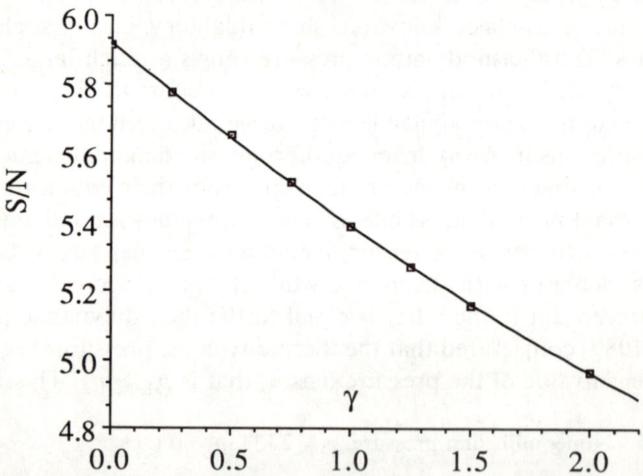

Fig. 10.15 The kinetic contribution to the system entropy as a function of strain rate. The system density is 0.1 and the energy per particle is 2.134. Within the accuracy of the data the entropy is essentially a linear function of strain rate. The derivative of the entropy with respect to strain rate gives ζ/T. ζ is positive but decreases with strain rate, mostly due to the decrease in the thermodynamic temperature with increasing strain rate.

can be given. A definition, based on (10.56) avoids the divergences inherent in the fine-grained entropy due to the contraction of the nonequilibrium phase space. At low densities this entropy reduces to the contraction of the nonequilibrium phase space. At low densities this entropy reduces to the Boltzmann entropy implicit in the Boltzmann H function. Our entropy is, for states of a specified energy and density, a maximum at equilibrium.

Defining a temperature on the basis of this entropy, indicates that far from equilibrium there is no reason to expect that the equipartition, or kinetic, temperature is equal to the thermodynamic temperature. Similarly, there seems to be no reason to expect that the average of the diagonal elements of the pressure tensor will be equal to the thermodynamic pressure far from equilibrium. The concept of minimum reversible virtual work, together with our numerical results suggests that the thermodynamic pressure might instead be equal to the minimum eigenvalue of the pressure tensor.

It remains to be seen whether the entropy, so defined, is a local maximum in nonequilibrium steady states. If this can be satisfactorily demonstrated then we will have for the first time a fundamental basis for a generalized thermodynamics of steady states far from equilibrium.

REFERENCES

Balatoni, J. and Renyi, A. (1956). *Publ. Math. Inst. Hungarian Acad. Sci.*, **1**, 9. *The Selected Papers of A. Renyi*, **1**, 558. Budapest: Akademia Budapest, 1976.

Baranyai, A. and Evans, D.J. (1989). *Phys. Rev., Ser. A*, **40**, 3817.

Benettin, G., Galgani, L. and Strelcyn, J.-M. (1976). *Phys. Rev., Ser. A*, **14**, 2338.

Boltzmann, L. (1872). See: *Lectures on Gas Theory* (translated by G. Brush). Berkeley: University of California Press, 1964.

Devaney, R.L. (1986). *An Introduction to Chaotic Dynamical Systems.* Menlo Park: Benjamin/Cummings.

Eckmann, J.-P. and Procaccia, I. (1986). *Phys. Rev., Ser. A*, **34**, 659.

Evans, D.J. (1983). *J. Chem. Phys.*, **78**, 3297.

Evans, D.J. (1985). *Phys. Rev., Ser. A*, **32**, 2923.

Evans, D.J. (1989). *J. Stat. Phys.*, **57**, 745.

Evans, D.J. and Hanley, H.J.M. (1980). *Phys. Lett.*, **80A**, 175.

Farmer, J.D. (1982). *Physica*, **4D**, 366.

Farmer, J.D., Ott, E. and Yorke, J.A. (1983). *Physica*, **7D**, 153.

Gibbs, J.W. (1902). *Elementary Principles in Statistical Mechanics.* Boston: Yale University Press.

Grassberger, P. (1983). *Phys. Lett.*, **97A**, 227.

Grassberger, P. and Procaccia, I. (1983a). *Physica*, **9D**, 189.

Grassberger, P. and Procaccia, I. (1983b). *Phys. Rev., Ser. A*, **28**, 2591.

Green, H.S. (1952). *The Molecular Theory of Fluids.* Amsterdam: North-Holland.

Haken, H. (1983). *Phys. Lett.*, **94A**, 71.

Halsey, T.C., Jensen, M.H., Kadanoff, L.P., Procaccia, I. and Shraiman, B.I. (1986). *Phys. Rev., Ser. A*, **33**, 1141.

Hardy, G.H., Littlewood, J.E. and Pólya, G. (1934). *Inequalities.* Cambridge: The University Press.

Hentschel, H.G.E. and Procaccia, I. (1983). *Physica*, **8D**, 435.

Holian, B.L., Hoover, W.G. and Posch, H.A. (1987). *Phys. Rev. Lett.*, **59**, 10.

Hoover, W.G. and Posch, H.A. (1985). *Phys. Lett.*, **113A**, 82.

Huang, K. (1963). *Statistical Mechanics.* New York: Wiley.

Hurewicz, W. and Wallman, H. (1948). *Dimension Theory.* Princeton, NJ: Princeton University Press.

Jensen, M.H., Kadanoff, L.P. and Procaccia, I. (1987). *Phys. Rev., Ser. A*, **36**, 1409.

Kadanoff, L.P. (1983). Roads to Chaos. *Physics Today*, December, pp. 46–53.

Kaplan, J. and Yorke, J.A. (1979). In: *Functional Differential Equations and Approximation of Fixed Points*, H.O. Peitgen and H.O. Walther (eds). Heidelberg: Springer-Verlag.

Kirkwood, J.G. (1942). *J. Chem. Phys.*, **10**, 394.

Ladd, A.J.C. and Hoover, W.G. (1985). *J. Stat. Phys.*, **38**, 973.

Lorenz, E.N. (1963). *J. Atmos. Sci.*, **20**, 130.

Mandelbrot, B. (1983). *The Fractal Geometry of Nature.* San Francisco: Freeman.

Mori, H. (1980). *Prog. Theor. Phys.*, **63**, 1044.

Morriss, G.P. (1985). *Phys. Lett.*, **A113**, 269.

Morriss, G.P. (1987). *Phys. Lett.*, **A122**, 236.

Morriss, G.P. (1988). *Phys. Rev., Ser. A*, **37**, 2118.

Morriss, G.P. (1989a). *Phys. Lett.*, **A143**, 307.

Morriss, G.P. (1989b). *Phys. Rev., Ser. A*, **39**, 4811.

Morriss, G.P., Isbister, D.J. and Hughes B.D. (1986). *J. Stat. Phys.*, **44**, 107.

Mountain, R.D. and Raveché, H.J. (1971). *J. Chem. Phys.*, **29**, 2250.

Posch, H.A. and Hoover, W.G. (1988). *Phys. Rev., Ser. A*, **38**, 473.

Posch, H.A. and Hoover, W.G. (1989). *Phys. Rev., Ser. A*, **39**, 2175.

Procaccia, I. (1985). *Phys. Script.*, **T9**, 40.

Saltzmann, B. (1961). *J. Atmos. Sci.*, **19**, 329.

Sarkovskii, (1964). *Ukr. Mat. Z.*, **16**, 61.

Schuster, H.G. (1988). *Deterministic Chaos, An Introduction*, 2nd edn. Weinheim: VCH.

Shimada, I. and Nagashima, T. (1979). *Prog. Theor. Phys.*, **61**, 1605.

Sparrow, C. (1982). *The Lorentz Equations: Bifurcations, Chaos, and Strange Attractors.* New York: Springer-Verlag.

Takahashi, Y. and Oono, Y. (1984). *Prog. Theor. Phys.*, **71**, 851.

Wallace, D.C. (1987). *J. Chem. Phys.*, **87**, 2282.

Index